Advanced Technologies of Welding, Surfacing, and Thermal Spraying of Modern Materials

Advanced Technologies of Welding, Surfacing, and Thermal Spraying of Modern Materials

Artur Czupryński
Marcin Adamiak
Antonín Kříž
Tünde Anna Kovács

MDPI

Basel • Beijing • Wuhan • Barcelona • Belgrade • Novi Sad • Cluj • Manchester

Editors

Artur Czupryński
Department of Welding Engineering
Silesian University of Technology
Gliwice
Poland

Marcin Adamiak
Laboratory of Materials Reserch
Silesian University of Technology
Gliwice
Poland

Antonín Kříž
Department of Manufacturing Technology
Czech Technical University
Prague
Czech Republic

Tünde Anna Kovács
Department of Materials Technology
Óbuda University
Budapest
Hungary

Editorial Office
MDPI AG
Grosspeteranlage 5
4052 Basel, Switzerland

This is a reprint of articles from the Special Issue published online in the open access journal *Materials* (ISSN 1996-1944) (available at: www.mdpi.com/journal/materials/special_issues/adv_technol_mat).

For citation purposes, cite each article independently as indicated on the article page online and as indicated below:

Lastname, A.A.; Lastname, B.B. Article Title. *Journal Name* **Year**, *Volume Number*, Page Range.

ISBN 978-3-7258-2260-7 (Hbk)
ISBN 978-3-7258-2259-1 (PDF)
doi.org/10.3390/books978-3-7258-2259-1

Contents

About the Editors

Artur Czupryński

Dr. Artur Czupryński is an Associate Professor at the Silesian University of Technology in Gliwice. He has been working in the Department of Welding since 1997, and, in 2001, he completed the European Welding Engineer (EWE) course. He has completed scientific internships at the Welding Institute in Gliwice (Poland), University of Warwick (UK), VSB Technical University of Ostrava (Czech Republic), Obuda University (Hungary), University of Ruse (Bulgaria), and University of West Bohemia (Czech Republic). He is a member of the Polish Welding Society and a member of the Programme Council of the Persons Certification Body TÜV Rheinland Polska. In his scientific endeavours, he deals with technologies of welding, brazing, thermal cutting, surfacing, and thermal spraying and oversees quality control in welding. Dr. Artur Czupryski is the author and coauthor of 3 books and academic textbooks, as well as more than 200 articles and 2 patents. In 2018, he received the Stanislaw Olszewski Medal awarded by the SIMP Welding Section to the most outstanding Polish and foreign welding engineers "in recognition of their contribution to Polish welding". For his scientific, didactic, and organisational achievements, he has been repeatedly awarded individual and team awards by the Rector of the Silesian University of Technology and decorated by the President of the Republic of Poland.

Marcin Adamiak

Prof. Marcin Adamiak, DSc., PhD., Eng., was born in 1968. He graduated from the Faculty of Mechanical Engineering at the Silesian University of Technology in 1993. He obtained his doctoral degree in materials engineering in 1997 and his postdoctoral degree in 2013 at the same Alma Mater. He currently works as a professor and the Dean of the Faculty of Mechanical Engineering at the Silesian University of Technology. In 2013–2014, as part of an industrial internship at Jeppesen Poland Sp. z o.o. a Boeing Company participated in implementing the Green Aircraft Interiors project with Boeing Research and Technology Europe in Madrid. In addition, he completed many internships abroad, both in Europe and beyond its borders. He was the head of the Department of Biomedical Materials Engineering and deputy director of the Institute of Engineering and Biomedical Materials for General Affairs and Didactics. His scientific interests include materials engineering, specializing in composite materials, manufacturing and processing technologies, biomaterials, and surface engineering. His scientific contributions include more than 180 scientific publications in international journals and scientific conference proceedings worldwide. Since 1 October 2019, he has been the Head of the Laboratory of Materials Research of the Faculty of Mechanical Engineering.

Antonín Kříž

Prof. Dr. Antonín Kříž is a full professor at the Department of Manufacturing Technology at the Czech Technical University in Prague. He is an International Welder Engineer (IWE) and lecturer of International Welder Engineer courses. He is a member of the International Advisory Board of Manufacturing Technology and a member of the committee of the Association for Heat Treatment of Metals. His research interests are unique in the field of materials science and technology. His unique research interests are in materials sciences and technologies, specifically in thin films, surface integrity, and bulk materials (e.g., tool steels and sintered carbide). His technological interests include heat treatment, polymer welding, heterogeneous welds, laser cladding, and laser metal deposition.

At universities in the Czech Republic and Slovakia, he lectures to PhD students in the field of materials, surface treatments, and welding and heat treatment technology. He is a member of the Research Board (University of West Bohemia) and Subject Area Boards (Production and Material Engineering) at Jan Evangelista Purkyně University in Ústí nad Labem and Czech Technical University in Prague.

Tünde Anna Kovács

Dr Kovács is an Associate Professor in the Department of Materials Technology at Óbuda University, Bánki Donát Faculty of Mechanical and Safety Engineering, Hungary. She is a member of the editorial board of the Acta Materialia Transylvania, Safety and Security Sciences Review, and Security Engineering of Anthropogenic Objects. Her research interests are in the field of materials science and technologies, unique welding processes (ultrasonic and explosive welding), welder robots, and collaborative welding robotics. She is an International Welder Engineer (IWE) and lecturer of the International Welder Engineer courses. She is a supervisor in the Doctoral School on Safety and Security Sciences and the Doctoral School on Materials and Technologies of Óbuda University, Hungary, and an author and co-author of numerous scientific publications.

Preface

Technological advancements have significantly changed metal joining and cladding processes. As a result, current and future welding technology trends are gradually shifting from traditional metals, techniques, and manual welding to advanced arc welding, automated welding, and the latest metals and alloys. Additionally, welding safety has also been elevated to a new level to ensure a safer working environment. In terms of advanced arc welding processes, there is increasing use of friction, plasma, laser beam, hybrid, magnetic arc, explosive, ultrasonic, friction stir, and electron beam welding. In the machinery industry, surface modification processes for workpieces primarily include surfacing, thermal spraying, plasma spraying, chrome plating, and nitriding. As the industry evolves, the demands for traditional surface modification processes are becoming more stringent, rendering some traditional methods inadequate to meet production requirements. Laser cladding technology, a new type of green remanufacturing surface modification technology, is gradually replacing traditional surface treatment processes. This reprint is a compilation of scientific content that provides an in-depth understanding of the relationships between structure, properties, and functions in welded joints, as well as in surface layers produced using advanced welding technologies and innovative engineering materials. It includes an overview of various welding techniques, including solid-state welding processes, used to join new metal alloys, composites, polymers, and ceramics from different perspectives. Furthermore, this book presents the latest developments in the research on innovative technologies and materials for producing surface layers and coatings that are resistant to mechanical wear, thermal degradation, and corrosion. It focuses on the influence of various factors in anti-wear surface engineering, the mechanisms of wear and wear resistance, the characteristics of different alloy categories, and the effects of layer and coating deposition. This reprint also covers recent developments in advanced materials and processes, including thermal spraying, plasma and laser surface treatments, surface modification techniques, multilayer structural composites, and nanosurface engineering (with the advancement of nanoscience and technology). This reprint includes original scientific papers that shed light on recent advancements in welding and surface technologies. The Editors recommend this reprint to scientists in the fields of materials science and production engineering.

Artur Czupryński, Marcin Adamiak, Antonín Kříž, and Tünde Anna Kovács
Editors

materials

Article

High-Temperature Corrosion of Flame-Sprayed Power Boiler Components with Nickel Alloy Powders

Artur Czupryński [1,*], Janusz Adamiec [2], Marcin Adamiak [3], Marcin Żuk [1], Antonin Kříž [4], Claudio Mele [5] and Monika Kciuk [6]

1 Department of Welding, Silesian University of Technology, Konarskiego Street 18A, 44-100 Gliwice, Poland
2 Department of Metallurgy and Recycling, Silesian University of Technology, Krasińskiego Street 8, 40-019 Katowice, Poland
3 Materials Research Laboratory, Silesian University of Technology, Konarskiego Street 18A, 44-100 Gliwice, Poland
4 Faculty of Mechanical Engineering, Institute of Technology and Materials, Jan Evangelista Purkyně University in Ústí nad Labem, Pasteurova Street 1, 400 96 Ústí nad Labem, Czech Republic
5 Department of Engineering for Innovation, University of Salento, via per Monteroni, 73100 Lecce, Italy
6 Department of Engineering Materials and Biomaterials, Silesian University of Technology, Konarskiego Street 18A, 44-100 Gliwice, Poland
* Correspondence: artur.czuprynski@polsl.pl

Abstract: The development trends in the energy sector clearly indicate an increase in the share of biomass and alternative fuels fed for combustion in power boilers, which results in the imposition of many unfavourable factors and a demanding working environment. During the operation of the energy system, this means a sharp increase in corrosion of the gas-tight pipe walls and coils by the destructive action of chlorine and sulphur. Implementing advanced surface protection in addition to the selection of materials of better quality and resistance to difficult working conditions would significantly reduce their wear by high temperature corrosion. Thermally sprayed coatings offer a great opportunity to protect machine components and energy systems against corrosion, erosion, impact load and abrasive wear. This article presents the test results of high-temperature corrosion resistance of coatings made with Ni-Cr-B-Si and Ni-B-Si alloy powders on a boiler steel substrate. Samples with sprayed coatings were exposed to an atmosphere with a composition of N_2 + 9% O_2 + 0.08% SO_2 + 0.15% HCl at 800 °C for 250, 500, 750 and 1000 h. Tests results of coatings made of Ni-Cr-B-Si alloys subjected to the influence of a corrosive environment showed the formation of a layer of scale on the surface, composed mainly of Cr_2O_3 oxide, which was a passive layer, reducing the rate of corrosion. Coatings sprayed with Ni-B-Si alloys showed significantly lower corrosion resistance. It was found that the developed technology of subsonic flame spraying with powders of the Ni-Cr-B-Si type allows the production of coatings compliant with the requirements of the energy industry, which allows their use as anti-corrosion protection on boiler elements intended for waste disposal and biomass combustion.

Keywords: flame powder spray process; coating; nickel alloy; high temperature corrosion; sheet pile walls; power boiler

check for updates

Citation: Czupryński, A.; Adamiec, J.; Adamiak, M.; Żuk, M.; Kříž, A.; Mele, C.; Kciuk, M. High-Temperature Corrosion of Flame-Sprayed Power Boiler Components with Nickel Alloy Powders. *Materials* **2023**, *16*, 1658. https://doi.org/10.3390/ma16041658

Academic Editor: Luigi Calabrese

Received: 15 January 2023
Revised: 7 February 2023
Accepted: 14 February 2023
Published: 16 February 2023

1. Introduction

High costs of energy produced from energy resources (natural gas, crude oil, fossil coals) and the need to protect the natural environment more and more often encourage world economies to use renewable energy sources obtained from wind, sunlight or water. Biomass is a natural energy resource, which, due to its origin, is divided into: wood biomass from trees and shrubs; green biomass from plants with non-lignified stems and whose above-ground part dies after the cultivation period; fruit biomass from plant parts containing seeds; and a mixture resulting from the deliberate mixing of biofuels [1]. The increase in the share of biomass and alternative fuels fed for combustion in power boilers is associated with surface phenomena that may accompany the contact of waste gases with boiler elements. Corrosion processes in waste incineration installations are mainly caused by the presence of moisture (surface corrosion, crevice corrosion, pitting corrosion, stress corrosion, intercrystalline corrosion) and high operating temperature (sulfiding, carbonization, dusting, nitriding corrosion, sulfiding). In the combustion chamber of a power boiler, corrosion processes may be of the following nature:

- oxidizing due to the presence of nitric and chromic acids, iron and copper salts, nitrites, chromates and urea;
- reducing due to the presence of sulphuric, hydrochloric and phosphoric acids, organic acids, a solution of alkaline salts and halides;
- mixed (oxidizing-reducing) related to the presence of a mixture of the above-mentioned compounds.

High-temperature corrosion, also known as low NOx corrosion, occurs most often in boilers with high steam parameters and mainly attacks the boiler structure elements heated to high temperatures. It is a chemical process of metal oxidation at high temperatures, exhaust gas atmospheres or environments containing sulphur or hydrogen sulphide [2,3]. The specificity of high-temperature corrosion consists in the fact that a homogenous, passive layer of oxides does not form on the surface of the metal, as is the case under normal oxidation conditions. In addition to the correct chemical composition of the exhaust gas, high temperature is required for this type of corrosion to occur.

According to Ostrowski et al. [4], with oxygen deficiency in the furnace of a power boiler, sulphur, chlorine and fluorine react with oxygen, forming a protective layer of oxides on the surface of the panel elements of the gas-tight pipe wall and coils. These elements then react with the iron, causing a rapid loss of wall thickness of these elements.

High-temperature corrosion is manifested by the destruction of the material and the loss of stress properties. This often causes failures related to leaks in pipes and coils and, consequently, the need to replace them, which leads to costs related to boiler downtime [5,6]. Resistance to aggressively corrosive environments is ensured by the modification of the structure and properties of construction materials and the use of protective coatings with the appropriate chemical composition of the alloy. In addition to high mechanical properties, these materials must also show heat resistance, creep resistance and thermal shock resistance, and be characterized by low susceptibility to corrosion at high temperatures and aggressive atmospheres containing, among others, sulphuric compounds, oxygen or water vapour [7]. The main group of materials resistant to high-temperature corrosion are nickel-based alloys. These alloys are characterized by very high wear resistance when used in particularly difficult working conditions. The characteristics of this group of materials are: the ability to work at temperatures up to 1250 °C; limited sensitivity to variable and dynamic loads; and resistance to highly aggressive corrosive environments of gases, sulphur compounds, nitrogen and carbon [8]. Incoloy 625, 686 and 825 superalloys are a popular group of nickel alloys. In these alloys, as in high-alloy austenitic stainless steels, additions of alloying elements such as: Mo, Si, Cu, Ti, Nb and low C content improve creep strength, as well as corrosion resistance and resistance to acids and alkalis [9–11]. According to Rozmus-Górnikowska et al. [12], it is related to the formation of passive oxide layers containing Cr_2O_3 and Fe_3O_4 on the surface of materials. With the increase in the chromium content in nickel alloys, the parabolic course of oxidation is extended due to changes in the oxide morphology, starting from Fe_2O_3/Fe_3O_4 through $Fe_3O_4/(Fe,Cr)_3O_4$ and $Fe_3O_4/(Fe,Cr)_3O_4/Cr_2O_3$ to the layer of pure Cr_2O_3, Figure 1c. The constant change of oxide morphology contributes to better protection of surfaces subjected to corrosion [7,13].

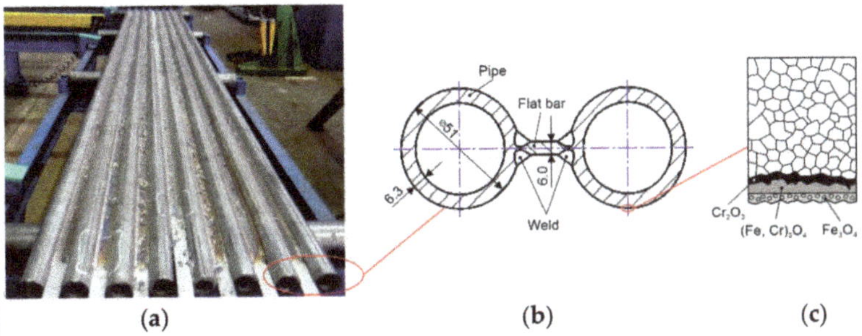

Figure 1. View of the gas-tight wall of a power boiler: (**a**) view of the panel before the surface protection process, (**b**) spectacle joint, (**c**) diagram of the formation of oxides on the surface of fine-grained austenitic steel at elevated temperature.

On the other hand, uninterrupted operation of the power unit is conducive to the development of corrosive processes of material destruction. Corrosion products are characterized by a layered structure with various properties, such as roughness, porosity or adhesion to the substrate. As a result, the oxide layer periodically peels off, which, with constant temperature jumps, affects the oxidation kinetics. Consequently, these processes lead to a thinning of the wall thickness and an increase in stress in the material of the gas-tight pipe wall and coils. The mechanism of the build-up of oxide layers on the surface of pipes is a beneficial phenomenon, as it inhibits further progress of corrosion phenomena. Nevertheless, as a result of the increase in the thickness of the oxide layer, heat exchange deteriorates, which causes an increase in the wall temperature, and thus promotes the degradation of the material structure [13,14].

Most often, the following methods are used to produce protective layers on gas-tight wall pipes and coils of power boilers: MIG (metal inert gas), PPTA (plasma powder transferred arc) or HPDDL (high power direct diode laser) [15–17]. However, these technologies do not always meet the high requirements related to the lack of welding imperfections, minimum depth of the heat affected zone and low iron content in the made cladding layer. Other methods of protecting boiler components include:

- thermal spraying of ceramic coatings, e.g., Al_2O_3, Cr_2O_3, TiO_2, ZrO_2 [18,19];
- use of composite pipes, e.g., 3R12/4L7 (304L/P265GH), Sanicro 38/HT8 (Incoloy 825/10CrMo9-10) and Sanicro 63/HT7 (Incoloy 625/X10CrMoVNb9-1) [20–22].

The use of thermally sprayed ceramic coatings is associated with many technical and economic challenges, such as the large dimensions of the power unit or the need to use a recirculation system to cool the boiler. This solution turned out to be technically complex and very expensive at the stage of implementation and operation. Currently, this type of anti-corrosion protection has been abandoned and can only be found in older power units. The structures of newly built power boilers are based on gas-tight walls and coils made of composite pipes or pipes protected with hardfacing layers [4,20,23].

Composite pipes with an outer layer of austenitic stainless steel, e.g., 3R12/4L7, did not fully meet the hopes placed in them, because various cracks or flaking of the entire layer often occurred in the cladding layer. According to the tests presented in [5,24], the cracks were caused by the difference between the coefficient of thermal expansion for austenitic steel and carbon structural steel, which in the temperature range of 30–500 °C is respectively, $\alpha_{ss} = 18.4 \times 10^{-6}$ 1/K and $\alpha_{cs} = 14.3 \times 10^{-6}$ 1/K, as well as thermal fatigue and, stress corrosion cracking. Sanicro 38/4L7 grade composite pipes with Incoloy 825 ($\alpha_{inc} = 16.2 \times 10^{-6}$ 1/K) superalloy outer layer turned out to be only slightly better than 3R12/4L7 grade composite pipes with 306L austenitic steel layer. In this case, cracks were also found in the plating and at the interface between the two layers [5]. On the other hand, composite pipes of the Sanicro 63/4L7 grade with an outer layer of Incoloy 625 ($\alpha_{inc} = 15.3 \times 10^{-6}$ 1/K) superalloy are characterized by very good properties, but they are not commonly used due to the high cost of production and the impossibility of making high-quality pipes of required lengths [5,13,24].

Welding technologies of applying layers and coatings resistant to corrosion, erosion and high temperatures, implemented on an industrial scale, still require comprehensive research, contributing to the production of high-quality protection of boiler elements that meet the increasingly demanding requirements of the power industry. Problems caused by periodic failures of power boilers that are already in operation require effective service actions using uncomplicated technologies. The high-temperature corrosion resistance tests undertaken were aimed at determining the resistance to high temperature and aggressive exhaust gas environment of coatings flame-sprayed with self-fluxing powders made of Ni-Cr-B-Si alloys and non-fluxing powders made of Ni-B-Si alloys. These alloys are considered to be easily weldable, and their properties are comparable to Incoloy superalloys [9–13,25]. The addition of boron causes the formation of eutectic with a lower melting point than pure metals, which allows the coating to be melted without melting the substrate. The silicon contained in the powder binds all impurities in the form of oxides that float to the surface during melting. In addition, these alloys are strengthened by the precipitation of carbides such as Ni_6C, $Cr_{23}C_6$ and B_4C.

The main aim of the conducted research was to confirm the ability of developed technology of subsonic flame spraying with powders of the Ni-Cr-B-Si and Ni-B-Si type to the production of coatings compliant with the requirements of standards and expectations of the energy industry, as anti-corrosion protection on boiler elements intended for waste disposal and biomass combustion.

2. Materials and Research Methods

The flame powder spraying process was carried out on a production stand equipped with a modern and precise, modular oxy-acetylene system for manual flame powder spraying CastoDyn DS 8000 (Castolin, Gliwice, Poland), Figure 2b. Thermally sprayed coatings were made on a fragment of a gas-tight pipe wall of a powder boiler consisting of a pipe ϕ 51.0 × 6.3 mm welded to flat bars 6.0 mm thick, Figure 1b.

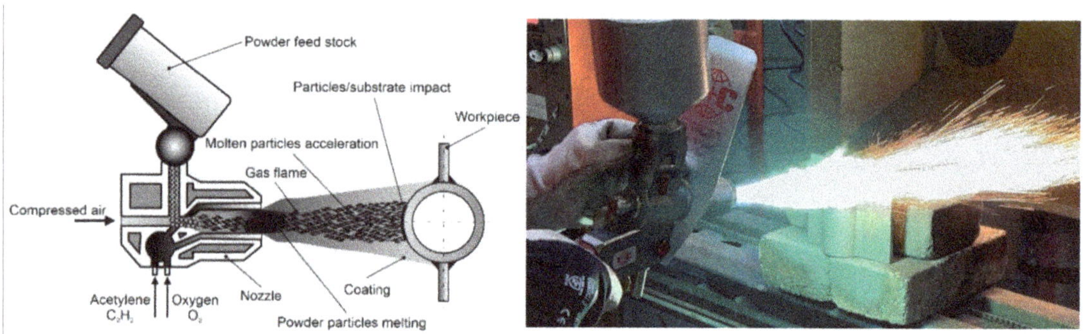

(a) (b)

Figure 2. *Cont.*

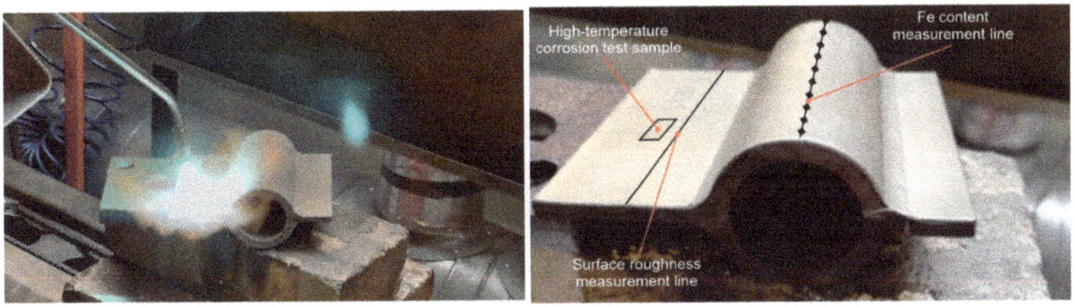

(c) (d)

Figure 2. Flame powder spraying process: (**a**) diagram of a manual flame powder spray burner, (**b**) view of a test of flame powder spraying of a remeltable Ni-Cr-B-Si alloy coating using a CastoDyn DS 8000 burner, (**c**) view of the remelting test of the coating sprayed with a self-fluxing powder made of the remeltable Ni-Cr-B-Si alloy, (**d**) view of the sample with the coating protecting against high-temperature corrosion.

All structural elements were made of low-alloy 16Mo3 (1.5415) boiler steel. Before the spraying process, the surface of the samples, in accordance with the requirements of ISO 2063-1 [26], was subjected to abrasive blasting in a cabin sandblaster, using an abrasive medium in the form of broken cast iron shot with G34 granulation (grain size 0.8–1.2 mm). This procedure was aimed at cleaning the external surface of the samples from impurities such as rust, scale or grease. After this stage of processing, the samples were additionally brushed to remove any traces of grit, and then chemically cleaned with tetrachloroethylene. The final surface roughness of the substrate after shot-blasting was Ra = 12 μm, Rz = 85 μm. As an additional material for flame spraying, two types of self-fluxing metallic powders made of Ni-Cr-B-Si alloys (Eutalloy® RW 17535 and Metco® 15E) were used and, for comparison, two types of non-fluxing metallic powders of the Ni-B-Si type (Eutalloy® BronzoChrom10185 and Eutalloy® NiTec 10224). Chemical composition and physical properties of low-alloy boiler steel according to EN 10273 [27] and powders for spraying according to manufacturers' data are given in Tables 1–5.

Table 1. Chemical composition and physical properties of 16Mo3 (1.5415) low-alloy boiler steel according to EN 10273.

Chemical Composition, wt.%										
C	Mn	Si	P	S	Cr	Mo	Ni	Cu	N	Fe
0.12–0.20	0.4–0.9	<0.35	<0.025	<0.010	<0.3	0.25–0.35	<0.3	<0.3	<0.012	rest
Mechanical properties										
Strength R_m, MPa		Hardness, HB 30			Melting range (liquidus/solidus), °C			Thermal resistance, °C		
440–590		130–170			1460/1420			530		

Table 2. Chemical composition of Ni-Cr-B-Si metal deposit according to the manufacturer's data.

Chemical Composition, wt.%					
C	Cr	Fe	B	Si	Ni
Eutalloy® RW 17535 (Castolin)					
0.8	26	1.0	3.0	3.7	rest
Metco® 15E (Sulzer)					
1.0	17	4.0	3.5	4.0	rest

Table 3. Physical properties of the Ni-Cr-B-Si metal deposit according to the manufacturer's data.

Mechanical Properties of the Coating Weld Metal [(1)]		
Hardness, HV30	Density, g/cm^3	Thermal resistance, °C
	Eutalloy® RW 17535 (Castolin)	
480	7.8-8.0	≤870
	Metco® 15E (Sulzer)	
745	7.8–8.1	≤840

Note: [(1)] alloy in the initial state, not heat-treated, test temperature 20 °C.

Table 4. Chemical composition of the Ni-B-Si metal deposit according to the manufacturer's data.

Chemical Composition, wt.%					
C	Cr	Fe	B	Si	Ni
		Eutalloy® BronzoChrom 10185 (Castolin)			
≤0.1	≤0.5	≤0.5	2.5	3.0	rest
		Eutalloy® NiTec 10224 (Castolin)			
≤0.1	≤0.5	≤0.5	1.5	2.0	rest

Table 5. Physical properties of the Ni-B-Si metal deposit according to the manufacturer's data.

Mechanical Properties of the Coating Metal Deposit [(1)]			
Hardness, HV30	Density, g/cm^3	Melting range (liquidus/solidus), °C	Thermal resistance, °C
	Eutalloy® BronzoChrom 10185 (Castolin)		
390	8.9	1050/1175	≤600
	Eutalloy® NiTec 10224 (Castolin)		
240	8.1	1050/1280	≤600

Note: [(1)] alloy in the initial state, not heat treated, test temperature 20 °C.

Each sample was preheated to 150 °C immediately before the spraying process. The burner was run in a wall-mounted position (PC) covering the entire surface of the protected element, and then the direction of spraying was changed several times by 90 degrees until a coating with a thickness of about 600 μm was obtained. After the spraying process, the coatings made with self-fluxing powders from remeltable alloys of the Ni-Cr-B-Si type were melted by heating with an oxy-acetylene flame of a gas burner to a temperature in the liquidus–solidus range of the alloy used. Coatings sprayed with Ni-B-Si alloys have not been remelted. Spray parameters are shown in Table 6.

Table 6. Parameters of flame-sprayed Ni-Cr-B-Si and Ni-B-Si coatings using CastoDyn DS 8000 torch.

Sample Number	Type of Powder	Oxygen Pressure [bar]	Acetylene Pressure [bar]	Air Pressure [bar]	Number of the Orifice for the Powder
1	Eutalloy® RW 17535	4	0.7	1	4
2	Metco® 15E	4	0.7	1	4
3	Eutalloy® BronzoChrom 10185	4	0.7	2	4
4	Eutalloy® NiTec 10224	4	0.7	2	4

Note: Standard modular nozzles regulating the flame outlet (SSM 20) were used. Slightly carburizing flame was used. The spraying distance was 200 mm.

Visual tests (VT) and penetration tests (PT) of coatings sprayed with self-fluxing powders made of Ni-Cr-B-Si alloys and powders made of Ni-B-Si non-fluxing alloys, carried out in accordance with the requirements of ISO 17637:2017 [28] and ISO 3452:2021 [29], did not show any non-conformity of coating surface and shape. No cracks, porosity or lack of adhesion to the substrate were found in the coatings. The established parameters of thermal spraying ensured the production of coatings with an acceptable level of quality that met the requirements of class "B" according to ISO 5817:2014 [30]. On the basis of preliminary tests, the influence of spraying parameters and the type of metal powder used on the depth of the heat affected zone (HAZ) "h", the degree of porosity of the coating "p" and the roughness of the coating surface "r" were determined. In addition, the following parameters were also determined: the thickness of the sprayed coating "g" and the iron content on the surface of the sprayed coating "Fe", Table 7.

The assessment of the surface microgeometry of the thermally sprayed coatings was performed using a surface roughness gauge SRT-200 (Sunpoc, Guiyang City, China). Surface roughness measurements were made on the length of the elementary section of 2.5 mm. Microscopic examinations were carried out on standardly prepared metallographic specimens. The samples to be observed with a light microscope were etched in two stages: the native material of the steel pipe was revealed using $FeCl_3$Et (Mi19Fe) solution, and the structure of the nickel alloy coatings was revealed by electrochemical etching with the following reagent: 20 cm^3 HCl, 10 g $FeCl_3$, 30 cm^3 of distilled water. Observation and recording of microstructure images were carried out using an Olympus SZX7 stereoscopic microscope (Olympus Corporation, Tokyo, Japan) and an Olympus GX 71 inverted metallographic microscope (Olympus Corporation, Tokyo, Japan). The porosity of the surface area fraction of the sprayed coatings was measured by computer image analysis using the MicroScan computer program for Windows.

The prerequisite for the recognition of the technology of corrosion protection of pipes used in the construction of a heat exchanger operating in biomass incineration plants is the assessment of the iron content on the surface of the thermally sprayed coating. In their recommendations, certification bodies require that the iron content in the surface layer does not exceed 7% [31]. The XRF X-MET8000 Expert portable spectrometer (Hitachi High-Technologies Corporation, Tokyo, Japan), commonly used in industrial conditions, was used to analyse the iron content on the surface of coatings sprayed with nickel alloys selected for testing. The tests were carried out at 10 points spaced 10 mm apart, Figure 2d. Sample results of microscopic examinations are shown in Figure 3, and all measurements of characteristic sizes are summarized in Table 7.

(a) (b)

Figure 3. *Cont.*

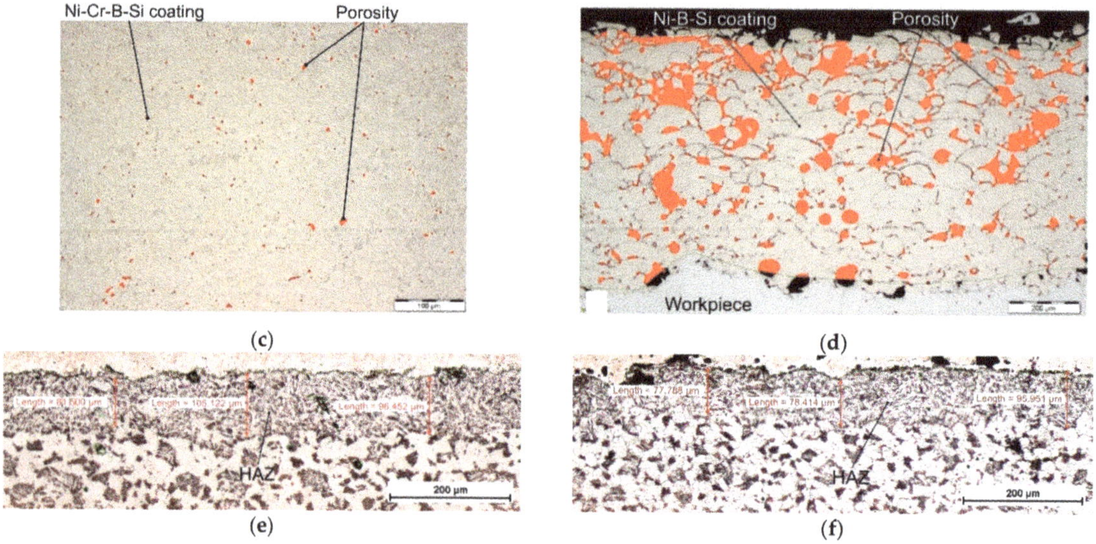

Figure 3. View of the microstructure of the cross-section of coatings sprayed with self-fluxing powder made of Ni-Cr-B-Si alloy (Eutalloy® RW 17535) and powder made of Ni-B-Si non-fluxing alloy (Eutalloy® BronzoChrom 10185) and base material of low-alloy boiler steel of the 16Mo3 grade: (**a**) microstructure of the Ni-Cr-B-Si alloy shell, (**b**) microstructure of the Ni-B-Si alloy shell, (**c**) image of porosity of the inner shell of Ni-Cr-B-Si alloy, (**d**) image of porosity of the inner shell made of Ni-B-Si alloy, (**e**) image of HAZ microstructure and base material under Ni-Cr-B-Si alloy coating, (**f**) image of HAZ microstructure and base material under Ni-B-Si alloy coating.

Table 7. Geometric and metallurgical properties of sprayed coatings.

Sample Number	Type of Coating	Geometrical Properties		Metallurgical Properties		
		g, μm	Ra, μm	h, μm	p, %	Fe, wt.%
1	Eutalloy® RW 17535	600	2.03	95	1.7	0.9
2	Metco® 15E	628	2.48	131	2.6	3.7
3	Eutalloy® BronzoChrom 10185	661	11.54	83	9.6	0.4
4	Eutalloy® NiTec 10224	634	10.93	74	8.3	0.3

Note: Average values are given in the table.

High-temperature corrosion resistance tests of the sprayed coatings were carried out on samples with an area of 250 mm^2 (25 mm × 10 mm) cut from the central part of the flat bar of the spectacle joint of a fragment of the gas-tight pipe wall, Figure 2b. Native material (steel 16Mo3) was removed from the samples taken before the corrosion test. The assessment of corrosion resistance concerned only coatings sprayed with metallic powders from Ni-Cr-B-Si and Ni-B-Si alloys. For comparative purposes, tests of high-temperature corrosion resistance of 16Mo3 steel (1.5415) were also carried out independently, from which the elements of the gas-tight pipe wall of the power boiler were made.

The diagram of the test stand consisting of tube furnaces with built-in quartz and ceramic tubes, reducers, gas intake points with rotameters, a neutralizing system and cylinders with gases included in the simulated gas mixture is shown in Figure 4.

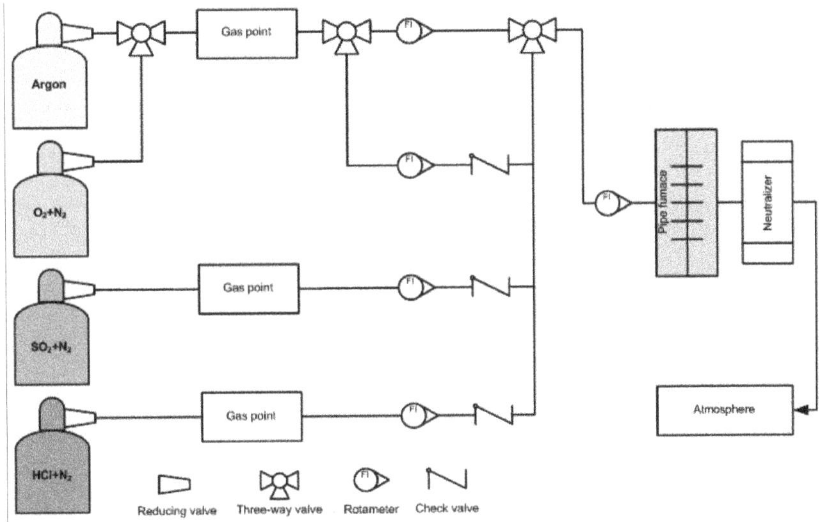

Figure 4. Scheme of equipment for corrosion tests.

The tested samples were placed in Al_2O_3 refractory crucibles and closed in the test chamber of the tube furnace. A mixture of gases with the chemical composition: N_2 + 9.0% O_2 + 0.08% SO_2 + 0.15% HCl was used as the working atmosphere. The chemical composition of the gas mixture was similar to the atmosphere present in power boilers. The tests were carried out at a temperature of 800 °C. The experiment procedure was divided into three stages:

- heating the furnace charge in an argon atmosphere up to the set temperature (800 °C) in order to avoid the oxidation of samples;
- withstanding of the samples in a corrosive atmosphere with the given concentrations of gases at a given temperature and the flow of the gas mixture ensuring a single exchange of the atmosphere within 4 h;
- cooling of the furnace charge in an argon atmosphere, up to 20 °C.

Corrosion resistance tests were carried out for up to 1000 h, with measurements of the mass increase of corrosion products every 250 h. Three weight measurements were made for each of the samples, which were averaged. The percentage relative standard deviation (RSD%) for samples with corrosion protective coatings was $\leq$1%. The change in mass (Δm) of the sample after the tests was adopted as a measure of the resistance to high-temperature corrosion of the coatings. The samples were weighed on an AS 60/220.R2 PLUS analytical balance (Radwag, Radom, Poland) with an accuracy of 0.0001 g.

Studies of the corrosion mechanisms of thermally sprayed coatings after corrosion tests were carried out in a scanning electron microscope Zeiss Supra 25 (Carl Zeiss AG, Oberkochen, Germany) using the EDS method—energy dispersive spectrometry.

X-ray diffraction tests enabling the phase analysis of coatings were performed using a Panalytical X'Pert Pro MPD diffractometer (Malvern Panalytical Ltd., Malvern, UK), using filtered radiation (Kβ Fe filter) of a lamp with a cobalt anode (λKα = 0.179 nm). The diffractograms were recorded in the Bragg—Brentano geometry, using a PIXcell 3D detector on the axis of the diffracted beam, in the range of angles 20–110 [2θ] (step = 0.05°, counting time per step = 100 s). The obtained diffraction patterns were analysed using the dedicated Panalytical High Score Plus software together with the PAN-ICSD structural database. The tests of chemical and phase composition made it possible to define the types of corrosion products formed after the corrosion tests.

Electrochemical tests on corrosion resistance were also performed on all tested coatings. Electrochemical study of corrosion resistance was conducted in accordance with the standard ISO 17475:2010 [32]. Corrosion resistance of the material was examined by Atlas 0531 EU (Atlas-Sollich ZSE, Rębiechowo, Poland) unit potentiostat/galvanostat and results were generated by the AtlasCorr05 software. The base steel for corrosion examinations was cut into 10 mm × 10 mm plates. Before starting the corrosion tests, specimens were cleaned in ethanol and dried in air. All electrochemical tests were performed in 3.5% NaCl solution, which was freshly prepared prior to the testing.

A typical three-electrode system was employed in this test set-up where the Ag/AgCl electrode was used as the reference electrode, the tested steel was the anode (0.384 cm^2 exposed area) and stainless steel was the counter electrode for current measurements. The potential was changed in the anodic direction at the rate of 1 $m \cdot V \cdot s^{-1}$. All the electrochemical tests were performed at a constant room temperature.

3. Results and Discussion

3.1. Flame Spraying with Nickel-Based Powders

Technological tests of anti-corrosion protection on a fragment of a gas-tight pipe wall were carried out by thermal spraying with the use of self-fluxing powders made of remeltable Ni-Cr-B-Si alloys and powders of non-fluxing Ni-B-Si alloys. These types of tubular walls are used for the production of heat exchangers in waste incineration boilers. Thermally sprayed coatings with nickel-based alloys significantly increase the resistance of pipes to corrosion and erosion [33–35]. They can be manufactured on new tubular structures or in the event of the need to repair elements of a panel damaged during operation. In many studies, the authors pointed to technological difficulties in the process of surfacing and thermal spraying with nickel-based alloys. Most often, they resulted from incorrectly selected padding parameters, which may cause an increase in the Fe content in the pad layer above the required value of 7% [12,31] and unacceptable quality and adhesion of thermally sprayed coatings.

The Fe content in the overly layer is exponentially dependent on the amount of heat supplied to the base material. During arc surfacing, the requirement to obtain the desired value below 7% Fe in the overlay requires limiting the line energy of the surfacing to approx. 3 kJ/cm [15,17]. The increase in the value of the linear energy of the surfacing also affects the increase in the thickness of the nickel oxide (NiO) layer, the presence of which on the surface of the gas-tight pipe wall is unacceptable. After the surfacing process, these oxides must be removed from the surface of the pipe wall. In the case of thermally sprayed coatings, the critical parameters are proper surface preparation before spraying and ensuring the diffusive nature of the coating-substrate connection.

In the proposed technology of flame powder spraying of coatings with a thickness of about 600 μm, such difficulties do not occur. No external inconsistencies were found in all the coatings made. The average value of the Ra roughness parameter for coatings made of Ni-Cr-B-Si alloys ranged from 2.03 μm to 2.48 μm, and for coatings made of Ni-B-Si alloys it ranged from 10.93 μm to 11.54 μm. These results are consistent with the data presented in [34,35]. A significant internal inconsistency of the coatings made of Ni-B-Si alloys is the porosity of the coatings, which reached 9.6%. According to Czupryński et al. [35], it is possible to flame spray powders made of Ni-B-Si non-meltable alloy with porosity even below 6%. In the case of coatings flame sprayed with self-fluxing powders made of remeltable Ni-Cr-B-Si alloys, the degree of porosity of the coatings can be significantly reduced by remelting the coating in the liquidus–solidus temperature range of the alloy used [34].

The obtained test results confirmed that it is possible to produce coatings from Ni-Cr-B-Si alloys with a porosity of 1.7%, Table 7. The coatings were characterized by proper adhesion, indicating the mechanical-diffusion nature of the connection with the substrate. The smallest amount of intermediate phases in the coating was observed in the case of coatings made of Ni-Cr-B-Si alloys. Intermediate phases were located in the metal deposit at the line of connection of the coating with the base material. In each case, a clearly separated heat affected zone (HAZ) could be observed in the native material. A maximum HAZ depth of more than 105 μm was recorded for a coating sprayed with Eutalloy® RW 17535 powder (Ni-Cr-B-Si) and a minimum HAZ depth of approximately 77 μm for a coating sprayed with Eutalloy® BronzoChrom 10185 powder (Ni-B-Si). The increase in HAZ depth in samples with coatings of remeltable Ni-Cr-B-Si alloys resulted from the additional heat supplied to the base material in the remelting process. The absence of the base material in the coating provides better control of the Fe content in the metal deposit. The highest average Fe content = 3.7% was determined on the surface of the coating sprayed with Metco® 15 remeltable alloy powder. The iron content on the surface of the other coatings was less than 1%, Table 7.

It was found that the factors determining the quality of coatings flame-sprayed with powders based on nickel alloys on the surface of elements of a gas-tight pipe wall are:

- proper preparation and cleaning of the substrate surface before spraying in order to obtain good adhesion of the coating,
- limiting the amount of heat supplied to the base material by maintaining the proper distance between the burner and the sprayed surface and the use of a slightly carburizing flame ($\lambda = 1.2$), ensuring low content of Fe and NiO oxides on the surface of the coating,
- the use of remelting the sprayed coating by heating it with an oxy-acetylene flame of a gas burner to a temperature in the liquidus–solidus range of the alloy used in order to minimize the internal porosity of the metal deposit and obtain the appropriate smoothness of the coating surface.

It was also shown that there is no need for forced internal cooling of the heat exchanger tube in order to limit the depth of the diffusion zone and the heat affected zone in the base material and to ensure the appropriate quality of the sprayed coatings.

3.2. Evaluation of High-Temperature Corrosion Resistance of Flame-Sprayed Coatings Made of Ni-Cr-B-Si and Ni-B-Si Powders

The results of testing the high-temperature corrosion resistance of coatings sprayed with flame-sprayed powders made of Ni-Cr-B-Si and Ni-B-Si alloys selected for testing, as well as the construction material from which the elements of the gas-tight pipe wall were made, i.e., 16Mo3 (1.5415) steel, are presented in Table 8. The course of changes in the mass of samples with anti-corrosion coatings is also illustrated in Figures 5 and 6. The curves in the graph show changes in the mass of the sample per unit area as a function of the duration of the corrosion test.

Table 8. Change in the weight of samples protected with protective coatings and the base material as a function of exposure to a corrosive atmosphere at a temperature of 800 °C.

Time, T [h]	Mass Change Δm, [mg]														
	Eutalloy® RW 17535			Metco® 15E			Eutalloy® BronzoChrom 10185			Eutalloy® NiTec 10224			Steel 16Mo3 (1.5415)		
	$\bar{x}$	S(x)	RSD%	$\bar{x}$	S(x)	RSD%	$\bar{x}$	S(x)	RSD%	$\bar{x}$	S(x)	RSD%	$\bar{x}$	S(x)	RSD%
0	0	0	0	0	0	0	0	0	0	0	0	0	0	0	0
250	0.012	0.17	0.29	0.010	0.20	0.33	0.021	0.30	0.50	0.018	0.20	0.33	1.559	0.40	0.66
500	0.024	0.20	0.33	0.026	0.10	0.17	0.048	0.30	0.51	0.044	0.30	0.50	1.946	0.50	0.83
750	0.034	0.35	0.58	0.038	0.56	0.93	0.068	0.40	0.66	0.062	0.40	0.66	3.208	0.70	1.17
1000	0.041	0.30	0.50	0.046	0.30	0.50	0.082	0.60	1.00	0.074	0.50	0.83	4.288	1.2	2.0

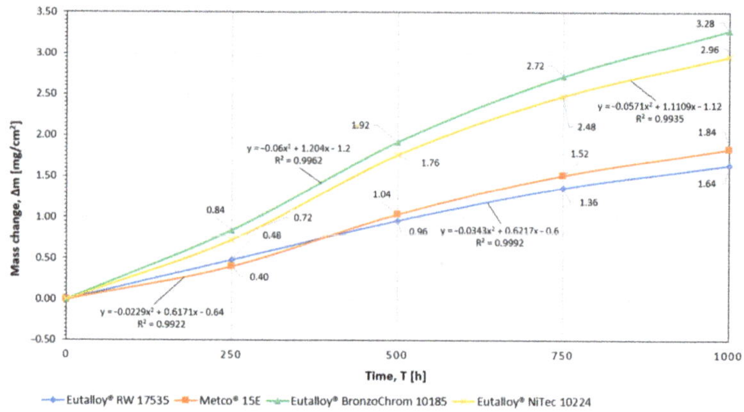

Figure 5. Graph of the unit mass change of samples protected with protective coatings as a function of exposure to a corrosive atmosphere.

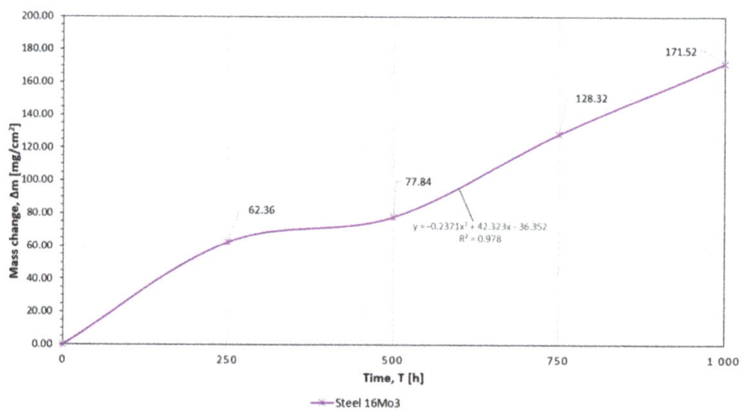

Figure 6. Graph of the unit weight change of 16Mo3 (1.5415) steel base material samples as a function of exposure to corrosive atmosphere.

The conducted analysis of the course of the corrosion process, based on the empirical relationship between the scale mass increase or metal loss and the time of the oxidation reaction, indicates the parabolic nature of the sample mass changes during the course of the experiment. Flame-sprayed coatings with self-fluxing powders made of remeltable alloys Ni-Cr-B-Si of the Eutalloy® RW 17535 and Metco® 15E grades showed a negligible tendency to corrosive wear, Figure 5. The corrosion rate of these coatings under the conditions of the experiment was 0.0016 ((mg/cm^2)/h) and 0.0018 ((mg/cm^2)/h), Figure 7. These results indicate that the Fe content in the coating sprayed with Metco® 15E powder of about 4% does not cause the formation of a large amount of iron oxide Fe_2O_3 on the surface of the coating, which has much lower adhesion to the substrate than chromium oxide Cr_2O_3, which in the conditions of corrosive and erosive wear could intensify the processes of coating degradation. Coatings flame-sprayed with powders made of Ni-B-Si non-meltable alloys of Eutalloy® BronzoChrom 10185 and Eutalloy® NiTec 10224 alloys, containing about 0.5% Fe, performed slightly worse in terms of corrosive wear. The corrosion rate of these coatings in given temperature-time conditions was, respectively, 0.0033 (mg/cm^2)/h) and 0.0030 (mg/cm^2)/h), Figure 7. However, in the case of these coatings, it can be seen that the mass increase of corrosion products is the most increased in the time interval of 250–500 h. After this period, the corrosion processes progressed more slowly. In addition, it was observed that after 500 h of exposure to high temperature and corrosive atmosphere, the wear resistance of coatings made of non-meltable alloys decreased, which was manifested by strong overheating of the coating structure. On the other hand, the unprotected construction material (16Mo3 steel) was characterized by the highest corrosion rate, which after 1000 h of the test was 0.1715 ($(mg/cm^2)/h$), Figure 6.

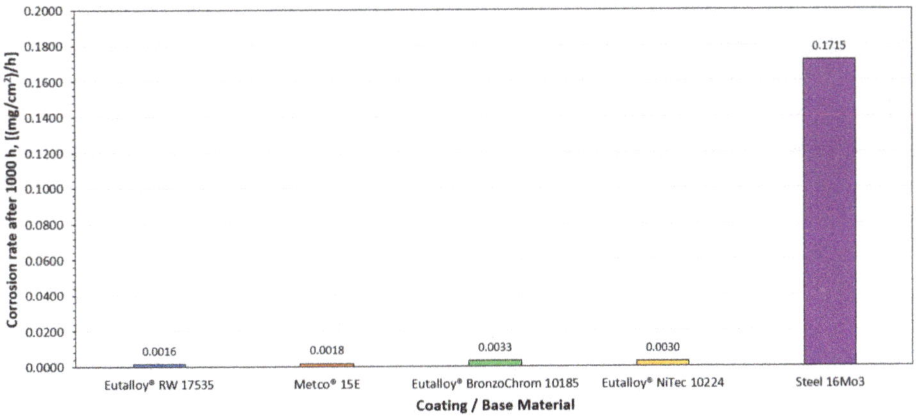

Figure 7. Graph of corrosion rate of protective coatings and base material exposed to corrosive atmosphere at 800 °C for 1000 h.

The general results of susceptibility to high-temperature corrosion for the adopted temperature-time conditions indicate a higher corrosion resistance of coatings made of Ni-Cr-B-Si alloys than coatings of Ni-Si-B alloys, which was slightly affected by the Fe content up to 4%, Figures 5 and 7.

The high-temperature corrosion resistance tests were supplemented by metallographic tests. Exemplary surfaces of samples after corrosion tests and the surface of the joint observed in the scanning electron microscope are shown in Figure 8. The tests were carried out in the technique of recording secondary electrons (SE) and backscattered secondary electrons (BSE). The SE images show the surface topography, while the BSE images inform about the varied chemical composition. In further studies, the corrosion products were identified by microanalysis of the EDS chemical composition (Figure 9) and the identification of the phase composition by XRD (Figure 10).

Microscopic examinations on cross-sections of corroded samples after 1000 h showed that in the case of a 16Mo3 steel sample, the thickness of the oxide layer, mainly Fe_3O_4, reaches even 1.25 mm.

On the basis of SEM microscopic observations and tests of the chemical composition of the surface of coatings sprayed with self-fluxing powders made of Ni-Cr-B-Si alloys after 1000 h of corrosion tests, the presence of places with high Cr content was found, Figure 9a,b. The analysis of the phase composition showed that the main corrosion product of all alloy sprayed samples was Cr_2O_3 chromium oxide. On the surface of the sample containing about 4% wt. of Fe, sites with a relatively high iron content were found, which would indicate the presence of this element oxides, Figure 9b. This was confirmed by the study of the phase composition, Figure 10. On the other hand, the high content of silicon in the layer of corrosion products reaching even more than 15% wt. is presumably related to the furnace lining.

The kinetics and mechanism of nickel oxidation are the subject of numerous studies due to the formation of NiO oxide on the surface of this metal, which, due to its transport properties, is treated as a model semiconductor. Despite the huge number of works devoted to this subject, there is still no full agreement in the literature regarding both the mechanism of high-temperature corrosion of nickel and the structure of nickel oxide defects. Some authors assume that in the process of oxide scale growth on nickel at temperatures lower than 1000 °C, apart from the core network diffusion of cations, the core diffusion of oxygen across grain boundaries is involved [36]. At higher temperatures, it is assumed that the core diffusion of the metal plays a decisive role, but there is no agreement as to the degree of ionization of defects [37].

In the case of both coatings flame sprayed with Ni-B-Si non-meltable alloy powders, after 1000 h of corrosion tests carried out at 800 °C and in a highly corrosive gas atmosphere N_2 + 9.0% O_2 + 0.08% SO_2 + 0.15% HCl, the oxidation process can be described with the kinetic parabolic law. The oxide scale formed on the surface of the coating was mainly composed of the NiO phase, Figure 10c,d. Nickel oxide is a p-type semiconductor with a metal deficiency, so its growth was due to core diffusion of Ni ions. The incorporation into the NiO lattice of cations with a valency higher than the valence of nickel ions causes an increase in the vacancy concentration in the cation sublattice, and thus the rate of oxidation increases. The presence of monovalent metals improves the protective properties of the scale, e.g., the presence of metals such as chromium or manganese increases the rate of nickel oxidation. According to Rivoaland et al. [36], each alloy addition has a certain concentration limit value at which, under certain reaction conditions, a saturated MeO–NiO solid solution is formed, and then the scale is two- or multi-phase. Most often, however, there is a two-layer scale, which consists of an outer layer made of the NiO phase and an inner layer consisting of a mixture of both MeO + NiO alloy components. If the alloying element has a greater affinity with oxygen than nickel, then a protective, single-phase scale layer composed only of the alloying element oxide may form on the surface of the alloy. Such additives may be aluminium, chromium and silicon.

In addition, nickel borides Ni_3B, which are characterized by high hardness and thermal resistance, were identified on the surface of the coatings. The melting point of Ni_3B nickel borides is 1478 °C, and their hardness is in the range of 10.49–10.75 GPa. According to González et al. [38], the presence of boron in the nickel coating reduces the diffusion movement of atoms and vacancies in the crystal lattice, increases the resistance to plastic deformation and reduces the size of crystallites during electro-crystallization and their growth during recrystallization, and also strengthens the nickel matrix.

On the surface of 16Mo3 grade steel, the results of the analysis of chemical and phase composition confirmed the presence of mainly Fe_3O_4 iron oxide with a small amount of Cr_2O_3 chromium oxide.

Examination of corrosion process products produced on coatings sprayed with self-fluxing powders from remeltable Ni-Cr-B-Si alloys showed their laminar morphology. On the surface of coatings sprayed with Eutalloy® RW 17535 and Metco® 15E powders, a Cr_2O_3 oxide film is formed, which is a barrier to oxygen access and significantly reduces the intensity of corrosion. The presence of Fe in sprayed coatings can lead to the formation of Fe_3O_4 oxide, the adhesion of which is much lower. In the iron oxide layer, unfavourable cracks are observed, which cause flaking of the layer, and thus a decrease in the corrosion resistance of the coating, Figure 8e,f. Therefore, limiting the Fe content in powders for flame spraying of coatings protecting against high-temperature corrosion in the range of 0.5–4% is reasonable, although the results of corrosion kinetics studies would indicate the possibility of increasing the Fe content up to about 7%.

The analysis of the obtained results indicates a high corrosion resistance of coatings sprayed with self-fluxing powders made of Ni-Cr-B-Si alloys in a gas atmosphere simulating flue gases in coal-fired boilers. A uniform layer of corrosion products was observed on the surface of all samples. This layer causes a slight reduction in heat exchange between the flue gases and the 16Mo3 steel sample material, at the same time it also protects well against further corrosion progress. The results of the microanalysis of the chemical composition of gas corrosion products indicate that they are rich in chromium, nickel and oxygen, which probably indicates the presence of a layer of chromium oxide Cr_2O_3 on the surface of the samples. Its presence protects samples against further oxidation, ensuring adequate corrosion resistance at high temperatures. This is advantageous due to the fact that the

Cr_2O_3 layer is continuous and adheres well to the coating, and under high-temperature corrosion conditions it quickly rebuilds, which ensures permanent corrosion protection. The analysis of changes in the weight of the samples during the experiment also confirms that a Cr_2O_3 film is formed on the surface, which covers the surface and hinders the access of the corrosive atmosphere to the surface. This is indicated by the initial increase in the mass of the samples (for 750 h) and then the slowdown of the growth (up to 1000 h), which confirms the formation of a continuous and tight layer of chromium oxide (Figure 5).

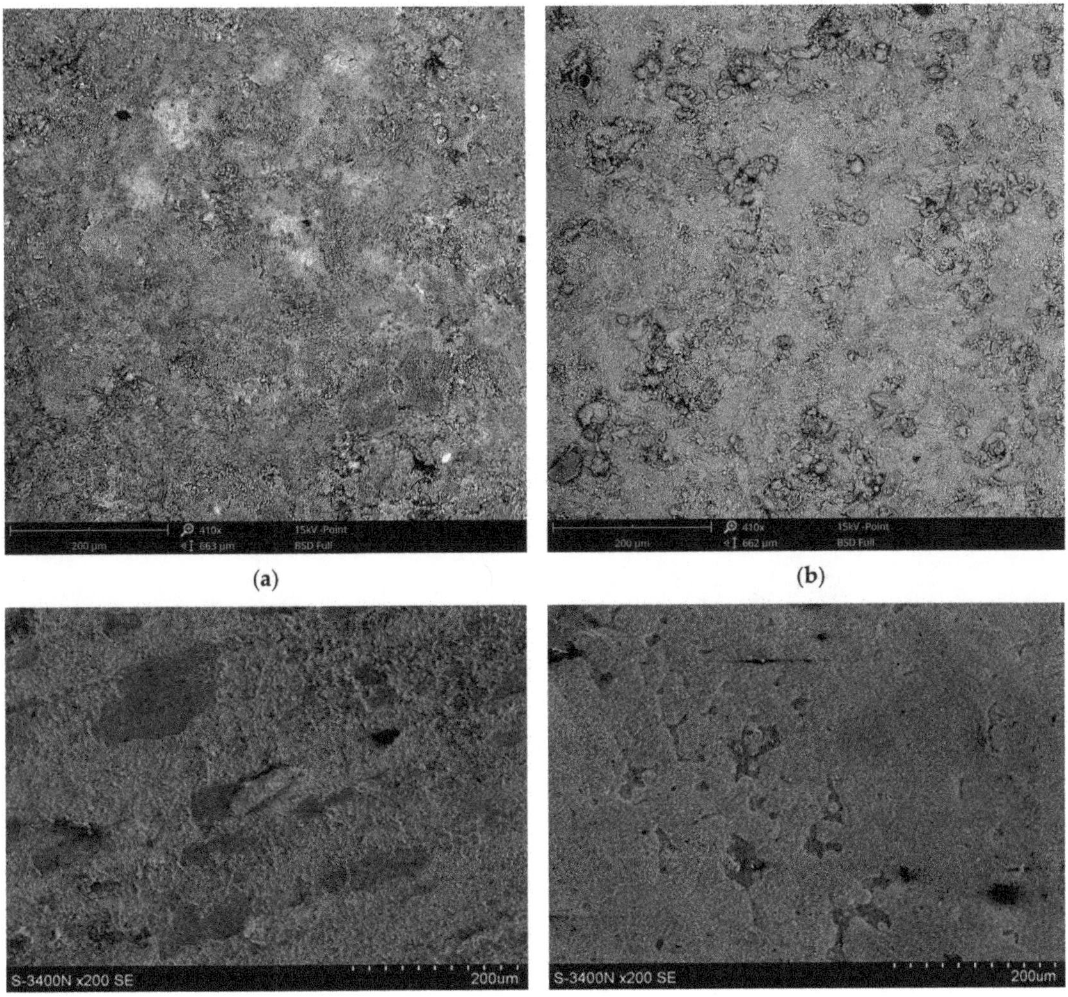

Figure 8. *Cont.*

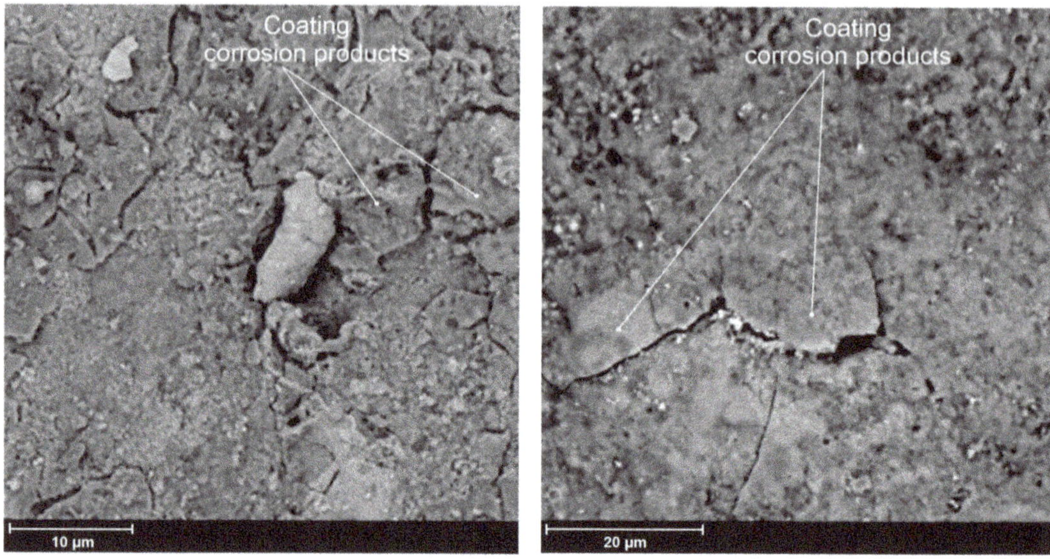

(e) (f)

Figure 8. Surface morphology after high-temperature corrosion resistance tests at 800 °C for 1000 h: (**a**) flame-sprayed coating with Eutalloy® RW 17535 powder, (**b**) flame-sprayed coating with Metco® 15E powder, (**c**) flame-sprayed coating with Eutalloy® BronzoChrom 10185, (**d**) flame-sprayed coating with 10185 Eutalloy® NiTec 10224 powder, (**e**) cracks in flame-sprayed coating with Eutalloy® RW 17535 powder, (**f**) cracks in flame-sprayed coating with Metco® 15E powder.

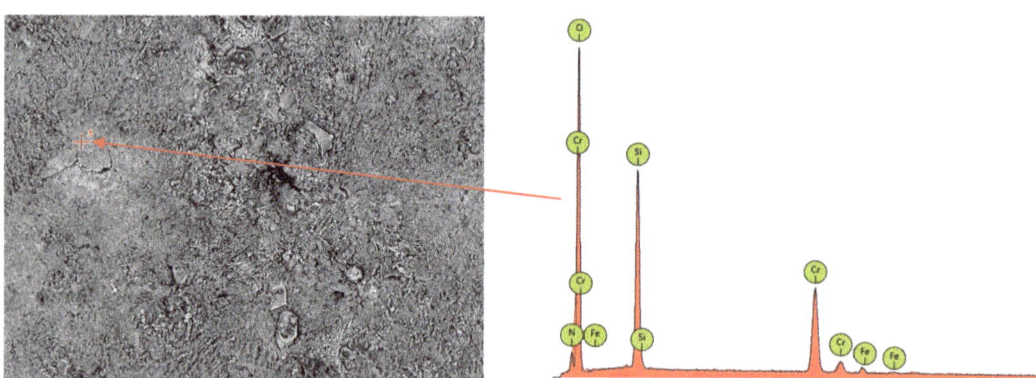

Percent	O	Cr	Si	N	Fe
Weight (%)	50.27	26.58	15.44	6.86	0.84
Atom (%)	66.74	10.86	11.68	10.40	0.32

(a)

Figure 9. *Cont.*

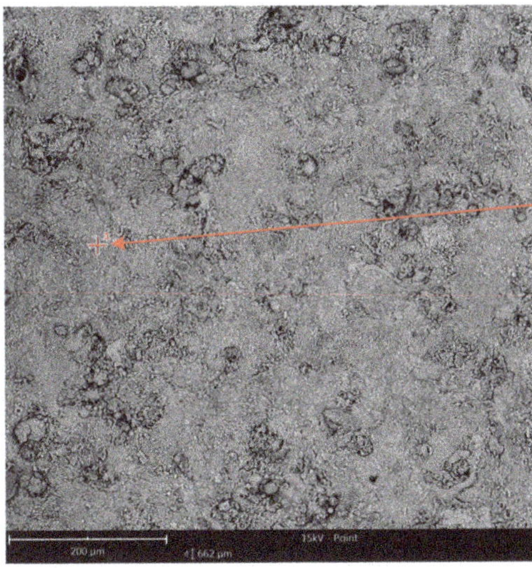

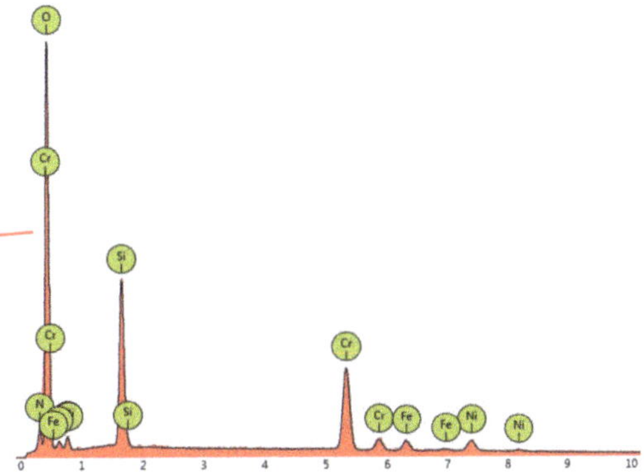

Percent	O	Cr	Si	Ni	N	Fe
Weight (%)	49.44	21.52	11.74	7.27	5.84	4.19
Atom (%)	68.10	9.12	9.21	2.73	9.19	1.65

(b)

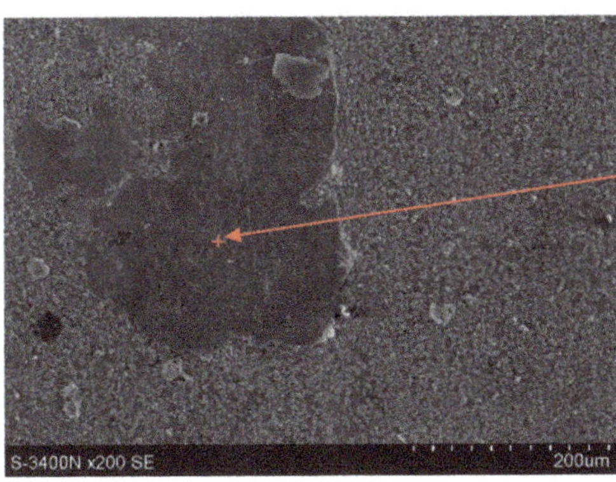

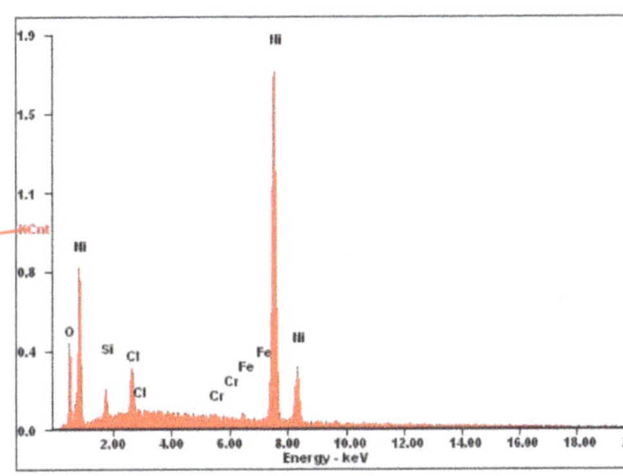

Percent	O	Si	Cl	Cr	Fe	Ni
Weight (%)	10.55	2.89	3.66	0.93	0.80	81.17
Atom (%)	28.92	4.51	4.53	0.78	0.63	60.63

(c)

Figure 9. *Cont.*

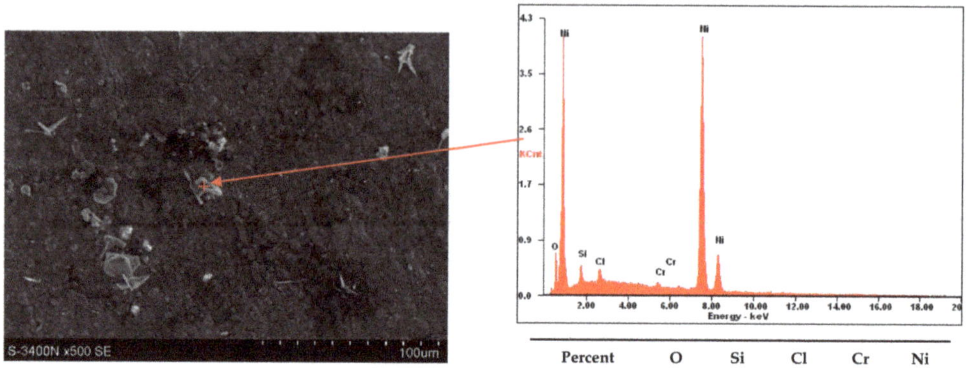

Percent	O	Si	Cl	Cr	Ni
Weight (%)	9.38	3.02	1.55	0.85	84.44
Atom (%)	26.58	4.88	1.98	0.74	65.19

(**d**)

Figure 9. Results of microanalysis of the chemical composition of corrosion products after high-temperature corrosion resistance tests at 800 °C for 1000 h of flame-sprayed coatings: (**a**) Eutalloy® RW 17535, (**b**) Metco® 15E, (**c**) Eutalloy® BronzoChrom 10185, (**d**) Eutalloy® NiTec 10224.

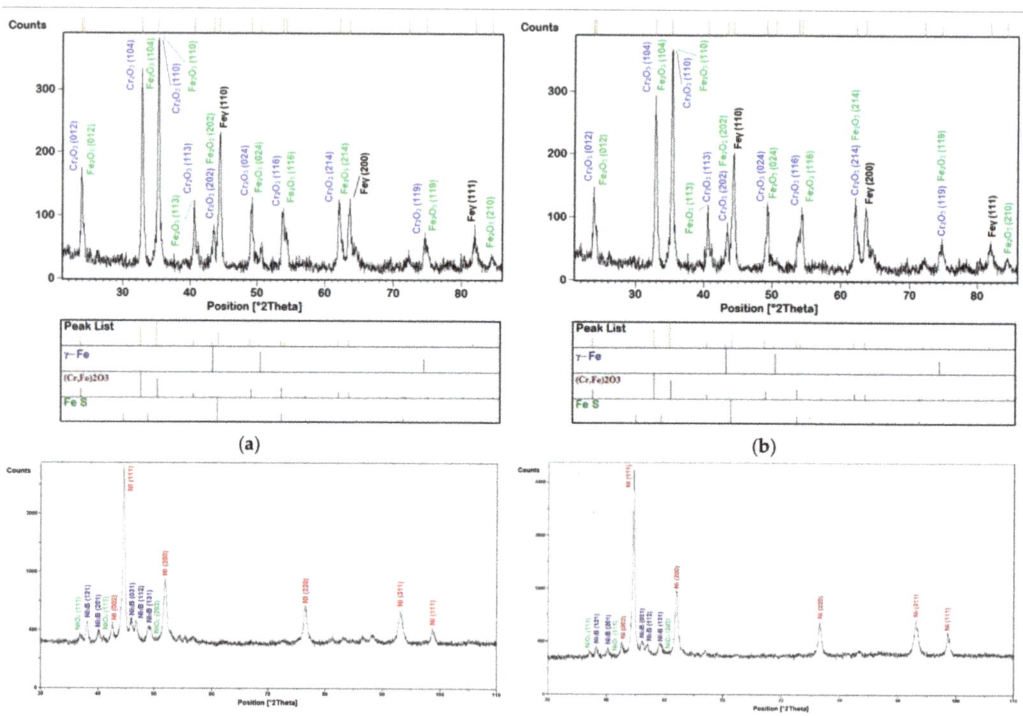

(**a**) (**b**)

(**c**) (**d**)

Figure 10. Results of analysis of the phase composition of corrosion products after high-temperature corrosion resistance tests at 800 °C for 1000 h of flame-sprayed coatings: (**a**) Eutalloy® RW 17535, (**b**) Metco® 15E, (**c**) Eutalloy® BronzoChrom 10185, (**d**) Eutalloy® NiTec 10224.

The results of potentiodynamic experiments conducted in water solution 3.5% NaCl are shown in Table 9 and the course of changes of current density as a function of potential are presented in Figure 11.

Table 9. Results of potentiodynamic tests for examined specimens.

Material	I_{cor}, [A/cm^2]	E_{cor}, [mV]	R_p, [k$\Omega \times$ cm^2]
Eutalloy® RW 17535	283×10^{-9}	−376	110
Metco® 15E	674×10^{-9}	−481	41.91
Eutalloy® BronzoChrom 10185	3.285×10^{-6}	−607	5.39
Eutalloy® NiTec 10224	2.586×10^{-6}	−619	6.06

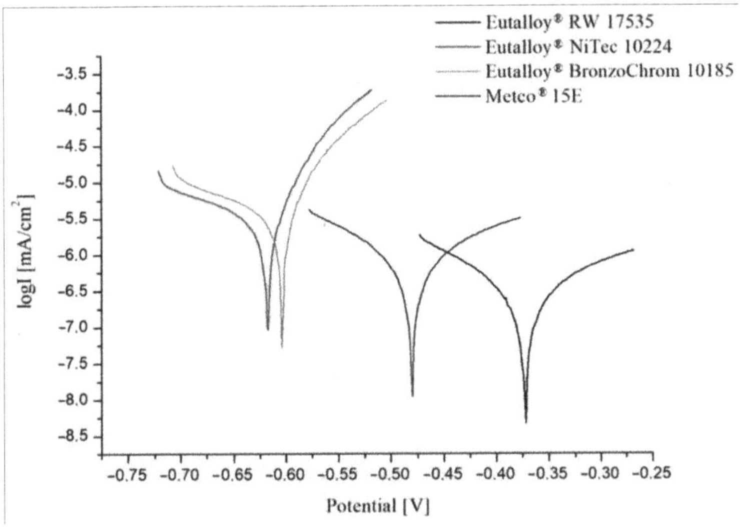

Figure 11. Potentiodynamic polarization curves of analyzed specimens.

In each case, a characteristic course of the polarization curve is observed. It was found that the analyzed samples had varied resistance to the corrosion.

In terms of the polarization resistance Rp and corrosion current density Icor, Eutalloy® RW 17535 seems to have a higher degree of corrosion resistance than other tested alloys due to the high content of chromium, which determines good corrosion resistance in the aggressive environment of chloride ions. However, Metco® 15E alloy display also excellent corrosion resistance properties ($I_{cor} = 674 \times 10^{-9}$ A/cm^2, $R_p = 41.91$ k$\Omega \times$ cm^2).

Alloys Eutalloy® NiTec 10224 and Eutalloy® BronzoChrom 10185 showed similar the lowest corrosion resistance ($I_{cor} = 2.586 \times 10^{-6}$ A/cm^2, $R_p = 6.06$ k$\Omega \times$ cm^2 and $I_{cor} = 3.285 \times 10^{-6}$ A/cm^2, $R_p = 5.39$ k$\Omega \times$ cm^2 respectivly).

The tests have shown that use of flame-sprayed coatings with self-fluxing powders made of remeltable Ni-Cr-B-Si alloys on the surfaces of gas-tight pipe wall panel elements and coils can effectively protect them against high-temperature corrosion and extend the lifetime of the power installation several times.

Compared to previously used surfacing technologies and the main advantage of flame spraying with self-fluxing powders from meltable alloys is the high quality of the coatings and the lack of substrate material in the metal deposit. In addition, this technology makes it possible to spray complex shapes even in hard-to-reach places. Further research will be devoted to the comparison of high temperature corrosion resistance of coatings sprayed with FPS and HVOF powders made of Ni-Cr-B-Si alloys.

4. Conclusions

Based on the conducted research and analysis of the results, the following conclusions were formulated:

1. During flame powder spraying of boiler elements with nickel alloys, the amount of heat supplied to the base material should be limited by maintaining the proper distance between the burner and the sprayed surface and using a slightly carburizing flame ($\lambda = 1.2$), ensuring low Fe content on the coating surface and avoiding the need to the use of internal cooling of the element to limit overheating of the base material.
2. Coatings flame sprayed with self-fluxing powders from remeltable Ni-Cr-B-Si alloys and non-fluxing powders from Ni-B-Si alloys are resistant to high-temperature corrosion in the exhaust gas atmosphere (N_2 + 9.0% O_2 + 0.08% SO_2 + 0.15% HCl) typical for power boiler installations. At temperatures up to 800 °C, high-temperature corrosion in the range from 0 to 1000 h runs in a parabolic manner. Coatings made of Ni-Cr-B-Si alloys show about 50% higher resistance to high-temperature corrosion than coatings made of Ni-B-Si alloys and much lower susceptibility to corrosion than low-alloy steel intended for boiler structures. The corrosion rate of coatings made of Ni-Cr-B-Si alloys in the conditions of the experiment was 0.0030–0.0033 (mg/cm^2)/h and was 52–57 times lower than the corrosion rate of low-alloy 16Mo3 (1.5415) boiler steel.
3. The results of potentiodynamic tests conducted in an aqueous solution of 3.5% NaCl confirmed the high corrosion resistance of coatings sprayed with self-fluxing powders from remeltable Ni-Cr-B-Si alloys.
4. The main corrosion product of coatings made of remeltable alloys Ni-Cr-B-Si is Cr_2O_3 oxide, which forms a thin passivation layer on the surface of the coating, reducing the rate of corrosion. On the surface of coatings made of non-smelting alloys Ni-B-Si, NiO_2 and NiB_3 phases are formed, which protect the surface less effectively against the aggressive exhaust gas environment and create an unfavourable thermal barrier limiting heat transfer. In the case of 16Mo3 boiler steel, a thicker layer of oxides is formed on its surface, mainly Fe_3O_4, which cracks and does not provide anti-corrosion protection.

Author Contributions: Conceptualization, A.C., M.A. and J.A.; methodology, A.C., J.A., M.Ż., M.A., A.K., C.M. and M.K.; formal analysis, A.C., J.A., M.Ż., A.K., C.M. and M.K.; investigation, A.C., J.A., M.A., A.K. and M.Ż.; resources, A.C., J.A. and M.Ż.; writing—original draft preparation, A.C. and M.Ż.; writing—review and editing, A.C. and A.K.; visualization, A.C., M.Ż. and J.A.; supervision, A.C. and C.M.; project administration, A.C. and M.Ż.; funding acquisition, A.C. and M.Ż. All authors have read and agreed to the published version of the manuscript.

Funding: The research was founded by Silesian University of Technology Rector's pro-quality grants 10/050/RGJ22/1029 and 10/050/BKM22/1038.

Institutional Review Board Statement: Not applicable.

Informed Consent Statement: Not applicable.

Data Availability Statement: Not applicable.

Conflicts of Interest: The authors declare no conflict of interest.

References

1. Ling, C.; Hamilton, J.; Khandelwal, B. Chapter 1—Feedstock and pathways for alternative aviation fuels. *Aviat. Fuels* **2021**, 1–22. [CrossRef]
2. Zhou, H.; Yang, Y.; Dong, K.; Liu, H.; Shen, Y.; Cen, K. Influence of the gas particle flow characteristics of a low-NOx swirl burner on the formation of high temperature corrosion. *Fuel* **2014**, *135*, 595–602. [CrossRef]
3. Xiong, X.; Liu, X.; Tan, H.; Deng, S. Investigation on high temperature corrosion of water-cooled wall tubes at a 300 MW boiler. *J. Energy Inst.* **2020**, *93*, 377–386. [CrossRef]
4. Adamiec, J. High temperature corrosion of power boiler components surfaced with nickel alloys. *Mater. Charact.* **2009**, *60*, 1093–1099. [CrossRef]

5. Young-Ho, L.; In-Sup, K. The effect of subsurface deformation on the wear behavior of steam generator tube materials. *Wear* **2002**, *253*, 438–447.
6. Vakkilainen, E.K. 11—Recovery Boiler. Steam Generation from Biomass. *Constr. Des. Large Boil.* **2017**, 237–259. [CrossRef]
7. Lazić, V.; Arsić, D.; Nikolić, R.R.; Rakić, D.; Aleksandrović, S.; Djordjević, M.; Hadzima, B. Selection and Analysis of Material for Boiler Pipes in a Steam Plant. *Procedia Eng.* **2016**, *149*, 216–223. [CrossRef]
8. Czupryński, A.; Wyględacz, B. Comparative Analysis of Laser and Plasma Surfacing by Nickel-Based Superalloy of Heat Resistant Steel. *Materials* **2020**, *13*, 2367. [CrossRef]
9. Naghiyan Fesharaki, M.; Shoja-Razavi, R.; Mansouri, H.A.; Jamali, H. Evaluation of the Hot Corrosion Behavior of Inconel 625 Coatings on the Inconel 738 Substrate by Laser and TIG Cladding Techniques. *Opt. Laser Technol.* **2019**, *111*, 744–753. [CrossRef]
10. Rozmus-Górnikowska, M.; Cieniek, Ł.; Blicharski, M.; Kusiński, J. Microstructure and Microsegregation of an Inconel 625 Weld Overlay Produced on Steel Pipes by the Cold Metal Transfer Technique. *Arch. Metall. Mater.* **2014**, *59*, 1081–1084. [CrossRef]
11. Czupryński, A. Research on 16Mo3 steel pipe overlaid with superalloys Inconel 625 using robotized PPTAW. *Weld. Tech. Rev.* **2019**, *91*, 9–16. [CrossRef]
12. Rozmus-Górnikowska, M.; Blicharski, M.; Kusiński, J. Influence of weld overlaying methods on microstructure and chemical composition of Inconel 625 boiler pipe coatings. *Kov. Mater.* **2014**, *52*, 141–147. [CrossRef]
13. Jordan, D.E. Welding of high molybdenum nickel-base alloys. *Weld. World* **1998**, *41*, 1–9.
14. Jian, L.; Yuh, C.Y.; Farooque, M. Oxidation behavior of superalloys in oxidizing and reducing environments. *Corros. Sci.* **2000**, *42*, 1573–1585. [CrossRef]
15. Golański, G.; Lachowicz, M.; Słania, J.; Jasak, J.; Marszałek, P. Research on 16Mo3 (16M) steel pipes overlaid with Haynes NiCr625 alloy using MIG (131) method. *Arch. Metal. Mater.* **2015**, *60*, 2521–2524. [CrossRef]
16. Janicki, D. Laser surfacing of Inconel 625-based composite coatings reinforced by porous chromium carbide particles. *Opt. Laser Technol.* **2017**, *94*, 6–14. [CrossRef]
17. Appiah, A.N.S.; Bialas, O.; Czupryński, A.; Adamiak, M. Powder Plasma Transferred Arc Welding of Ni-Si-B+60 wt%WC and Ni-Cr-Si-B+45 wt%WC for Surface Cladding of Structural Steel. *Materials* **2022**, *15*, 4956. [CrossRef]
18. Vargas, F.; Ageorges, H.; Fournier, P.; Fauchais, P.; López, M.E. Mechanical and Tribological Performance of Al_2O_3-TiO_2 Coatings Elaborated by Flame and Plasma Spraying. *Surf. Coat. Technol.* **2010**, *205*, 1132–1136. [CrossRef]
19. Czupryński, A. Properties of Al_2O_3/TiO_2 and ZrO_2/CaO Flame-Sprayed Coatings. *Mater. Tehnol.* **2017**, *51*, 205–212. [CrossRef]
20. Ciokan, R.; Urbańczyk, M. The Development of a Technology of the Laser-Based Welding of Butt Joints in Composite Tubes. *Biul. Inst. Spaw.* **2021**, *1*, 23–31. [CrossRef]
21. Montgomery, M.; Karlsson, A. In-situ corrosion investigation at Masnedø CHP plant—A straw-fired power plant. *Mater. Corros.* **1999**, *50*, 579–584. [CrossRef]
22. Kawahara, Y. Application of High Temperature Corrosion-Resistant Materials and Coatings Under Severe Corrosive Environment in Waste-to-Energy Boilers. *J. Therm. Spray. Tech.* **2007**, *16*, 202–213. [CrossRef]
23. Rajkumar, V.; Arjunan, T.V.; Rajesh Kannan, A. Metallurgical and mechanical investigations of Inconel 625 overlay welds produced by GMAW-hardfacing process on AISI 347 pipes. *Mater. Res. Express* **2019**, *6*, 076534. [CrossRef]
24. Madsen, O.H. New technologies for waste to energy plants. In *4th International Symposium on Waste Treatment Technologies*; Babcock & Wilcox Vølund: Sheffield, UK, 2003.
25. Lippold, J.C.; Kiser, S.D.; DuPont, J.N. *Welding Metallurgy and Weldability of Nickel-Base Alloys*, 1st ed.; Wiley: Hoboken, NJ, USA, 2011.
26. *ISO 2063*; Thermal Spraying—Zinc, Aluminium and Their Alloys—Part 1: Design Considerations and Quality Requirements for Corrosion Protection Systems. ISO: Geneva, Switzerland, 2017.
27. *EN 10273*; Hot rolled Weldable Steel Bars for Pressure Purposes with Specified Elevated Temperature Properties. CEN: Brussels, Belgium, 2016.
28. *ISO 17637*; Non-Destructive Testing of Welds—Visual Testing of Fusion-Welded Joints. ISO: Geneva, Switzerland, 2016.
29. *ISO 3452*; Non-Destructive Testing—Penetrant Testing—Part 2: Testing of Penetrant Materials. ISO: Geneva, Switzerland, 2021.
30. *ISO 5817*; Welding—Fusion-Welded Joints in Steel, Nickel, Titanium and Their Alloys (Beam Welding Excluded)—Quality Levels for Imperfections. ISO: Geneva, Switzerland, 2014.
31. Vd. TÜV-Mbl 1166. *Instruction of Preparing and Testing of Padded Membrane Walls for Power Boilers*; Schweisstechnik 1166; Verband der Ueberwachungs-Verinee: Essen, Germany, 2001.
32. *ISO 17475*; Corrosion of Metal and Alloys-Electrochemical Test Methods-Guidelines for Conducting Potentiostatic and Potentiodynamic, Polarization Measurements. ISO: Geneva, Switzerland, 2010.
33. Czupryński, A.; Mele, C. Properties of Flame Spraying Coatings Reinforced with Particles of Carbon Nanotubes. *Adv. Mater. Sci.* **2021**, *21*, 57–76. [CrossRef]
34. Kondej, A. The Ni-Cr-B-Si coatings obtained in surfacing and thermal spraying processes. *Przegląd Spaw.* **2017**, *89*, 28–31.
35. Czupryński, A.; Adamiak, M.; Bayraktar, E.; Wyględacz, B. Comparison of tribological properties and structure of coatings produced in powder flame spraying process on grey cast iron. *Weld. Tech. Rev.* **2020**, *92*, 7–21. [CrossRef]
36. Rivoaland, L.; Maurice, V.; Josso, P.; Bacos, M.-P.; Marcus, P. The Effect of Sulfur Segregation on the Adherence of the Thermally-Grown Oxide on NiAl—I: Sulfur Segregation on the Metallic Surface of NiAl(001) Single-Crystals and at NiAl(001)/Al_2O_3 Interfaces. *Oxid. Met.* **2003**, *60*, 137–157. [CrossRef]

37. Jeong, S.; Lee, D. The effect of Fe on high temperature oxidation of NiAl. *Met. Mater.* **1998**, *4*, 1077–1083. [CrossRef]
38. González, R.; García, M.A.; Peñuelas, I.; Cadenas, M.; del Rocío Fernández, M.; Hernández Battez, A.; Felgueroso, D. Microstructural study of NiCrBSi coatings obtained by different processes. *Wear* **2007**, *263*, 619–624. [CrossRef]

materials

MDPI

Article

Effect of Surface Modifications on Surface Roughness of Ti6Al4V Alloy Manufactured by 3D Printing, Casting, and Wrought

János Kónya [1,2], Hajnalka Hargitai [3], Hassanen Jaber [4], Péter Pinke [4] and Tünde Anna Kovács [4,*]

1 Doctoral School on Materials Sciences and Technologies, Óbuda University, Bécsi út. 96/B., H-1034 Budapest, Hungary; janos@dentarttechnik.hu
2 Dent-Art Technik Ltd., Csokonai u. 10., H-9024 Győr, Hungary
3 Department of Materials Science and Technology, Széchenyi István University, Egyetem tér 1., H-9026 Győr, Hungary; hargitai@sze.hu
4 Bánki Donát Faculty of Mechanical and Safety Engineering, Óbuda University, Népszínház u. 8., H-1081 Budapest, Hungary; hassen.jaber@bgk.uni-obuda.hu (H.J.); pinke.peter@bgk.uni-obuda.hu (P.P.)
* Correspondence: kovacs.tunde@bgk.uni-obuda.hu; Tel.: +36-16665317

Abstract: This work aimed to comprehensively evaluate the influence of different surface modifications on the surface roughness of Ti6Al4V alloys produced by selective laser melting (SLM), casting and wrought. The Ti6Al4V surface was treated using blasting with Al_2O_3 (70–100 μm) and ZrO_2 (50–130 μm) particles, acid etching with 0.017 mol/dm^3 hydrofluoric acids (HF) for 120 s, and a combination of blasting and acid etching (SLA). It was found that the optimization of the surface roughness of Ti6Al4V parts produced by SLM differs significantly from those produced by casting or wrought processes. Experimental results showed that Ti6Al4V alloys produced by SLM and blasting with Al_2O_3 followed by HF etching had a higher surface roughness (Ra = 2.043 μm, Rz = 11.742 μm), whereas cast and wrought Ti6Al4V components had surface roughness values of (Ra = 1.466, Rz = 9.428 m) and (Ra = 0.940, Rz = 7.963 m), respectively. For Ti6Al4V parts blasted with ZrO_2 and then etched by HF, the wrought Ti6Al4V parts exhibited higher surface roughness (Ra = 1.631 μm, Rz = 10.953 μm) than the SLM Ti6Al4V parts (Ra = 1.336 μm, Rz = 10.353 μm) and the cast Ti6Al4V parts (Ra = 1.075 μm, Rz = 8.904 μm).

Keywords: surface roughness; 3D printing; Ti6Al4V; selective laser melting (SLM)

check for updates

Citation: Kónya, J.; Hargitai, H.; Jaber, H.; Pinke, P.; Kovács, T.A. Effect of Surface Modifications on Surface Roughness of Ti6Al4V Alloy Manufactured by 3D Printing, Casting, and Wrought. *Materials* **2023**, *16*, 3989. https://doi.org/10.3390/ma16113989

Academic Editor: Thomas Niendorf

Received: 9 May 2023
Revised: 22 May 2023
Accepted: 23 May 2023
Published: 26 May 2023

1. Introduction

Osseointegration is a crucial factor in the success of dental and bone implants [1]. The term is generally understood to mean the formation of good interaction and functional connection between the surface of an implant and living bone tissue. As a result, the osseointegration process is strongly influenced by implant surface conditions such as surface roughness, chemical composition, charge, and energy [2]. The surface roughness is recognized as being the most important parameter influencing the speed and quality of osseointegration [3]. There are three categories of surface roughness based on size: macro-rough (10–30 μm), micro-rough (1–10 μm), and nano-rough (less than 1 μm). It is shown that an increase in the macro-, micro-, and nano-structured surface morphologies can improve osseointegration and enhance bone fixation [4,5]. Therefore, dental implant quality is directly dependent on surface conditions. To improve the osseointegration of dental implants, surface modification technologies are often used, such as blasting, polishing, chemicals (acid etching), and blasting plus acid etching (SLA) [6]. In blasting, hard ceramic particles are shot through a nozzle into the surface of implants utilizing compressed air at high speed. Acid etching treatment involves immersing the implants in a strong acid such as hydrofluoric acid (HF), nitric acid (HNO_3), and/or sulphuric acid (H_2SO_4). The

SLA is blasting followed by acid etching. Souza et al. [6] investigated the effect of blasting followed by acid etching (SLA) on the proteomic profile of layers of proteins adsorbed from saliva and blood plasma on the surface of a Ti-Zr alloy. Wang et al. [7] studied the impact of the processing parameters of electron beam melting (EBM) on the surface roughness of manufactured parts. Szymczyk-Zi'ołkowska et al. [8] investigated the influence of surface modifications (polishing, sandblasting, and acid-polishing) of Ti6Al4V implants produced by EBM on essential biological properties (wettability, cytotoxicity, and biofilm formation). They concluded that surface modification has a very strong influence on biological properties.

Titanium alloys, especially Ti6Al4V, are an important alloy for dental and orthopaedic implants owing to their excellent mechanical and biological properties [9]. In recent years, there has been increased interest in the use of 3D printing technology (selective laser melting, SLM) for the manufacture of Ti64 medical implants in place of powder metallurgy, wrought and casting processes [10,11]. In this work, the surface roughness in terms of arithmetic mean roughness (Ra) and mean depth of roughness (Rz) of Ti6Al4V samples manufactured by SLM, casting, and wrought were measured and compared. It was found that the surface roughness was different for each process.

2. Materials and Methods

2.1. Preparation of Ti6Al4V Samples

Polished cylindrical specimens of Ti6Al4V produced by three methods, SLM, casting and wrought, were used as the base material to study the surface roughness (Figure 1). The SLM Ti6Al4V specimens (Figure 1a) were fabricated using an SLM machine (Sisma MYSINT 100, Via dell'Industria, Vicenza, Italia) equipped with a 200 W fibre laser and a 55 μm laser spot. The dimensions of the SLM Ti6Al4V samples were 9 mm in diameter, and 50 mm in height. The optimal settings consisted of a continuous laser power of 125 W, a scanning speed of 1000 mm/s, and a layer thickness of 20 μm. A constant flow of 35 L of argon gas per minute was used for protection. The starting material for SLM Ti6Al4V specimens was Ti6Al4V plasma-atomized spherical powder (Gr.5) provided by LPW Technology (Runcorn UK), as shown in Figure 2. The chemical composition of Ti64 powder is shown in Table 1. The size distribution ranged from 15 to 45 μm.

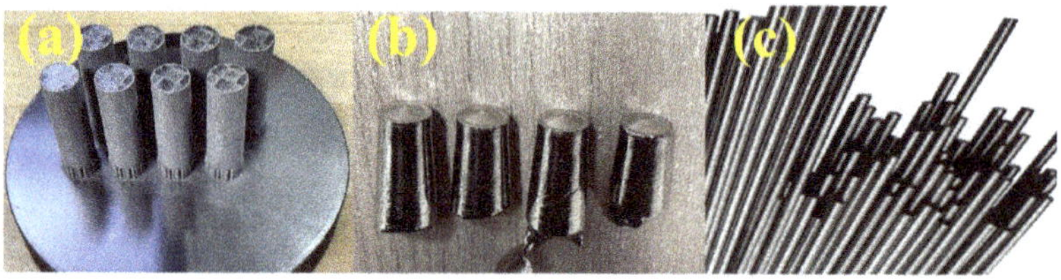

Figure 1. (**a**) 3D-printed, (**b**) Casting, and (**c**) wrought specimens.

Table 1. Chemical analysis of Ti64 powder and ASTM specification.

(Mass%)		Al	V	Fe	O	N	C	H	Ti
Ti6Al4V powder		6.11	4.02	0.17	0.090	0.023	0.01	0.003	Bal
ASTM B348 Gr.23 [12]	Max	6.75	4.50	0.40	0.20	0.05	0.08	0.015	Bal
	Min	5.50	3.50	-	-	-	-	-	-

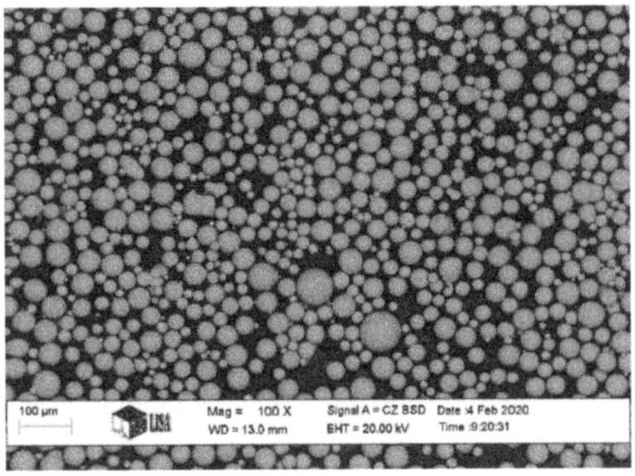

Figure 2. SEM micrograph shows the morphology of Ti6Al4V powder.

The casting Ti6Al4V specimens (Figure 1b) were fabricated using a vacuum-pressure, plasma jet-heated casting machine. To ensure better chemical homogeneity, the 20 g ingots were remelted three times. A red copper mould was used to cast the experimental alloys used in this investigation. The mould was a truncated cone which had a 10 mm top diameter and 14 mm base diameter and a height of 50 mm (Figure 1b). In addition, Ti6Al4V drawn-rolled specimens (Figure 1c), in its wrought condition, were used as the base metal, and was 9 mm in diameter. After the manufacturing process, all samples were subjected to a polishing process. The polishing process was performed by a WP-EX 2000 machine (Wassermann, Hamburg, Germany) equipped with rag polishing discs. The samples were polished with a #1200 grit SiC foil.

2.2. *Surface Modification Technologies*

The manufactured and polished samples were divided into three groups: Casting, wrought, and 3D Printing. Each group was subjected to five types of surface modification (see below).

1. Etched in 0.017 mol/dm^3 of hydrofluoric acid (HF) for 120 s at room temperature.
2. Blasted with Al_2O_3 particles (70–100 µm) with 4 bar blasting pressure. The blasting was performed with a Renfert Basic Quattro IS.
3. Blasted with Al_2O_3 particles and etched in 0.017 mol/dm^3 of hydrofluoric acid (HF) for 120 s at room temperature.
4. Blasted with ZrO_2 particles (50–130 µm) with 4 bar blasting pressure.
5. Blasted with ZrO_2 and etched in 0.017 mol/dm^3 of hydrofluoric acid (HF) for 120 s at room temperature.

2.3. *Surface Roughness and Topography*

The Ra and Rz surface roughness were determined using an ALICONA Infinite Focus equipped with Vision software. For each surface, five measurements were performed.

3. Results and Discussion

The values of surface roughness, Ra and Rz, for all specimens, are detailed in Table 2. It is revealed that 3D-printed (SLM) Ti6Al4V components are significantly different from cast and wrought Ti6Al4V parts when it comes to optimizing surface roughness by surface treatments such as Al_2O_3 blasting + HF etching. Ti6Al4V alloys produced by 3D printing and blasting with Al_2O_3 followed by HF etching exhibit the highest surface roughness compared to cast and wrought Ti6Al4V parts. The surface roughness of the 3D-printed samples is the roughest (Ra = 2.043, Rz = 11.742 µm), followed by the surface of the cast samples (Ra = 1.466, Rz = 9.428 µm), and the surface of the wrought samples (Ra = 0.940,

Rz = 7.963 μm). The increase in surface roughness and the change in surface morphology of Ti alloys have been reported in sandblasting and acid etching processes [13].

Table 2. The Ra and Rz values of the surface of the Ti6Al4V alloys manufactured by 3D printing, casting, and wrought after surface modifications.

Production Technology	Surface Treatment	Surface Roughness, Ra (μm)				Roughness Height, Rz (μm)			
		Specimen Number			Average	Specimen Number			Average
		1	2	3		1	2	3	
Casting	Polishing	0.479	0.531	0.499	0.503	3.397	3.506	3.816	3.573
Casting	Etching	0.312	0.353	0.367	0.344	2.317	2.889	2.964	2.723
Casting	Al_2O_3 Blasting	1.208	1.140	1.360	1.236	9.329	7.482	8.265	8.359
Casting	Al_2O_3 Blasting + etching	1.423	1.468	1.508	1.466	11.703	8.000	8.580	9.428
Casting	ZrO_2 Blasting	0.734	1.049	0.916	0.900	7.499	9.777	6.417	7.898
Casting	ZrO_2 Blasting + etching	0.804	1.307	1.061	1.075	8.198	9.843	8.670	8.904
Wrought	Polishing	0.360	0.475	0.555	0.463	2.497	3.083	3.679	3.086
Wrought	Etching	0.332	0.499	0.556	0.462	2.416	3.310	3.610	3.112
Wrought	Al_2O_3 Blasting	0.493	0.634	0.823	0.650	3.515	4.415	6.222	4.171
Wrought	Al_2O_3 Blasting + etching	0.729	0.877	1.215	0.940	6.144	5.982	10.953	7.693
Wrought	ZrO_2 Blasting	1.329	1.460	1.415	1.401	9.017	7.951	8.965	8.644
Wrought	ZrO_2 Blasting + etching	1.519	1.636	1.738	1.631	10.001	10.462	12.397	10.953
3D printing	Polishing	0.474	0.701	0.502	0.559	4.079	4.915	3.4516	4.149
3D printing	Etching	0.755	0.995	0.579	0.776	4.787	5.974	2.923	4.561
3D printing	Al_2O_3 Blasting	1.328	1.239	1.444	1.377	9.091	7.507	9.183	8.594
3D printing	Al_2O_3 Blasting + etching	2.623	1.763	1.743	2.043	12.625	11.252	11.349	11.742
3D printing	ZrO_2 Blasting	0.715	0.677	0.786	0.726	6.103	5.024	5.473	5.533
3D printing	ZrO_2 Blasting + etching	1.549	1.557	0.903	1.336	11.618	12.693	6.748	10.353

It is interesting to note that the surface treatment (ZrO_2 blasting + HF etching) of the wrought Ti6Al4V parts has a higher surface roughness (Ra = 1.631, Rz = 10.953 μm) than the cast parts (Ra = 1.075, Rz = 8.904 μm) and the 3D-printed ones (Ra = 1.336, Rz = 10.353 μm). The reason for this is probably the difference in the surface properties of the manufactured samples, which leads to different inclusion of ejected particles on the surface of the samples. In addition, as can be seen from Table 2, the HF etching process leads to a reduction in the surface roughness of the polished cast specimen (polishing and then etching) from (Ra = 0.503, Rz = 3.573 μm) to (Ra = 0.344, Rz = 2.723 μm), as the surface oxidation removes material, resulting in the ionization of atoms. In the wrought sample (polishing then etching), the roughness remains the same without an increase or decrease. On the other hand, the HF etching process leads to an increase in the surface roughness of the polished 3D sample (polishing and then etching). This is due to the high hardness of the 3D-printed sample, which reduces the oxidation process. It has been reported that the hardness of specimens manufactured by 3D printing (SLM) (377 HV) [10] is higher than those manufactured by casting (340 HV) [14] or wrought (306 HV) [15].

3.1. Casting

Figure 3 compares the surface roughness of the as-polished casted samples with samples after etching, blasting with Al_2O_3 or ZrO_2, or a combination of these methods. As can be seen, the surface roughness was reduced from (Ra = 0.503, Rz = 3.573 μm) to (Ra = 0.344, Rz = 2.723 μm) by etching, and increased to (Ra = 1.236, Rz = 8.359 μm) by blasting with Al_2O_3, (Ra = 1.466, Rz = 9.428 μm) by blasting with Al_2O_3 and etching, (Ra = 0.900, Rz = 7.898 μm) by blasting with ZrO_2, and (Ra = 1.075, Rz = 8.904 μm) by blasting with ZrO_2 and etching. The highest surface roughness was achieved after a combination of blasting with Al_2O_3 and etching. This is confirmed by the surface roughness profile (Figure 4a) of the sample after Al_2O_3 blasting and etching. The surface is rougher than that of the other samples (Figure 4b–d). In addition, Figure 4a shows the alternation of sharp peaks with a height of 4 μm and sharp valleys with a depth of 6 μm. Figure 5 shows the surfaces of the samples when (a) blasting with ZrO_2, (b) blasting with ZrO_2 and etching with HF, (c) blasting with Al_2O_3, and (d) blasting with Al_2O_3 and etching with HF. In the blasting with ZrO_2 (Figure 5a) and blasting with Al_2O_3 (Figure 5c) conditions, it can be seen that the sandblasted surface displayed an anisotropic structure of craters, valleys and peaks due to plastic deformation caused by the impact of Al_2O_3 and ZrO_2 particles, and there may well be some particles embedded in the surface. During the plastic deformation process some materials can be removed from the surface [16]. The SEM images of the surface blasted with Al_2O_3 (Figure 5c) are identical to the images of the surface blasted with ZrO_2 (Figure 5a). It is also possible to see the disordered position of the valleys and peaks produced. Figure 5b and d show the surfaces of the sample blasted with ZrO_2 and etched with HF and the sample blasted with Al_2O_3 and etched with HF. As can be seen, the etching process produced a very rough surface due to surface cleaning as well as material removal from the surface due to oxidation. The preferential dissolution of the alpha phase of the Ti6Al4V alloys has been reported in an etching by HF [13]. After HF etching, the Al_2O_3-blasted surface becomes sharper in appearance (Figure 4a). Rounded peaks (Figure 4b) become sharp, which is confirmed by the surface roughness profile (Figure 4a).

3.2. *Wrought*

Figure 6 shows the surface roughness of the as-polished wrought manufactured Ti6Al4V components after various surface modifications. As can be seen, there is a remarkable increase in surface roughness from (Ra = 0.463, Rz = 3.086 μm) (in the as-polished condition) to (Ra = 0.650, Rz = 4.171 μm) after blasting with Al_2O_3, (Ra = 0.940, Rz = 7.693 μm) after blasting with Al_2O_3 and etching, (Ra = 1.401, Rz = 8.644 μm) after blasting with ZrO_2, and (Ra = 1.631, Rz = 10.953 μm) after blasting with ZrO_2 and etching. It should be noted that the surface roughness of the etched sample (Ra = 0.462, Rz = 3.122 μm) is the same as in the polished state (Ra = 0.463, Rz = 3.086 μm), without any change. The surface of Ti6Al4V after ZrO_2 blasting was characterized by the presence of several craters, as shown in Figure 7a. The formation of craters could be attributed to ZrO_2 abrasive particles. After ZrO_2 blasting and etching (Figure 7b), a change in the surface was noticeable. Etching cleans the surface and removes material, resulting in a very rough surface. Figure 8 shows the roughness profile of each condition. After ZrO_2 blasting and etching (Figure 8b), the surface shows several peaks (6 μm)-to-valley (6 μm) relationships, indicating that the surface became rougher after acid etching (Ra = 1.631, Rz = 10.953 μm) compared to the blasted ZrO_2 sample (Ra = 1.401, Rz = 8.644 μm) and the other conditions. Figure 7c,d consists of SEM micrographs of the blasting with Al_2O_3 and blasting with Al_2O_3 and etching, indicating some small valleys and peaks, which are confirmed by Figure 8c,d to be of less roughness than the sample after ZrO_2 blasting and etching. The surface blasted with Al_2O_3 exhibited regular and homogeneous pore features. After blasting with Al_2O_3, more uniform and smaller micro-rough valleys (average 7 μm in diameter) formed on the surface than in the other conditions. Similar surface characteristics were observed in previous results [17]. A distinct surface change can be observed on the rolled specimen after Al_2O_3 blasting. The surface topography consists of valleys (3 μm) and peaks (2 μm), as shown in Figure 8c. In addition, the peaks and valleys are present in approximately equal proportions.

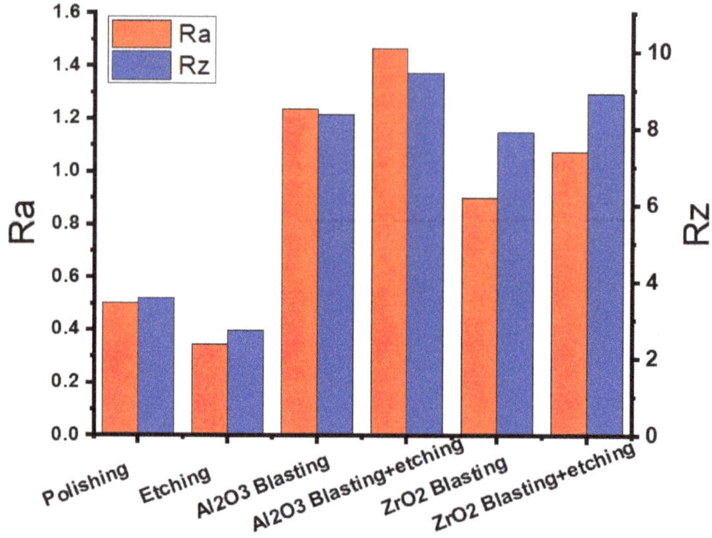

Figure 3. Results of the surface roughness of the cast-polished and surface-modified specimens.

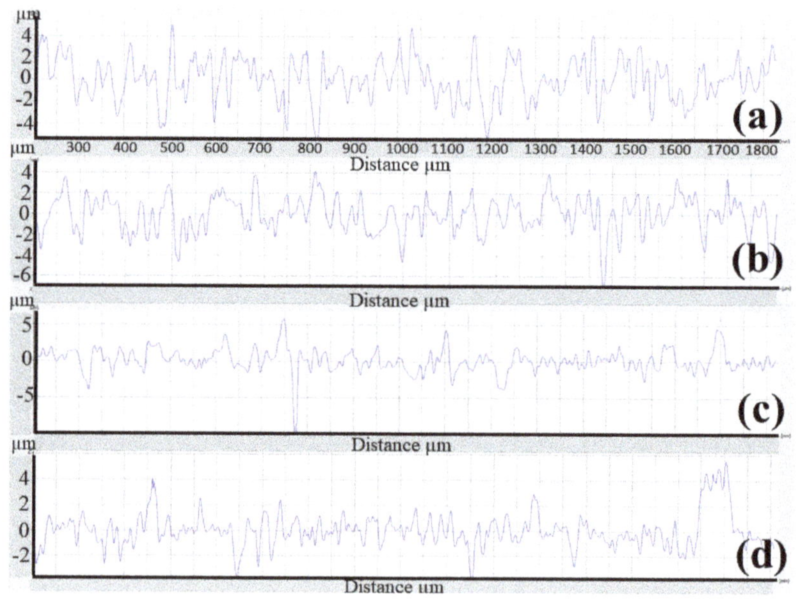

Figure 4. Surface roughness profile for each condition of the cast specimen. (**a**) Blasting with Al_2O_3 and etching with HF, (**b**) blasting with Al_2O_3, (**c**) blasting with ZrO_2 and etching with HF, and (**d**) blasting with ZrO_2.

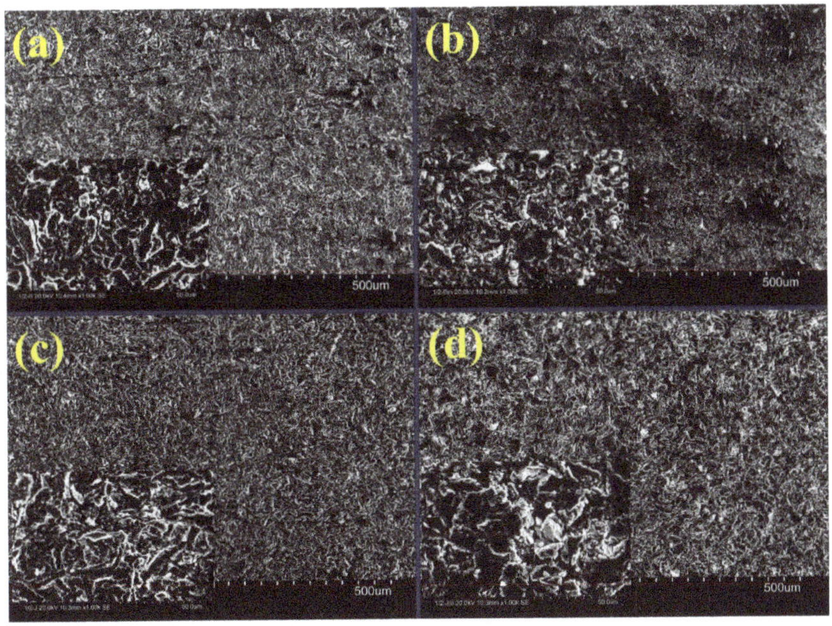

Figure 5. SEM micrographs for each condition of the cast specimen. (**a**) Blasting with ZrO_2, (**b**) blasting with ZrO_2 and etching with HF, (**c**) blasting with Al_2O_3, and (**d**) blasting with Al_2O_3 and etching with HF.

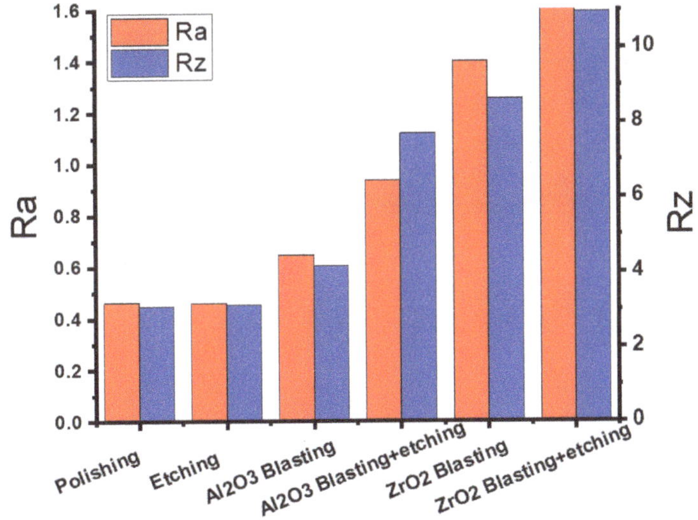

Figure 6. Results of the surface roughness of the wrought-polished and surface-modified specimens.

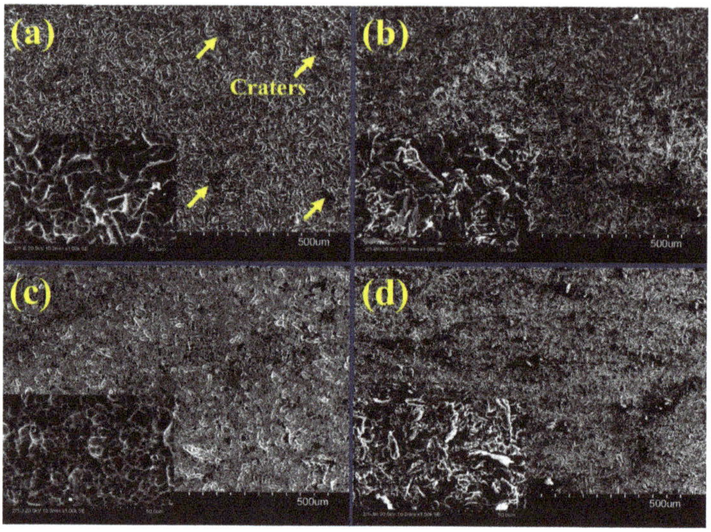

Figure 7. SEM micrographs for each condition of the wrought specimen. (**a**) Blasting with ZrO_2, (**b**) blasting with ZrO_2 and etching with HF, (**c**) blasting with Al_2O_3, and (**d**) blasting with Al_2O_3 and etching with HF.

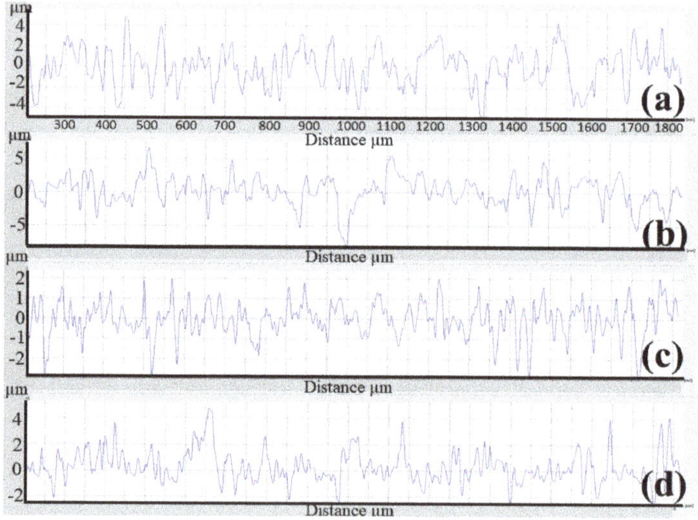

Figure 8. Surface roughness profile for each condition of the wrought specimen. (**a**) Blasting with ZrO_2, (**b**) blasting with ZrO_2 and etching with HF, (**c**) blasting with Al_2O_3, and (**d**) blasting with Al_2O_3 and etching with HF.

3.3. 3D Printing

The as-polished 3D-printed sample had Ra and Rz values of 0.559 and 4.149 μm, respectively. Etching with HF, blasting with Al_2O_3, blasting with Al_2O_3 and etching, blasting with ZrO_2, and blasting with ZrO_2 and etching, increased Ra and Rz to 0.776 and 4.561 μm, 1.377 and 8.594 μm, 2.043 and 11.742 μm, 0.726 and 5.533 μm, and 1.336 and 10.353 μm, respectively, as shown in Figure 9. Figure 10a–d highlights the corresponding

SEM micrographs after blasting with ZrO_2, blasting with ZrO_2 and etching, blasting with Al_2O_3, and blasting with Al_2O_3 and etching. It was noted that Figure 10d appears rougher because there are more valleys and cavities on the surface. Figure 11 shows the roughness profile difference between each condition. The absolute difference between the roughness profile of blasting with Al_2O_3 and etching (Figure 11d) and other samples (Figure 11a–c) was the surface topography, consisting of deep valleys (6 μm) and sharp peaks (6 μm). These results show that the surface modification process (blasting with Al_2O_3 and etching) is a suitable process to obtain the highest surface roughness of the produced titanium alloys. Furthermore, this surface roughness is described as a hierarchical structure composed of three different types of surface roughness based on dimensions: macro-rough (10–30 μm), micro-rough (1–10 μm), and nano-rough (less than 1 μm), all of which are advantageous to the osseointegration process [18].

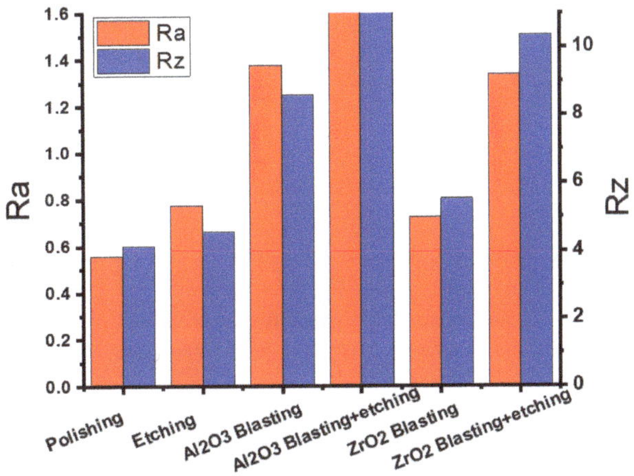

Figure 9. Results of the surface roughness of the 3D-polished and surface-modified specimens.

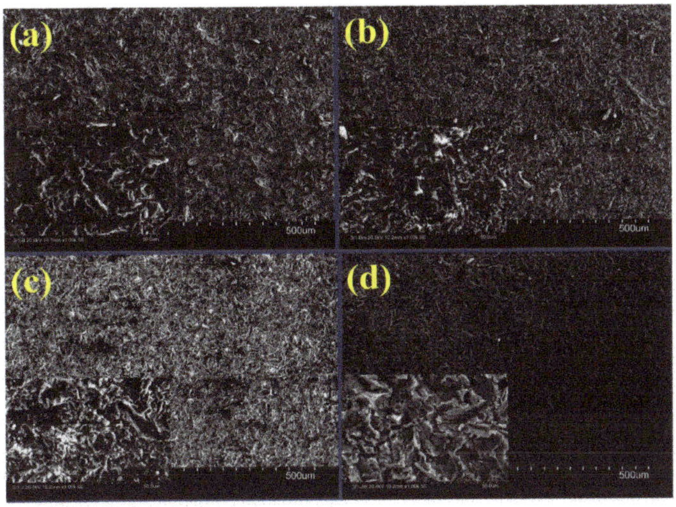

Figure 10. SEM micrographs for each condition of the 3D specimen. (**a**) Blasting with ZrO_2, (**b**) blasting with ZrO_2 and etching with HF, (**c**) blasting with Al_2O_3, and (**d**) blasting with Al_2O_3 and etching with HF.

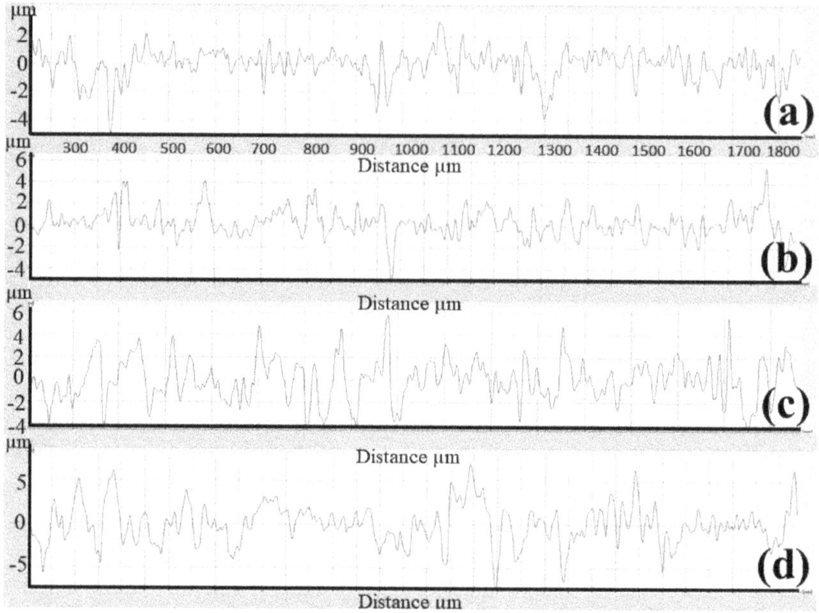

Figure 11. Surface roughness profile for each condition of the SLM specimen. (**a**) Blasting with ZrO_2, (**b**) blasting with ZrO_2 and etching with HF, (**c**) blasting with Al_2O_3, and (**d**) blasting with Al_2O_3 and etching with HF.

4. Conclusions

The effects of surface modifications on the surface roughness of Ti6Al4V alloy components produced by 3D printing, casting, and wrought have been studied in detail. The following conclusions can be drawn from the results.

1. Significant differences were found in the surface roughness of specimens produced by 3D printing compared to those produced by casting and wrought after surface modifications were performed. This can be attributed to the difference in the surface properties of the manufactured samples, which leads to different inclusion of ejected particles on the surface of the samples.
2. The highest outcomes were collected for Ti6Al4V alloys fabricated using SLM and blasting with Al_2O_3, followed by HF etching (Ra = 2.043, Rz = 11.742 μm), or with Ti6Al4V fabricated using wrought and blasting with ZrO_2, followed by HF etching (Ra = 1.631, Rz = 10.953 μm).
3. The surface roughness of the SLM-fabricated samples and blasting with Al_2O_3 or ZrO_2 was considerably influenced by HF etching. In the case of the specimens with Al_2O_3 blasting + HF etching, the surface roughness increased from (Ra = 1.337, Rz = 8.594 μm) to (Ra = 2.043, Rz = 11.742 μm). For the specimens of ZrO_2 blasting + HF etching, the surface roughness increased from (Ra = 0.726, Rz = 5.533 μm) to (Ra = 1.336, Rz = 10.353 μm).

Author Contributions: Conceptualization, J.K. and H.J.; methodology, H.H.; validation, H.H. and T.A.K.; formal analysis, T.A.K.; investigation, J.K.; data curation, J.K. and P.P.; writing—original draft preparation, H.J.; writing—review and editing, T.A.K.; visualization, J.K.; supervision, H.H. and P.P.; project administration, T.A.K. All authors have read and agreed to the published version of the manuscript.

Funding: The research was supported by the NKFIH from the project 'Research on the health application of artificial intelligence, digital imaging, employment and material technology developments by linking the scientific results of Széchenyi István University and Semmelweis University' under grant number TKP2021-EGA-21.

Institutional Review Board Statement: Not applicable.

Informed Consent Statement: Not applicable.

Data Availability Statement: The data used for the research are available upon request.

Acknowledgments: The authors acknowledge the support by the NKFIH from the project 'Research on the health application of artificial intelligence, digital imaging, employment and material technology developments by linking the scientific results of Széchenyi István University and Semmelweis University' under grant number TKP2021-EGA-21. The authors would like to express their gratitude to Dent-Arc Technik Ltd. for the technical support.

Conflicts of Interest: The authors declare no conflict of interest.

References

1. Jaber, H.; Kovacs, T. Selective laser melting of Ti alloys and hydroxyapatite for tissue engineering: Progress and challenges. *Mater. Res. Express* **2019**, *6*, 082003. [CrossRef]
2. Jemat, A.; Ghazali, M.J.; Razali, M.; Otsuka, Y. Surface Modifications and Their Effects on Titanium Dental Implants. *BioMed Res. Int.* **2015**, *2015*, 791725. [CrossRef] [PubMed]
3. Monetta, T.; Bellucci, F. The effect of sand-blasting and hydrofluoric acid etching on Ti CP2 and Ti CP4 surface topography. *Open J. Regen. Med.* **2012**, *1*, 41–50. [CrossRef]
4. Li, J.; Zhou, P.; Attarilar, S.; Shi, H. Innovative Surface Modification Procedures to Achieve Micro/Nano-Graded Ti-Based Biomedical Alloys and Implants. *Coatings* **2021**, *11*, 647. [CrossRef]
5. Liu, X.; Chu, P.K.; Ding, C. Surface modification of titanium, titanium alloys, and related materials for biomedical applications. *Mater. Sci. Eng. R Rep.* **2004**, *47*, 49–121. [CrossRef]
6. Souza, J.G.S.; Bertolini, M.; Costa, R.C.; Lima, C.V.; Barão, V.A.R. Proteomic profile of the saliva and plasma protein layer adsorbed on Ti–Zr alloy: The effect of sandblasted and acid-etched surface treatment. *Biofouling* **2020**, *36*, 428–441. [CrossRef] [PubMed]
7. Wang, P.; Sin, W.J.; Nai, M.L.S.; Wei, J. Effects of Processing Parameters on Surface Roughness of Additive Manufactured Ti-6Al-4V via Electron Beam Melting. *Materials* **2017**, *10*, 1121. [CrossRef] [PubMed]
8. Szymczyk-Ziółkowska, P.; Hoppe, V.; Rusińska, M.; Gąsiorek, J.; Ziółkowski, G.; Dydak, K.; Czajkowska, J.; Junka, A. The Impact of EBM-Manufactured Ti6Al4V ELI Alloy Surface Modifications on Cytotoxicity toward Eukaryotic Cells and Microbial Biofilm Formation. *Materials* **2020**, *13*, 2822. [CrossRef] [PubMed]
9. Jaber, H.; Kónya, J.; Kovács, T.A. Selective Laser Melting of Ti6Al4V-2%Hydroxyapatite Composites: Manufacturing Behavior and Microstructure Evolution. *Metals* **2021**, *11*, 1295. [CrossRef]
10. Jaber, H.; Kónya, J.; Kulcsár, K.; Kovács, T. Effects of Annealing and Solution Treatments on the Microstructure and Mechanical Properties of Ti6Al4V Manufactured by Selective Laser Melting. *Materials* **2022**, *15*, 1978. [CrossRef] [PubMed]
11. Jaber, H.; Kovacs, T.; János, K. Investigating the impact of a selective laser melting process on Ti6Al4V alloy hybrid powders with spherical and irregular shapes. *Adv. Mater. Process. Technol.* **2020**, *8*, 715–731. [CrossRef]
12. *ASTM Standard B348/B348M*; Standard Specification for Titanium and Titanium Alloy Bars and Billets. ASTM International: West Conshohocken, PA, USA, 2019; pp. 1–5. [CrossRef]
13. Ferraris, S.; Spriano, S.; Pan, G.; Venturello, A.; Bianchi, C.L.; Chiesa, R.; Faga, M.G.; Maina, G.; Verné, E. Surface modification of Ti–6Al–4V alloy for biomineralization and specific biological response: Part I, inorganic modification. *J. Mater. Sci. Mater. Med.* **2011**, *22*, 533–545. [CrossRef] [PubMed]
14. Da Rocha, S.S.; Adabo, G.L.; Henriques, G.E.P.; Nóbilo, M.A.D.A. Vickers hardness of cast commercially pure titanium and Ti-6Al-4V alloy submitted to heat treatments. *Braz. Dent. J.* **2006**, *17*, 126–129. [CrossRef] [PubMed]
15. Shunmugavel, M.; Polishetty, A.; Goldberg, M.; Singh, R.; Littlefair, G. A comparative study of mechanical properties and machinability of wrought and additive manufactured (selective laser melting) titanium alloy—Ti-6Al-4V. *Rapid Prototyp. J.* **2017**, *23*, 1051–1056. [CrossRef]
16. Reshadi, F.; Khorasani, S.; Faraji, G. Surface characterization of nanostructured commercially pure titanium modified by sandblasting and acid-etching for implant applications. *Proc. Inst. Mech. Eng. Part J J. Eng. Tribol.* **2019**, *234*, 414–423. [CrossRef]
17. Demirci, S.; Dikici, T.; Güllüoğlu, A.N. Micro/Nanoscale Surface Modification of Ti6Al4V Alloy for Implant Applications. *J. Mater. Eng. Perform.* **2021**, *31*, 1503–1511. [CrossRef]
18. Souza, J.C.M.; Sordi, M.B.; Kanazawa, M.; Ravindran, S.; Henriques, B.; Silva, F.S.; Aparicio, C.; Cooper, L.F. Nano-scale modification of titanium implant surfaces to enhance osseointegration. *Acta Biomater.* **2019**, *94*, 112–131. [CrossRef] [PubMed]

materials

MDPI

Article

Experimental Comparison of Laser Cladding and Powder Plasma Transferred Arc Welding Methods for Depositing Wear-Resistant NiSiB + 60% WC Composite on a Structural-Steel Substrate

Marcin Adamiak [1,*], Augustine Nana Sekyi Appiah [1], Radosław Żelazny [2], Gilmar Ferreira Batalha [3] and Artur Czupryński [4]

[1] Materials Research Laboratory, Faculty of Mechanical Engineering, Silesian University of Technology, 18A Konarskiego Street, 44-100 Gliwice, Poland; augustine.appiah@polsl.pl
[2] Autoneum Poland Sp. Z o.o., Owsiana 60A, 40-780 Katowice, Poland; radoslawzelazny96@gmail.com
[3] Department of Mechatronics and Mechanical Systems Engineering, Polytechnic School of Engineering of the University of Sao Paulo (USP), São Paulo 05508-900, Brazil; gfbatalh@usp.br
[4] Welding Department, Faculty of Mechanical Engineering, Silesian University of Technology, 18A Konarskiego Street, 44-100 Gliwice, Poland; artur.czuprynski@polsl.pl
* Correspondence: marcin.adamiak@polsl.pl; Tel.: +48-502-155-968

Abstract: A Ni-based powder composed of NiSiB + 60% WC was deposited onto a structural-steel substrate using two methods: laser cladding (LC) and plasma powder transferred arc welding (PPTAW). The resulting surface layers were analyzed and compared. Both methods resulted in the precipitation of secondary WC phases in the solidified matrix, but the PPTAW clad exhibited a dendritic microstructure. The microhardness of the clads prepared by both methods was similar, but the PPTAW clad showed higher resistance to abrasive wear compared to the LC clad. The thickness of the transition zone (TZ) was thin for both methods, with a coarse-grain heat-affected zone (CGHAZ) and peninsula-like macrosegregations observed in clads from both methods. The PPTAW clad showed a unique cellular–dendritic growth solidification (CDGS) and a type-II boundary at the TZ attributed to its thermal cycles. While both methods resulted in metallurgical bonding of the clad to the substrate, the LC method exhibited a lower dilution coefficient. The LC method also resulted in a larger HAZ with higher hardness compared to the HAZ of the PPTAW clad. The findings of this study indicate that both methods are promising for antiwear applications due to their wear-resistant properties and metallurgical bonding to the substrate. The PPTAW clad may be particularly useful in applications that require higher resistance to abrasive wear, while the LC method may be advantageous in applications that require lower dilution and larger HAZ.

Keywords: dilution; heat-affected zone (HAZ); metal matrix composite (MMC); scanning electron microscopy (SEM); transition zone

Citation: Adamiak, M.; Appiah, A.N.S.; Żelazny, R.; Batalha, G.F.; Czupryński, A. Experimental Comparison of Laser Cladding and Powder Plasma Transferred Arc Welding Methods for Depositing Wear-Resistant NiSiB + 60% WC Composite on a Structural-Steel Substrate. *Materials* **2023**, *16*, 3912. https://doi.org/10.3390/ma16113912

Academic Editor: Zhenyu Zhang

Received: 8 May 2023
Revised: 19 May 2023
Accepted: 22 May 2023
Published: 23 May 2023

1. Introduction

Surface coatings play a crucial role in enhancing material properties and performance in various industries. For instance, in the rail industry, where structural steel is extensively used in the production of railway components such as wheels, axles, and rails, these coatings are essential due to the harsh operating conditions these components are exposed to, including high loads, abrasion, and corrosion that can lead to premature failure [1–3]. Surface coatings have been an integral part of the metals industry for many years, serving to improve the wear and corrosion resistance of components. With advancements in material science and engineering, there has been a rise in surface coating technologies that offer opportunities to enhance the performance and extend the service lives of metallic parts and components [4–6]. Studies aimed at improving the wear resistance of metals have explored various forms of coatings, including the use of wires [7] and powders [8]. One approach to coating these materials with wear-resistant powder metal matrix composites is through the use of powders consisting of matrices based on cobalt [9], nickel [10], and other materials reinforced with particles such as WC, TiC, Cr_3C_2, VC, BN, TiB_2, and diamond reinforcements [11–15].

Various methods have been utilized to apply wear-resistant coatings onto steel substrates, including flame spraying, thermal spraying, laser melt injection, and magnetron sputtering [16–18], among others. However, these conventional techniques often yield coatings with limited bonding strength and relatively higher porosity, and they are insufficiently thick for heavy-loaded structures such as bearings and gears. In contrast, cladding technology, which encompasses laser cladding (LC) and powder plasma transferred arc welding (PPTAW), presents a cutting-edge approach for coating preparation, enabling complete metallurgical bonding between the coating and substrate [19]. Cladding methods exhibit superior bonding strength, increased coating thickness, and reduced porosity. Moreover, the higher heating and cooling rates associated with cladding can refine the microstructure of the alloys, resulting in enhanced hardness and wear resistance [20,21]. The versatility of LC and PPTAW in working with diverse powders has rendered them popular techniques for surface modification, augmenting strength, hardness, corrosion resistance, and abrasive-wear resistance, as supported by existing literature [22–26].

Tungsten carbide (WC) is widely used in industrial applications due to its excellent wear resistance. However, WC is susceptible to dissolution in the melt pool generated during the cladding process, as it has a low free enthalpy of formation of 38.5 kJ/mol [27]. Therefore, extensive research has been conducted to determine optimal values for the cladding process to minimize carbide dissolution and ensure proper melting. Previous studies have suggested that in the case of LC, laser power ranging from 1.0 kW to 2.0 kW is optimal for cladding meta l= based powder blends containing WC reinforcements [28,29]. Similarly, for PPTAW, studies have reported that plasma gas flow rates (PGFR) ranging from 1.0 L/min to 2.0 L/min yield better cladding performance [30,31].

The performance and properties of a clad are strongly associated with its characteristics. Kim and Lee [32] recently reported that cladding processes, compared to other coating methods, have higher heat inputs and, as a result, lead to an oriented microstructure formation, which has a significant influence on the mechanical properties of the clad. In reported studies regarding Ni-based clads, Yoon et al. [33] and DuPont et al. [34] observed hot cracking. They explained this as resulting from the presence of eutectic phases that were topologically close-packed as a consequence of increased dilution during clad solidification. Additionally, Gittos et al. [35] reported a reduction in the corrosion resistance of their prepared clad after observing interdendritic precipitates and a high content of Fe in the clad. This was reported to have resulted from high dilution. It is therefore useful to apply surface clads using techniques that result in low dilution, such as LC and PPTAW. Such techniques have the potential to mitigate these cladding issues and to produce much more desirable clads.

The transition zone (TZ) at the clad/substrate interface is an area of much microstructural consideration, as the resulting microstructure in this zone impacts the performance of the clad. At the TZ, there is the possibility of hydrogen embrittlement (HE) occurring in the clad. The phenomenon of HE is described by Örnek et al. [36] as the vulnerability of a clad to becoming brittle due to high hardenability, elevated hardness, and the inclusion of partially diluted macrosegregation and a gradient in chemical composition in the TZ. The TZ can also possess a type-II boundary, which is a high-angle grain boundary that is usually in a parallel orientation to the TZ. In the work of Dong et al. [37], it is stipulated that the type-II boundary can serve as an avenue for sulfide stress corrosion cracking (SSCC). In the laser cladding on Ni-steel with stainless steel AISI 304 research conducted by Wu et al. [38], it was reported that the TZ exhibited a type-I boundary and had harder martensite phases. However, during the cladding of this same Ni-steel with a Ni-based alloy, by a different technique, hot-wire gas tungsten arc welding (GTAW-HW), Farias et al. [39] reported the observation of both type-I and type-II boundaries in the TZ without SSCC. It can be inferred from these studies that the resulting microstructure of the TZ and its characteristics vary depending on the cladding method used for material deposition.

Given the increasing use of Ni-based powder blends in diverse industrial applications [40–42] and the crucial role of microstructural features in relation to different cladding methods, along with the existing knowledge gap in this area, this study seeks to accomplish two primary goals: (1) to successfully clad a wear-resistant Ni-based powder reinforced with tungsten carbide (WC) with a composition of NiSiB + 60% WC onto a structural-steel substrate using LC and PPTAW techniques, and (2) to conduct a comprehensive comparative analysis of the mechanical properties, such as hardness and wear resistance, as well as the microstructural characteristics of the clads prepared by LC and PPTAW. To achieve these objectives, we used methodologies including scanning electron microscopy (SEM), energy-dispersive X-ray spectroscopy (EDS), light microscopy, digital microscopy, abrasive-wear resistance tests, Vickers microhardness tests, and macroanalytical dilution and penetrant tests. The findings of this research will contribute to knowledge and understanding of these cladding processes and their performance in antiwear applications, filling the existing gap in the literature and providing valuable insights for practical implementation.

2. Materials and Methods

2.1. Materials and Surface Cladding

2.1.1. Laser Cladding (LC)

The commercially available powder blend, which consists of Ni-Si-B + 60% WC, was acquired from Castolin Eutectic® in Gliwice, Poland. Structural steel grade EN S355 was utilized as the substrate material. The steel plate samples were prepared with dimensions of 90 mm × 30 mm × 10 mm.

The YLS 400 laser system (IPG Photonics, Oxford, MA, USA) was utilized for laser cladding in this study. The process incorporated a powder feeder and direct injection of powder into the melt pool. Argon gas was employed as both the powder transport and shielding gas to protect the clad. The clads were created through 7 multi-runs of single passes with a 50% overlap. Preheating was not employed during the laser processing, and the interpass temperature remained below 30 °C. To determine the optimal power for the clad's desired performance, the laser power was adjusted between 1.5 kW and 2.0 kW for samples L1 and L2, respectively. All other parameters were kept constant for both samples. The rate of cladding was 0.2 m/min, and the deposition speed was 8 g/min.

2.1.2. Powder Plasma Transferred Arc Welding (PPTAW)

The EuTronic Gap 3511 DC synergic PPTAW system, manufactured by Castolin Eutectic in Gliwice, Poland, was utilized for performing PPTAW. The cladding process involved the use of a plasma torch with a tungsten electrode placed in the center, acting as the cathode to generate a plasma arc, through which powders were fed. The anode was structural-steel-plate substrate material. The plasma gas utilized was argon 5.0 (99.999%) in accordance with ISO 14175-I1:2009 [43], while a shielding and carrier gas mixture of argon/hydrogen 5% H2, Ar (welding mixture ISO 14175-R1-ArH-5 [43]) was employed. The plasma transferred arc had a thermal efficiency coefficient of k = 0.6, with a tungsten electrode (WS-2) measuring 4 mm in diameter and a plasma nozzle size of 3 mm. The plasma gas flow rate (PGFR) was adjusted between 1.0 L/min and 1.2 L/min to prepare samples P1 and P2, respectively. All other PPTAW parameters, such as pilot arc current of 30 A, travel speed of 1.3 mm/s, power flow (cos phi) of 0.99, open-circuit voltage of 95 V (direct current), and plasma transferred arc current of 110 A, remained constant for both samples during the clad deposition.

A summary of the prepared samples is given in Table 1, with their corresponding variation in parameters for both cladding methods.

Table 1. Prepared samples with their respective parameter variations.

Sample ID	Cladding Method	Laser Power [kW]	PGFR * [L/min]
L1	LC *	1.5	–
L2	LC	2.0	–
P1	PPTAW *	–	1.0
P2	PPTAW	–	1.2

* LC–Laser cladding; PPTAW–Powder plasma transferred arc welding: PGFR–Plasma gas flow rate.

2.2. Characterization

Samples for metallographic analysis were obtained by cutting along the cross-section from the surface clad to the substrate material, as shown in Figure 1. The samples were then subjected to various analyses, including microhardness measurements, scanning electron microscopy (SEM), energy-dispersive X-ray spectroscopy (EDS), light microscopy (LM), and digital microscopy (DM), to further investigate their properties.

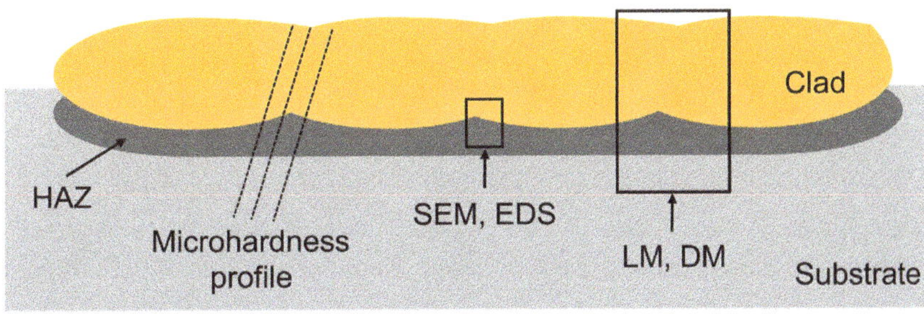

Figure 1. Cross-section of metallographic sample showing the areas examined via microhardness testing, SEM, EDS, LM, and DM.

The specimens were prepared for metallography by sequentially grinding with SiC papers of various grit sizes including 220, 500, 800, and 1200, followed by polishing with a coarse diamond suspension, and finally mirror polishing using colloidal silica with a particle size of 0.04 μm.

The Vickers microhardness of the surface layers on the cross-section of the clad was evaluated utilizing a 9.81 N load with the FM-ARS 9000 microhardness tester (Future Tech Corporation, Tokyo, Japan). The measurements were carried out subsequent to performing metallographic polishing.

The characterization of powder and evaluation of microstructural features in metallographic samples were conducted utilizing a combination of light microscopy with AxioVision (ZEISS, Jena, Germany), digital microscopy with Leica DVM6 (Leica Microsystems, Heerbrugg, Switzerland), and SEM using Zeiss Evo MA 15 series instrument equipped with an EDS system.

To identify cracks that may have developed in the prepared surface clads after the LC and PPTAW cladding processes, a penetration test was conducted in accordance with the PN-EN ISO 3452 standard [44], utilizing MR® 70 developer, MR® 79 remover (acetone), and MR® 68NF penetrant. Prior to the test, each specimen's surface was thoroughly cleaned with acetone to remove any impurities. The penetrant was then sprayed onto the surface and left to dry for approximately 10–15 min, after which it was removed using the remover and paper. Next, the developer was sprayed onto the sample's surface and allowed to settle. Finally, cracks were identified on the surface of the specimen where the penetrant had seeped in.

The prepared surface clads were subjected to tests to determine their resistance to abrasive wear. A reference material, AR400 steel, which is known for its resistance to abrasive wear, was also tested. The testing method used was the "rubber wheel" method, as described in ASTM G65 standard. This method involves placing abrasive materials between the surface being tested and a rubber-lined wheel, as depicted in Figure 2. The rubber-lined wheel is rotated while the surface being tested remains stationary for a certain duration. The testing parameters utilized in the study are detailed in Table 2.

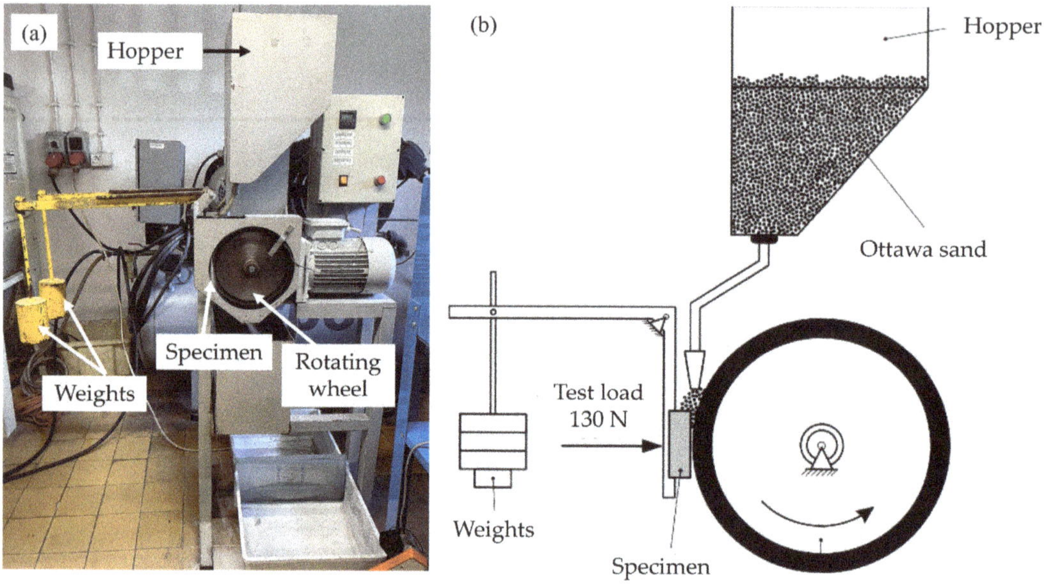

Figure 2. (**a**) Abrasive-wear resistance testing setup based on ASTM G65, (**b**) schematic diagram of the process.

Table 2. Abrasive-wear resistance testing parameters.

Parameter	Value	Unit
Abrasive particle grain size	210–297	μm
Feed rate	335	g/min
Pressure	130	Pa
Rubber wheel turns	6000	turns
Test time	30	min

Equations (1)–(3) were used to compute abrasive-wear test parameters. In the equations, M_B represents the initial mass of the specimen, M_A represents the mass of the specimen after the abrasive-wear test, ρ denotes the density of the material, V_{LR} signifies the average volume loss of the reference material, and V_{LS} denotes the average volume loss of the specimen being studied.

$$\text{Mass Loss, ML [g]} = \text{MB [g]} - \text{MA [g]} \tag{1}$$

$$\text{Volume Loss, VL}\left[\text{mm}^3\right] = \frac{M_L\,[g]}{\rho\left[\frac{g}{cm^3}\right]} \times 1000 \tag{2}$$

$$\text{Relative Abrasive Wear Resistance} = \frac{V_{LR}\left[\text{mm}^3\right]}{V_{LS}\left[\text{mm}^3\right]} \tag{3}$$

3. Results and Discussion

3.1. Powder Characterization

An SEM image of the NiSiB + 60% WC powder blend used for the cladding processes is shown in Figure 3a. The powder particles showed a mixture of spherical and angular morphologies. In Figure 3b, the result of a laser-assisted powder particle size analysis is presented, with statistical deductions (inset in Figure 3b) revealing a low span (indicating particle uniformity) of 1.3 for the 1475 particles analyzed. This low span value indicates that the powders have a low degree of particle non-uniformity, resulting in improved powder flow during PPTAW and LC processes. In the EDS characterization (Figure 3c–i and Table 3), the spherical powder particles are the Ni-based matrix phase particles, confirmed by the higher content of Ni, relative to other elements. The angular powder particles were found to be WC reinforcements. Boron was too light to be detected by the EDS technique.

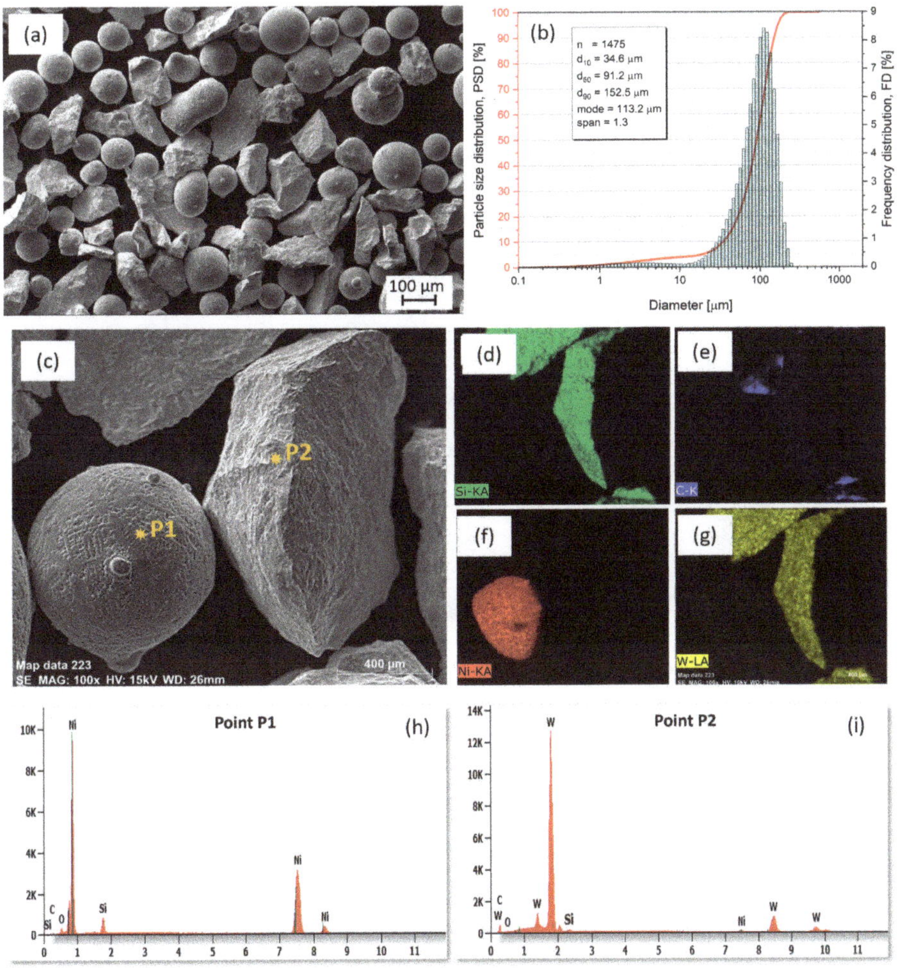

Figure 3. (**a**) NiSiB + 60% WC powder-blend particles as observed under the scanning electron microscope, (**b**) powder-blend particle size distribution (PSD) plot showing statistical parameters and powder particle span, (**c–g**) EDS elemental maps showing the position and amounts of elements present in the analyzed powder particles, (**h**,**i**) X-radiation energy diagrams of measured points from analyzed powder particles.

Table 3. Chemical composition in Weight % and Atom % of measured points from analyzed powder particles.

		C	O	Si	Ni	W
Point P1	Atom %	6.6	1.7	6.3	84.9	–
	Weigh t%	1.5	0.5	3.4	94.4	–
Point P2	Atom %	54.4	3.1	0.3	0.3	41.8
	Weight %	7.7	0.6	0.4	0.2	91.1

The morphology of the powder particles prior to cladding has been reported to affect the performance of the clad for various applications. ArunKumar et al. [45] have reported that angular powder particles perform better in wear-resistant applications, whereas spherical powder particles are more desirable for applications requiring high toughness. According to Li et al. [46], powders that have a mixture of spherical and angular particles result in enhanced stability and function better in applications requiring wear resistance, high hardness, and toughness.

3.2. Cladding

A micrograph of the substrate material, EN S355 structural steel grade, was obtained using a light microscope and is presented in Figure 4a. The microstructure mainly consists of allotriomorphic ferrite and pearlite, with traces of Widmanstätten ferrite, which is described as acicular ferrite plates emanating from a prior austenite grain boundary [47]. Microhardness measurements yielded an average of 405 HV with a standard deviation of 6.1 (Figure 4b), and equiaxed grains were observed in the through-thickness direction. The relatively high hardness of this steel grade is attributed to the presence of pearlite and ferrite zones with solute segregation, which can extend up to 400 μm [48], making it suitable for applications requiring high strength and hardness.

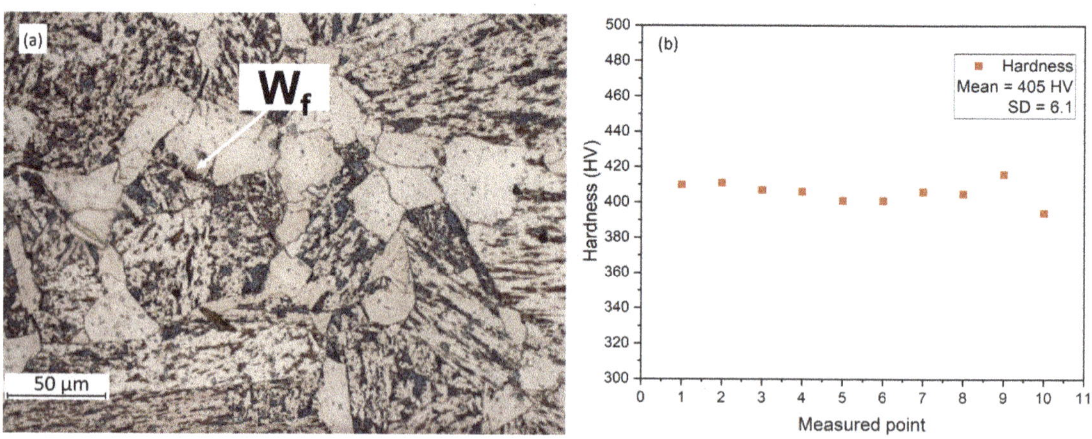

Figure 4. (**a**) Microstructure of substrate material, steel grade EN S335 showing the site of Widmanstätten ferrite, (**b**) microhardness (HV) of the base material.

Figure 5 shows the top view of the deposited surface clad onto the substrate material. Figure 5a,b are samples prepared using the LC method, and Figure 5c,d are samples prepared using the PPTAW method.

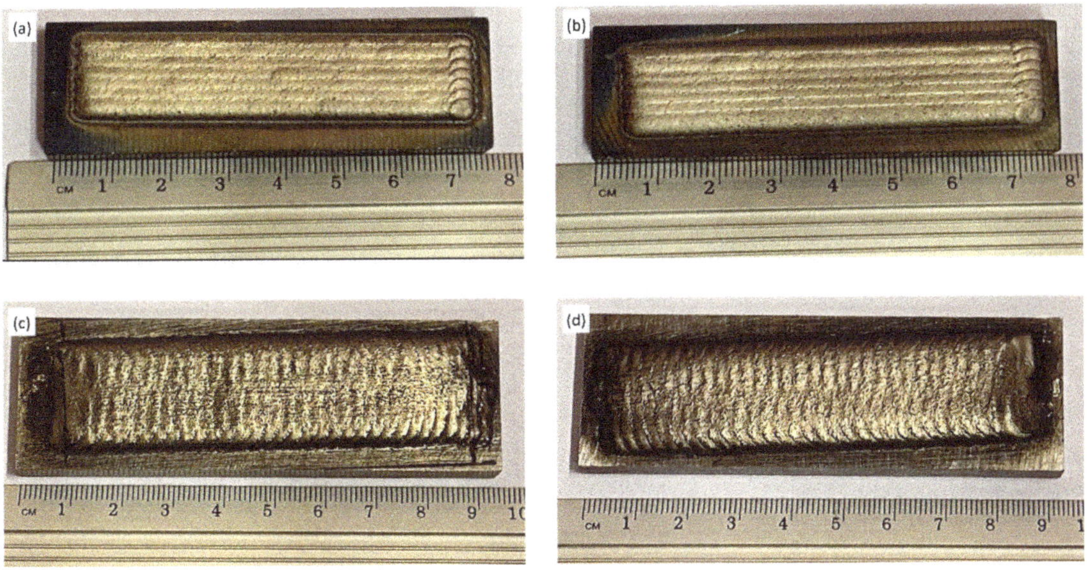

Figure 5. Top view of samples after cladding: (**a**,**b**) samples L1 and L2 prepared by laser cladding, (**c**,**d**) samples P1 and P2 prepared by PPTAW.

3.3. *Surface Crack Investigation*

The occurrence of surface cracks resulting from the cladding processes was investigated using penetrant testing, and the results are shown in Figure 6. The results reveal that the samples prepared via LC exhibited a significantly higher number of crack sites compared to those prepared via the PPTAW method. This discrepancy can be attributed to the considerably higher heat input from plasma during PPTAW, which promotes better melting and bonding of powder particles, resulting in fewer surface cracks. Sample P2, prepared with a PGFR of 1.2 L/min in the PPTAW method, demonstrated better resistance to surface crack formation compared to sample P1, prepared with a PGFR of 1.0 L/min. This observation has been previously explained by the entrapment of plasma gas in the plasma jet at lower PGFR, and an increase in the PGFR leads to increased heat in the system and expansion of the plasma jet, which fosters melting of powder particles and reduces pore formation and surface cracks [10].

On the other hand, it was noted that the number of crack sites relatively increased with an increase in laser power from 1.5 kW to 2.0 kW in the LC method. This phenomenon has been attributed to the differences in thermal expansion coefficients of powder particles and the higher hardness of WC, resulting in weaker interfacial bonding between powder particles at higher laser powers, leading to increased surface crack sites [49]. Sample L1, prepared with a laser power of 1.5 kW in the LC method, showed better resistance to surface crack formation compared to sample L2, prepared with a laser power of 2.0 kW.

It is worth noting that cladding surface cracks do not significantly impact the overall wear performance of the clad, as reported by earlier research [50]. In some cases, the crack sites can even be utilized as lubrication sites for applications that require surface lubricants [51].

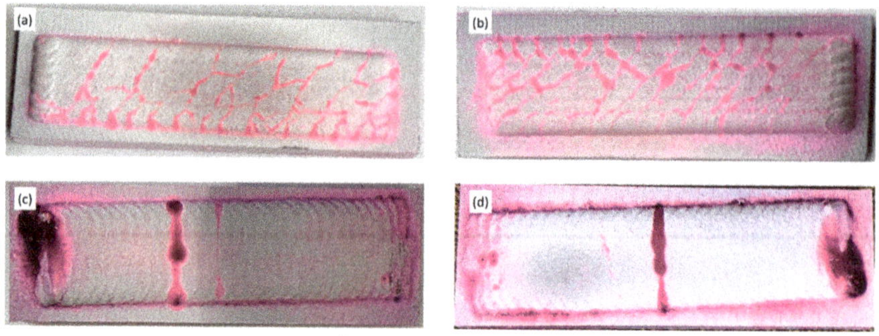

Figure 6. Top view of samples after penetrant test: (**a**,**b**) samples L1 and L2 prepared by laser cladding, (**c**,**d**) samples P1 and P2 prepared by PPTAW.

3.4. Microhardness Tests

The results of room-temperature Vicker's microhardness measurements of the prepared clads are visualized in the plot in Figure 7. It is observed that the mean hardness of the matrix phase in the clad prepared by the LC method is relatively higher than the mean of the matrix phase in the clad prepared by the PPTAW method. For the LC samples, L1 had matrix mean hardness of 695 HV and sample L2 had a mean of 634 HV. Samples P1 and P2 from the PPTAW method had matrix mean hardnesses of 600 HV and 620 HV respectively. Point measurements taken from the reinforcing carbide phases had contrasting mean values; PPTAW samples had higher values than LC samples. For LC samples, L1 had a carbide mean value of 2139 HV and L2 had a mean of 2120 HV. Samples P1 and P2 had average values of 2244 HV and 2261 HV, respectively. The standard deviations, as indicated by the bars in the plot, suggest that the mean values are not significantly further from each other. This outcome is expected because both processes are relatively high-energy-density processes, and the powder used is the same. However, it is observed in this hardness testing that sample L1 from LC showed higher hardness than sample L2. Although higher laser power introduces higher energy input in the cladding process, it is reported that increasing the laser power can also result in turbulent flow, resulting in reduced particle melting and a corresponding reduction in clad hardness [52]. It is also observed that sample P2 from the PPTAW method showed higher hardness than sample P1. As the PGFR was increased, the corresponding rise in plasma heat input led to higher particle temperatures, accelerating the dissolution of carbides into the matrix. This resulted in improved adhesion between the particles, leading to lower porosity and enhanced hardness, as reported by [11] in a similar PPTAW study.

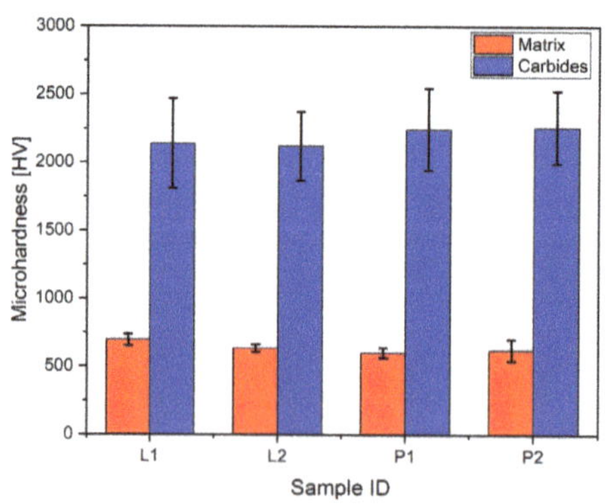

Figure 7. Average microhardness results of the matrix and carbide phases of the clads prepared by LC and PPTAW.

3.5. Abrasive-Wear Resistance Tests

To assess the abrasion resistance of the surface clads in metal-mineral abrasive-wear conditions, tests were conducted using Hardox 400 (AR400) as a reference material, which is a known abrasion-resistant material widely used in previous research [53–56]. The volume loss was determined by considering the density differences between WC, metallic matrix, and the reference material. The results of the abrasive-wear tests are presented in Table 4.

Table 4. Results of abrasive-wear resistance tests of surface clads composed of NiSiB + 60% WC prepared by laser cladding and powder plasma transferred arc welding methods in comparison to the abrasive-wear resistance of AR400 reference material.

Sample	Mass before Test, [g]	Mass after Test, [g]	Mass Loss, [g]	Material Density, [g/cm^3]	Volume Loss, [mm^3]	Relative Abrasive-Wear Resistance
			Surface clads			
L1	159.5765	159.2499	0.3266	11.1935	29.1776	4.5
L2	165.1587	164.6972	0.4615	11.1935	41.2292	3.2
P1	196.0594	195.6905	0.3689	11.1935	32.9566	4.0
P2	195.6418	195.3264	0.3154	11.1935	28.1771	4.7
			Reference Samples			
H1	104.6219	103.4971	1.0318	7.7836	132.5607	1.0
H2	111.7377	110.7989				

The findings reveal that the abrasive-wear performance of the surface clads varied depending on the process parameters for both LC and PPTAW methods. Specifically, the LC sample L1, prepared with a laser power of 1.5 kW, exhibited higher relative abrasive wear compared to sample L2, which had a laser power of 2.0 kW. Additionally, sample P2, fabricated with a PGFR of 1.2 L/min, showed superior relative abrasive-wear resistance compared to sample P1, which had a PGFR of 1.0 L/min.

The findings from the macroanalysis of surface cracks, microhardness measurement, and abrasive-wear resistance tests revealed appreciable results for sample L1 from the LC method and sample P2 from the PPTAW method. These samples showed superior performance compared to their counterparts that were prepared with variations in the process parameters. As a result, these select samples were chosen for further in-depth studies and investigations to gain a deeper understanding of the wear mechanism, microstructural evolution, and dilution mechanisms of the clads prepared by both methods in a comprehensive comparative analysis.

Digital micrographs in Figure 8 depict the surface of the clads before and after undergoing tests to determine their resistance to abrasive wear. The mechanism underlying the wear resistance was investigated and found to occur in two distinct stages. Initially, as the rubber wheel was set in motion with quartz sand in contact with the surface of the layer, the surface underwent wear. This process continued until the quartz sand had removed enough material, bringing the wear debris, which was rich in tungsten carbide (WC), closer to the quartz sand. When these WC particles came into contact with the quartz sand, smearing of the WC particles was observed. With increasing contact time, the WC particles reduced friction between the quartz sand and the revolution of the rubber wheel, resulting in a significant decrease in wear [57]. Micrographs in Figure 8 reveal that WC particles were exposed at the surface when further abrasion was hindered, and regions on the surface without WC particles (matrix) showed deeper abrasion compared to regions with a higher concentration of WC particles.

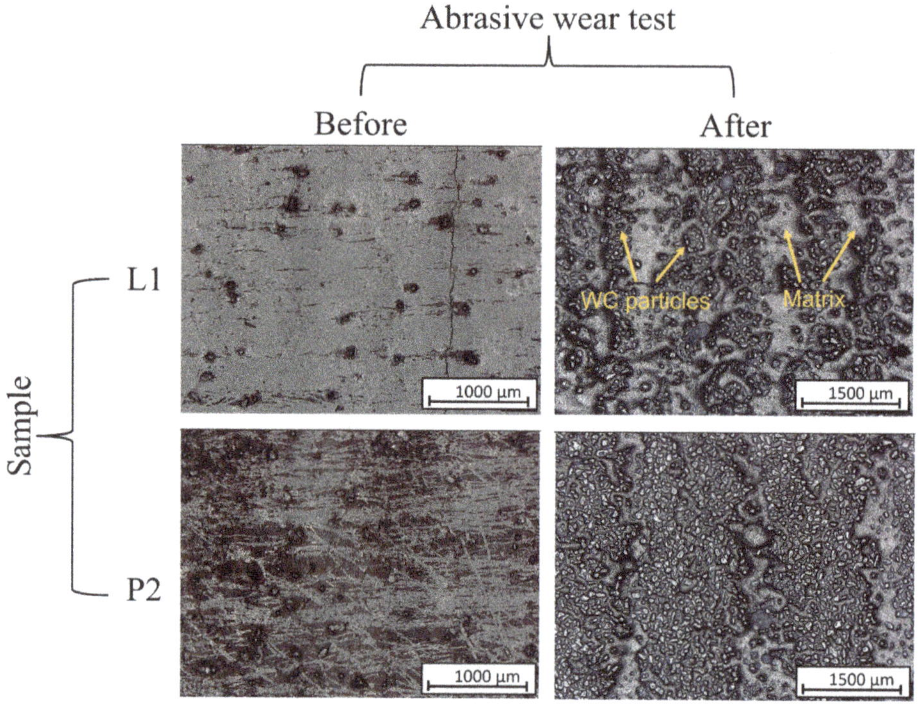

Figure 8. Digital micrographs of surface clads before and after abrasive-wear resistance tests for sample prepared by laser cladding (L1) and sample prepared by PPTAW (P2).

3.6. Microstructural Evolution

The metallographic cross-section of the clad and substrate material was examined in Figure 9 for samples fabricated using LC and PPTAW methods. The cross-section revealed distinct zones, including the clad subsurface, middle zone (MZ), transition zone (TZ), and heat-affected zone (HAZ). Microscopic analysis of the clad in both samples displayed varying degrees of particle dispersion, believed to be WC. Further magnification of the MZ in both samples obtained using a light microscope is shown in Figure 10.

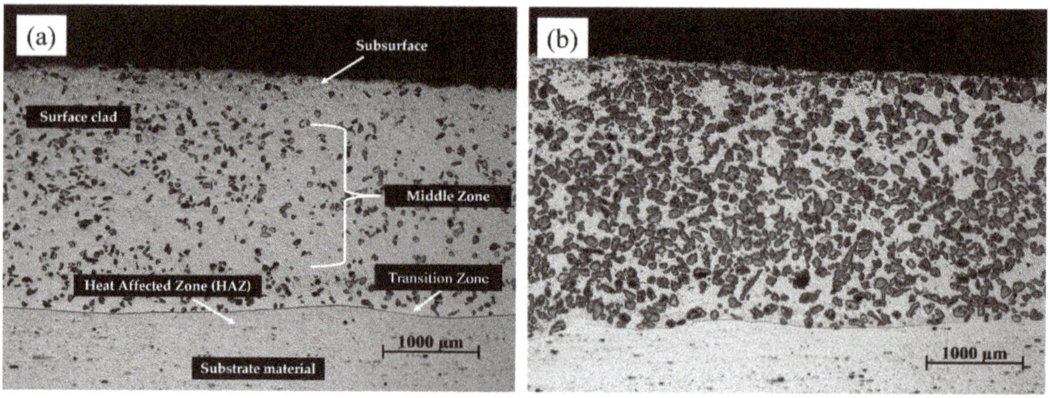

Figure 9. Light micrographs of prepared surface clads showing the observable zones across the cross-section: (**a**) sample L1 prepared by laser cladding, (**b**) sample P2 prepared by PPTAW.

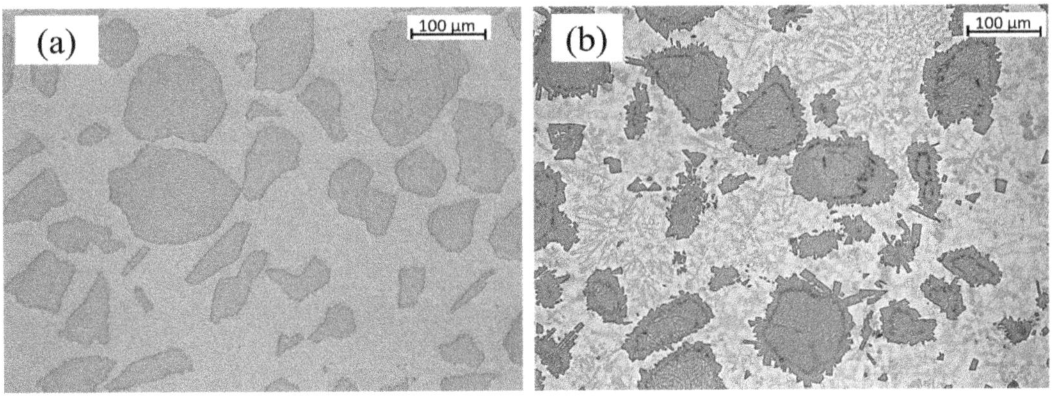

Figure 10. Light micrograph of the middle zone of surface clads at 200× magnification: (**a**) sample L1, (**b**) sample P2.

Upon solidification of the deposited clads, particle size analysis revealed different characteristics of WC particles in both the LC clad and the PPTAW clad. In the LC clad, the measurements showed a minimum size of 20 µm, a maximum size of 216 µm, and a mean particle size of 167.2 µm, with a standard deviation of 41.6. On the other hand, the PPTAW clad exhibited a minimum carbide particle size of 16 µm, a maximum size of 229 µm, and a mean size of 126.8 µm, with a standard deviation of 55.2. The mode particle sizes were determined to be 189 µm and 196 µm for the LC clad and PPTAW clad, respectively. These findings are further supported by the images in Figure 9, where it is evident that the carbide particles in the PPTAW clad appear larger compared to those in the LC clad. This disparity suggests that the PPTAW method resulted in a more even distribution of carbide particles within the clad, in contrast to the LC method. The discrepancy in particle size between the two methods can be attributed to the operational conditions. When employing the LC method, as described by Kotarska et al. [52], a laser power of 1.5 kW resulted in turbulent flow during the powder feeding process. This turbulence led to collisions among carbide particles, causing them to break down and resulting in relatively smaller sizes upon solidification of the clad. In contrast, the PPTAW method operated at a powder gas flow rate (PGFR) of 1.2 L/min, ensuring a consistent flow of powder particles. This continuous flow facilitated a more uniform dispersion of carbides within the clad, leading to the observed larger particle sizes in the PPTAW clad compared to the LC clad.

3.6.1. Middle Zone (MZ)

Figure 11 shows an SEM image of the MZ in the laser-clad sample, accompanied by EDS maps illustrating the distribution of elements across the examined MZ area. The EDS maps reveal non-uniform dispersion of primary WC particles within the Ni-based matrix, with precipitation of secondary WC phases observed, as shown in the inset in Figure 12. The MZ of the PPTAW sample, as observed under SEM, is presented in Figure 13, along with EDS maps illustrating the distribution of powder constituents in the MZ after solidification. The matrix exhibits a dendritic multi-directional microstructure upon solidification, characterized by the presence of partially dissolved primary carbides and precipitates of secondary carbides dispersed throughout.

EDS maps from Figures 11 and 13 confirm that the bright-grey regions in the SEM images correspond to the primary carbides, and the analyzed precipitates shown in Figures 12 and 14 are the secondary precipitates nucleated from the WC reinforcing phase during solidification. The EDS maps reveal consistent segregation of Si and W in the clads produced by both methods. The heat input required to fully melt the Ni-based matrix (with a melting point of 1555 °C) surpasses the melting temperature of Si (1410 °C). Consequently, the Si phases undergo melting prior to other materials. Due to the strong tendency to form a WSi_2 phase, the liquid Si forms bonds with W, leading to their observed segregation at the same locations according to EDS analysis. Döscher et al. [58] reported the formation of a WSi_2 phase at processing temperatures below 1414 °C, resulting from a solid-state reaction between W and Si. However, when processing temperatures exceed 1414 °C, the WS_2 phase forms from a reaction between molten Si and solid W. In typical MMC clads, ceramic reinforcement particles tend to agglomerate upon solidification due to density differences between the carbides and the matrix [59]. However, in the samples prepared in this study using LC and PPTAW methods, agglomeration of carbides was not observed, which can be attributed to the effective mixing of powder particles, resulting in the dispersion of carbides in the matrix phase.

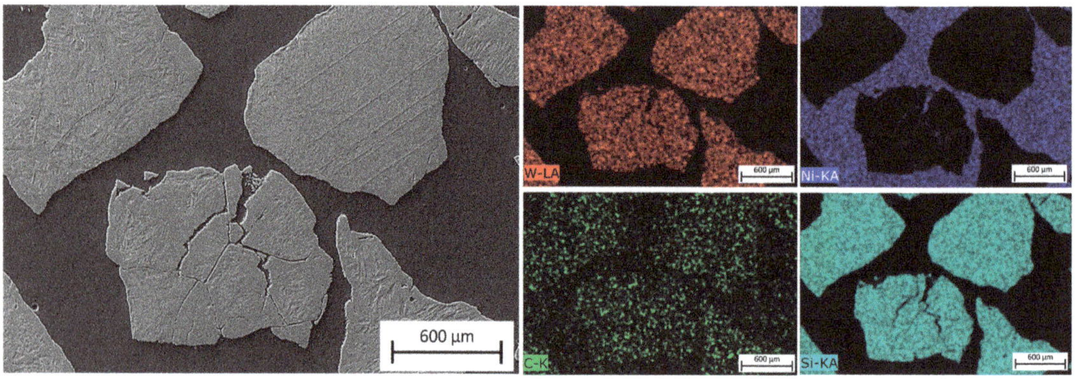

Figure 11. SEM image and maps from EDS microanalysis of the middle zone of sample L1 showing the distribution of the MMC clad constituent within the matrix.

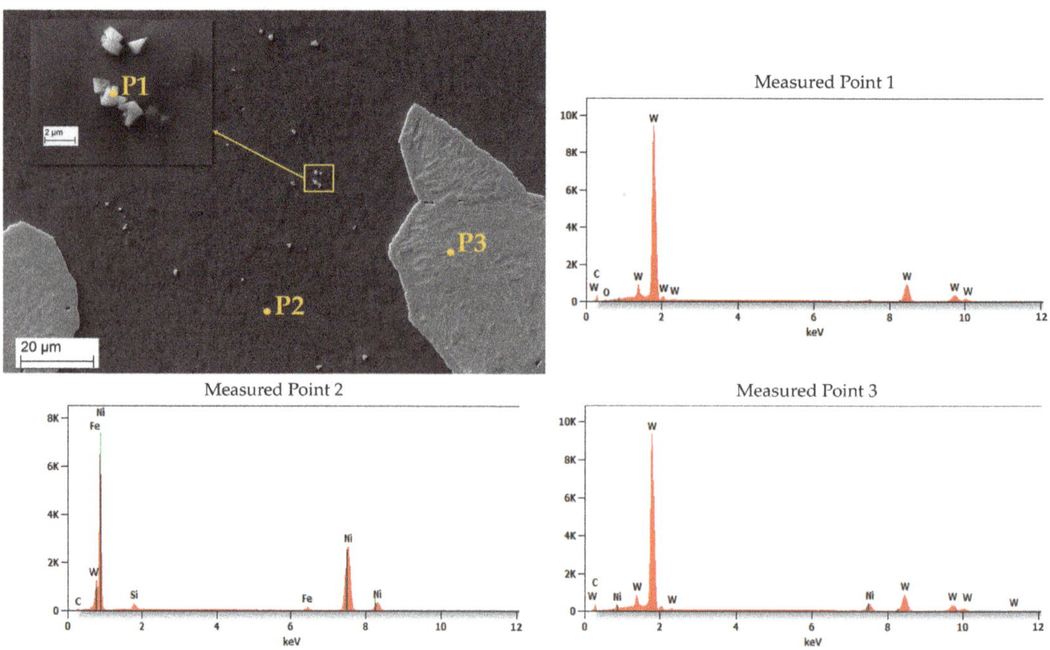

Figure 12. Energy-dispersive X-ray spectroscopy (EDS) microanalysis of the middle zone of surface clad for sample L1 prepared by laser cladding.

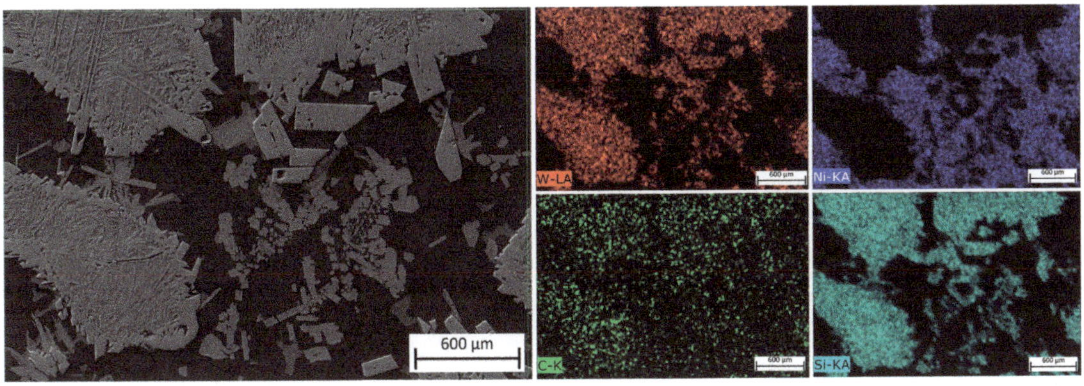

Figure 13. SEM image and maps from EDS microanalysis of the middle zone of sample P2 showing the distribution of the MMC clad constituent within the matrix.

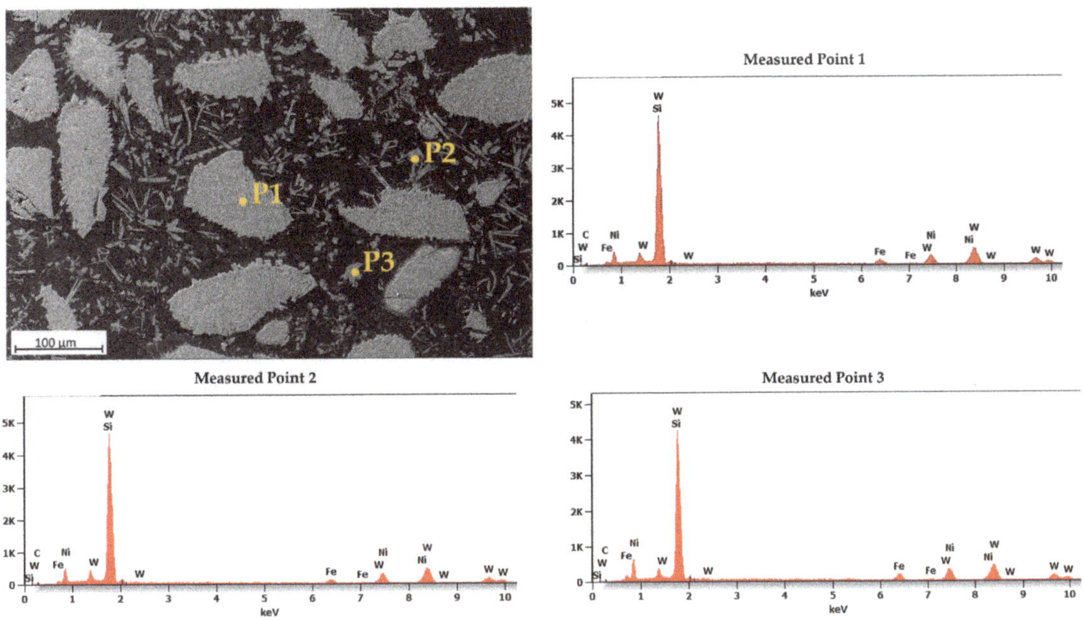

Figure 14. Energy-dispersive X-ray spectroscopy (EDS) microanalysis of the middle zone of surface clad for sample P2 prepared by PPTAW, showing the precipitates of secondary WC particles in the matrix.

During the cladding process, partial dissolution of the carbide particles (with a melting point of 2785 °C) in the completely dissolving Ni-based matrix phase is responsible for the observation of the dendritic structure formed in the MZ. Earlier research [60] has indicated that the bonding of these partially dissolving carbides in the matrix improves the stability of the clad, which is enhanced even further by the precipitation of secondary phases via accelerated diffusion.

3.6.2. Transition Zone (TZ)

Figure 15a provides an overview of the transition zone (TZ) for sample L1. Upon closer examination of the TZ (Figure 15b), the presence of precipitates with the same composition as those identified in the middle zone (MZ) can be observed, indicating nucleation of precipitates at the TZ. The heat-affected zone (HAZ) shown in Figure 15c is characterized by coarse grains composed of martensite laths and can be described as a coarse-grain heat-affected zone (CGHAZ). Additionally, a peninsula-like macrosegregation is observed at the TZ, as depicted in Figure 15d.

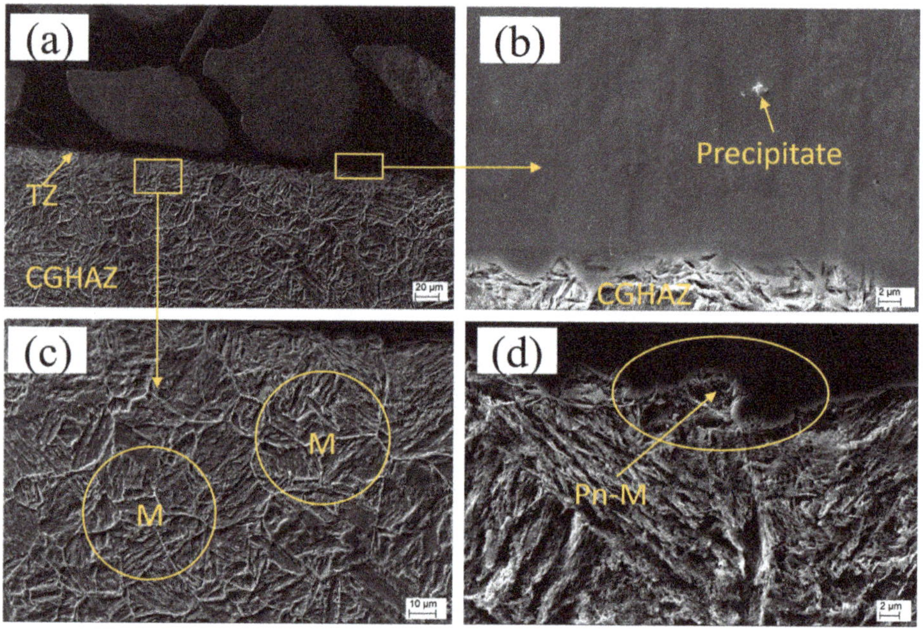

Figure 15. Scanning electron micrographs of sample L1: (**a**) overall view of the transition zone (TZ) showing the coarse-grain heat-affected zone (CGHAZ), (**b**) precipitation in the TZ, (**c**) CGHAZ martensite laths (M) in the substrate, (**d**) peninsula-like macrosegregation (Pn-M).

Similarly, the overview of the TZ for sample P2 is shown in Figure 16a. The TZ displays distinct regions, as shown in Figure 16b: a planar solidification growth region and a cellular–dendritic solidification growth (CDSG) region. The HAZ for this sample (Figure 16c) also exhibits characteristics of a CGHAZ, like that observed in sample L1. A type-II boundary is present at the TZ of sample P2, as shown in Figure 16d. Furthermore, peninsula-like macrosegregation is also observed at the TZ of sample P2, as depicted in Figure 16e. A magnified image of the CDSG region containing interdendritic precipitates is shown in Figure 16f.

The SEM images reveal that the transition zones in the samples prepared using LC and PPTAW methods are remarkably thin, measuring approximately 9 µm for LC and approximately 11 µm for PPTAW. Importantly, the TZs in both methods appear free from any discontinuities, such as partially dissolving powder particles, hot cracking, or absence of fusion. Additionally, the substrates of both LC and PPTAW samples exhibit clear evidence of metallurgical bonding with the clad material. In sample L1, the TZ shows a melt pool of the matrix phase that terminates in a planar mode upon solidification, which is a commonly observed phenomenon in cladding processes with relatively slower cooling rates, such as LC. However, the higher thermal cycles (heating and cooling) of the PPTAW method, which are approximately 10^8 K/s [61], and the associated high heat absorption by the substrate material, relative to the LC method, which has thermal cycles of approximately 10^4 K/s [62], resulted in a different solidification mode. The solidification advanced in three stages, starting with a planar-mode solidification. This was then followed by a cellular-mode solidification and terminated with a cellular–dendritic-mode solidification. The observed transitions in the solidification mode are linked to the solid–liquid interface instability. This is caused by a decrease in the rate of cooling, as the substrate also becomes heated during the cladding process. In addition, constitutional supercooling, which is described as the reduction in the solid solubility relative to the liquid [63], accounts for the observed changes in solidification mode. The PPTAW clad had a higher rate of nucleation close to the TZ, resulting from the substrate's heat sink and the presence of several sites for nucleation. This observation is consistent with the findings of Xibao et al. [64], who reported that some of these sites for nucleation emanate from partially melted powder particles.

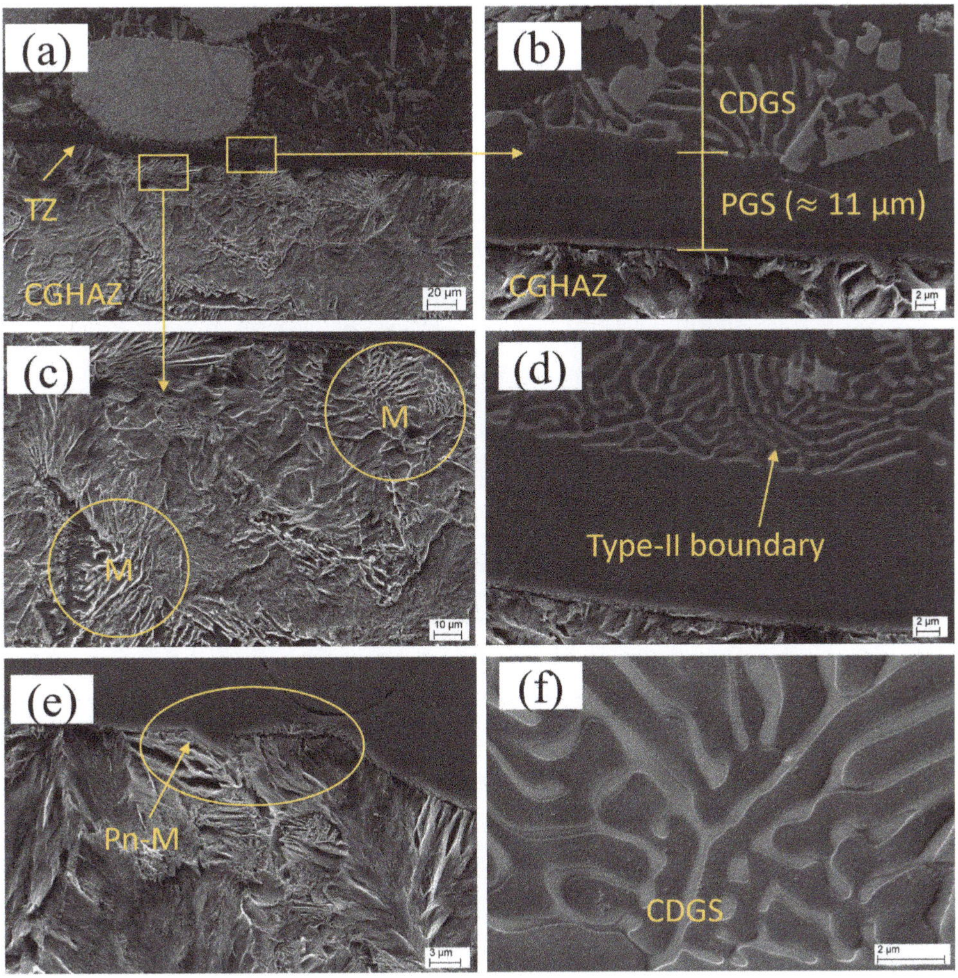

Figure 16. Scanning electron micrographs of PPTAW sample P2: (**a**) overall view of the transition zone (TZ) showing the coarse-grain heat-affected zone (CGHAZ), (**b**) planar growth solidification (PGS) and cellular–dendritic growth solidification (CDGS), (**c**) CGHAZ with martensite laths (M) in the substrate, (**d**) type-II boundary, (**e**) peninsula-like macrosegregation, (Pn-M), (**f**) oriented cellular/dendritic growth at the TZ containing interdendritic precipitates.

The occurrence of type-II boundaries at the transition zone (TZ) was observed exclusively in plasma powder transferred arc welding (PPTAW), while laser cladding (LC) did not exhibit this phenomenon. Previous research by Nelson et al. [65] has proposed that the transformation of ferrite (δ-γ) allotropic phases in the coarse-grain heat-affected zone (CGHAZ) during cooling is a critical prerequisite for the formation of type-II boundaries in structural steel. Recently, Farias et al. [39] postulated that the interface between the substrate and the clad is mobile and can induce projection of the CGHAZ into the clad during cooling, resulting in the formation of type-II boundaries. It is worth mentioning that the type-II boundary was only observed in a small section of the TZ. This is because the precipitation occurring in the TZ during the transition in the solidification mode impedes the advancement of the type-II boundary in the planar growth region [25].

Both PPTAW and LC samples exhibited tiny peninsula-like macrosegregations in the TZ of the clad, as observed in Figures 15d and 16e. The stagnation of the substrate liquid layer, which is very thin, accounted for this observation. It can be deduced from this that the dilution was low, resulting from the disparity in the substrate and MMC fusion temperatures, as well as the turbulence generated by the laser power and the PGFR. Consequently, the melted substrate was transported into the clad pool, where it solidified rapidly due to the presence of adjacent melted powder (liquid). According to the mechanism of the formation of type-I macrosegregation as proposed by Yang et al. [66], this type of macrosegregation limits the inter-liquid chemical diffusion. This observed phenomenon is an indication that LC and PPTAW used for these cladding operations are low-dilution processes and are able to impede the formation of macrosegregations. The absence or smaller size of macrosegregations is advantageous for improved clad performance. This is because, as proposed by Soysal et al. [67], the macrosegregations can potentially function as sites for hydrogen embrittlement because of their higher hardness.

The growth mode of cellular–dendritic solidification, as observed in the PPTAW sample, is a result of clad-pool grain-growth competition. This means that grains that possess higher growth rates impede the growth of their neighboring grains. This leads to the formation of an oriented microstructure. It is noteworthy that the fine interdendritic precipitates in the sample were not continuous, and no hot cracks were detected, indicating that the MMC clad and the structural-steel substrate exhibit good weldability. Furthermore, the PPTAW process, which has a higher cooling rate compared to LC, resulted in a greater presence of interdendritic precipitates.

Figure 17 depicts the microanalysis of the TZ in the LC sample using EDS. Similarly, Figure 18 presents the microanalysis of the TZ in the PPTAW sample. The EDS maps in these figures reveal the presence of a chemical composition gradient along the TZ, involving Si, Ni, W, C, and Fe, which can be attributed to the lower dilution during the cladding process. The gradient is not as steep as in mechanical bonding, indicating that the substrate and clad are metallurgically bonded through a diffusion layer. Additionally, the distribution of Fe content in close proximity to the TZ is found to be heterogeneous, with a decrease in Fe content observed with increasing distance from the TZ.

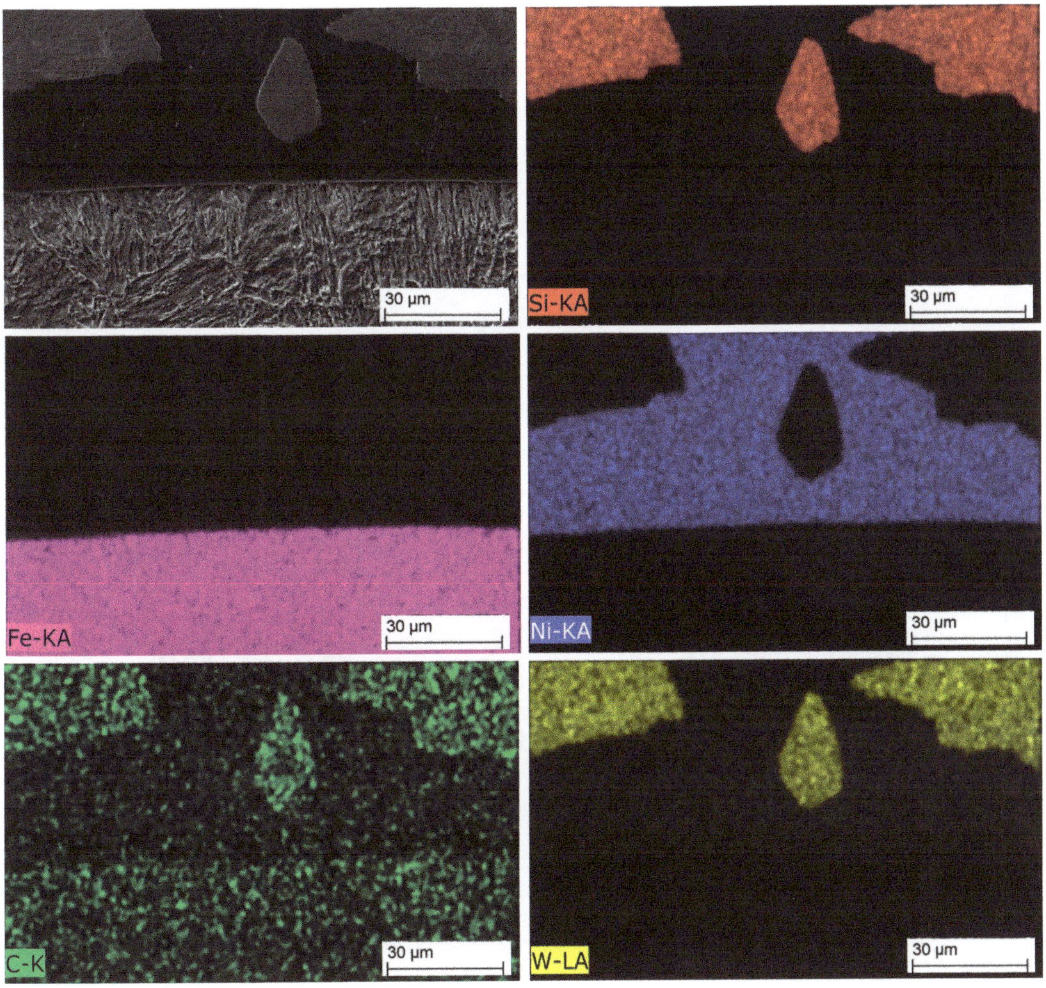

Figure 17. Energy-dispersive X-ray spectrometry mapping of the transition zone of laser-cladded sample L1.

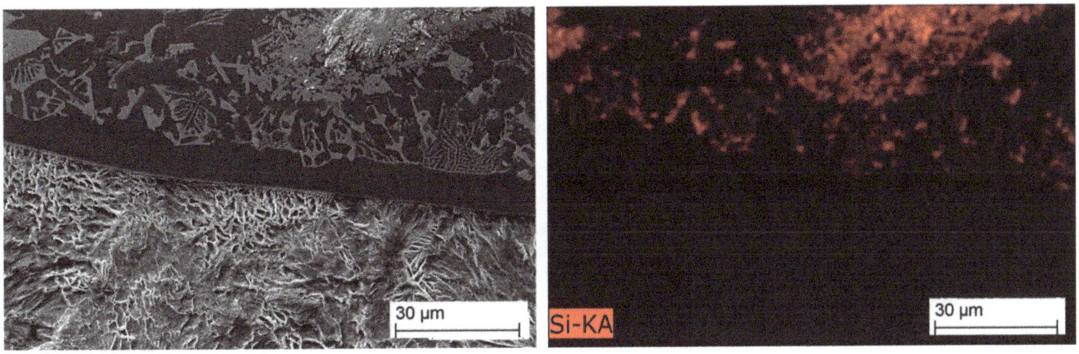

Figure 18. *Cont.*

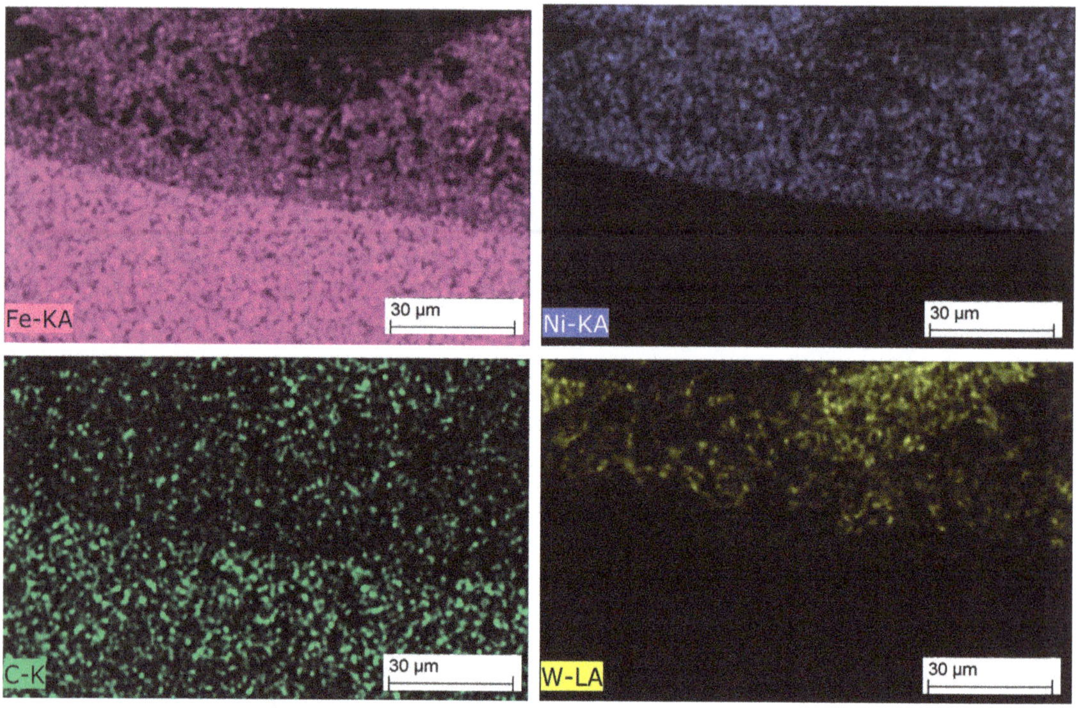

Figure 18. Energy-dispersive X-ray spectrometry mapping of the transition zone of PPTAW sample L1.

3.7. Dilution and Heat-Affected Zone (HAZ)

Figure 19 illustrates the relationship between the dilution ratio (D) and the area of dilution, as calculated using Equation 4. The reinforcement area (Fn) and the fusion zone area (Fw) were determined based on the geometrical schematic shown in Figure 19a. To obtain these values, measurements were taken using the AutoCAD software program (Autodesk, CA, USA) for the height of the clad on the substrate material (R), penetration depth of the clad into the substrate (P), and width of the clad across the x-plane (w). The results of the measurements and the computed values of the dilution coefficient D for the investigated clads are summarized in Table 5.

$$\text{Dilution, D} = \text{Fw}/(\text{Fw} + \text{Fn}) \times 100 \quad (4)$$

The findings indicate that LC demonstrated a significantly lower dilution ratio of 2.1% in comparison to PPTAW, which had a dilution ratio of 4.5%. These results are in line with previous literature that has consistently reported dilution ratios of less than 10% for both LC and PPTAW, contrasting with other surface-cladding technologies such as metal arc welding, gas tungsten arc welding, and gas metal arc welding, which typically exhibit dilution ratios ranging from 10% to 30%. Thus, the findings of this study support the existing literature's observations of low dilution ratios for LC and PPTAW [68,69].

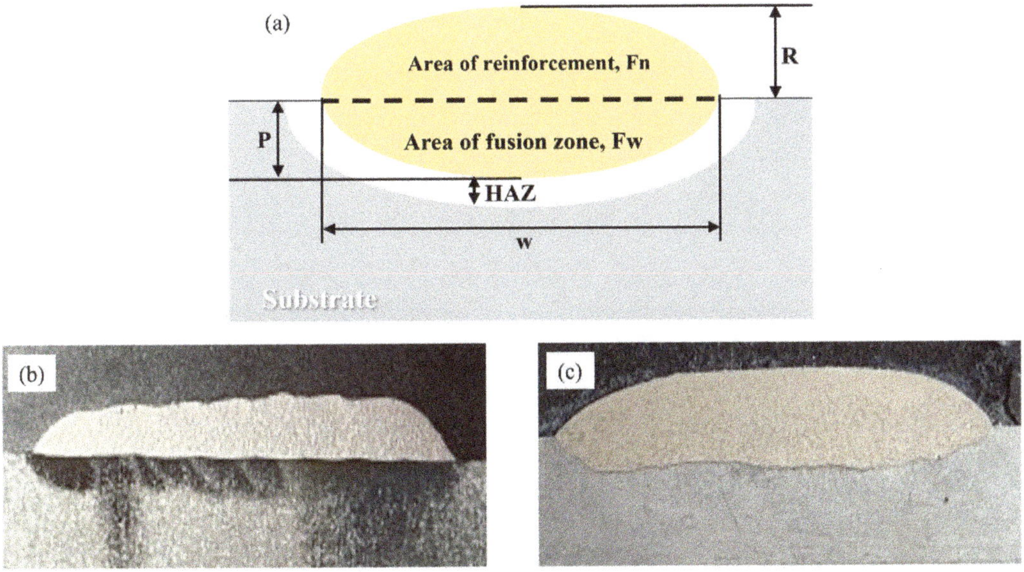

Figure 19. (**a**) Schematic representation of the geometric parameters of the cross-section of the final material after deposition of the clad onto the substrate material, depicting the area of reinforcement (Fn), area of fusion (Fw), heat-affected zone (HAZ), clad layer height on the substrate material (R), depth of clad penetration into the substrate material (P), and width of the clad across the x-plane (w). (**b**) Cross-sectional view of sample L1. (**c**) Cross-sectional view of sample P2.

Table 5. Geometrical properties and dilution ratio of prepared surface clads.

Sample	Cladding Method	Layer Height, R (mm)	Penetration Depth, P (mm)	Layer Width, w (mm)	Dilution, D, %
L1	LC	1.8	0.2	16	2.1
P2	PPTAW	2.7	0.5	24	4.5

In Figure 20, the microhardness measurement results of the heat-affected zone (HAZ) are presented, accompanied by a digital micrograph depicting the HAZ. The microhardness values of both samples L1 and P2 in this region were found to be higher than that of the substrate material. Notably, the LC sample exhibited a larger HAZ of approximately 283 μm (Figure 20a) compared to the PPTAW sample, with approximate HAZ of 35 μm (Figure 20c), attributed to the relatively higher heat retention of the LC process. The average microhardness value of the LC sample was measured to be 680 HV (standard deviation of 17.0), while the PPTAW sample had an average microhardness of 480 HV (standard deviation of 5.6). The HAZ in both samples predominantly consisted of martensite laths, forming what is commonly referred to as the coarse-grain heat-affected zone (CGHAZ), which is consistent with previous studies by Barick et al. [70] on ferritic steel. Furthermore, the increase in microhardness of the HAZ in both samples compared to the substrate material aligns with findings from earlier studies [71,72], which also reported higher microhardness in the coarse-grain structure of the HAZ.

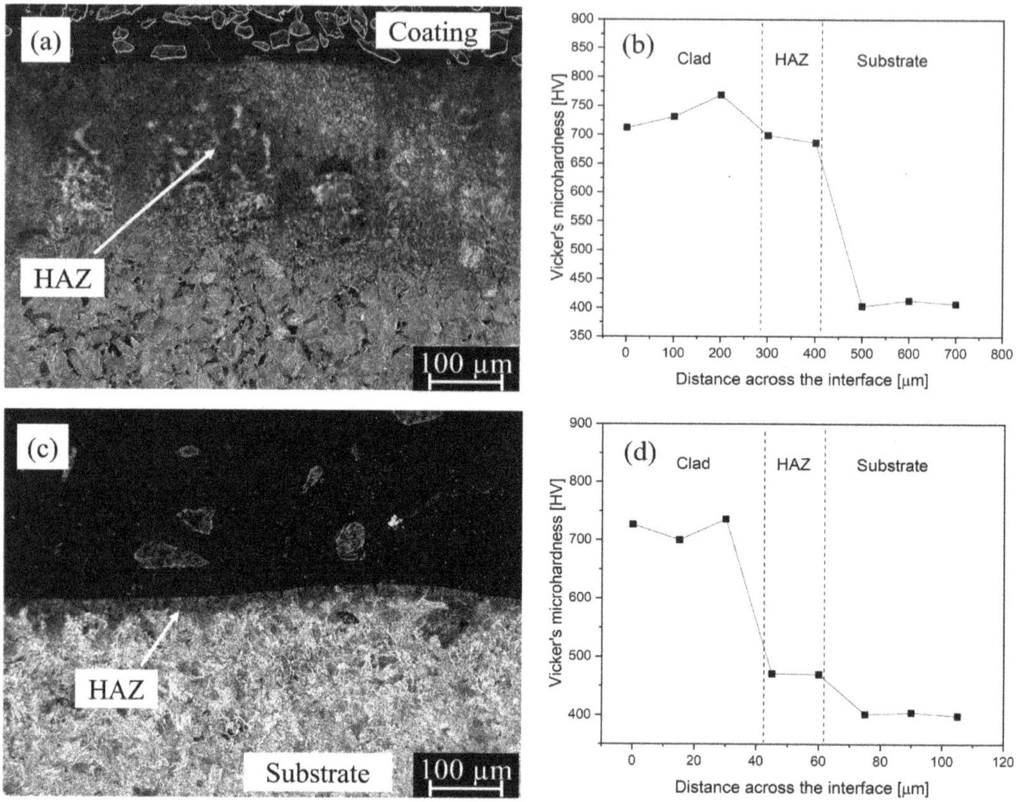

Figure 20. Digital micrograph of the cross-section of investigated sample showing the coating/cladding area, the heat-affected zone (HAZ) and the substrate area (Left), and microhardness profile of points taken along the indicated cross-section (Right). (**a**,**b**) Sample L1, (**c**,**d**) Sample P2.

4. Conclusions

1. A wear-resistant powder blend of composition NiSiB + 60% WC was successfully deposited on a structural-steel substrate using LC and PPTAW methods. There was precipitation of secondary WC phases in the matrix upon solidification for clads prepared by both methods. However, the clad from PPTAW had a dendritic microstructure, which is attributed to its relatively higher thermal cycles than those of the LC process.
2. Penetrant test macroanalysis revealed that the clad prepared by the LC method had more surface cracks than the PPTAW counterpart, owing to the lower laser heat input. The microhardness of the matrix and reinforcing WC phases of the clads prepared by both methods were comparatively similar. However, when compared to a reference abrasive-wear-resistant material, AR400, the relative abrasive-wear resistance of the PPTAW clad was relatively higher, at 4.7, than the clad prepared by LC, at 4.5. The wear mechanism was found to be the same for clads prepared by both methods.
3. The TZs for both clads were observed to be thin (approx. 9 µm for LC and approx. 11 µm for PPTAW), with a coarse-grain heat-affected zone (CGHAZ), made up of martensite laths, and a peninsula-like macrosegregation observed for clads from both methods. However, the PPTAW clad had a cellular–dendritic growth solidification (CDGS) and a type-II boundary at the TZ, which is explained as resulting from the thermal cycles of this method.

4. The LC clad had a lower dilution coefficient, at 2.1%, than the PPTAW clad, with dilution coefficient of 4.5%. The higher heat retention of the LC method resulted in a larger HAZ (≈283 μm) with higher hardness (average of 680 HV) than the HAZ of the PPTAW clad, which was ≈35 μm and had an average hardness of 480 HV. EDS analysis showed elemental diffusion at the clad/substrate interface with a diffusion gradient, which shows that the bonding of the clad by both methods was metallurgical, making them more desirable for industrial antiwear applications.

Author Contributions: Conceptualization, M.A. and A.C.; methodology, A.N.S.A., R.Ż. and A.C.; validation, G.F.B., M.A., A.N.S.A. and A.C.; formal analysis, A.N.S.A. and R.Ż.; investigation, A.N.S.A., R.Ż., A.C., G.F.B. and M.A.; resources, A.C. and M.A.; data curation, R.Ż. and A.N.S.A.; writing—original draft preparation, A.N.S.A. and M.A.; writing—review and editing, M.A., A.N.S.A., G.F.B., A.C. and R.Ż.; supervision, M.A. and A.C.; funding acquisition, M.A. All authors have read and agreed to the published version of the manuscript.

Funding: This research received no external funding.

Institutional Review Board Statement: Not applicable.

Informed Consent Statement: Not applicable.

Data Availability Statement: Not applicable.

Acknowledgments: Publication supported as part of the Excellence Initiative—Research University program implemented at the Silesian University of Technology, the year 2023.

Conflicts of Interest: The authors declare no conflict of interest.

References

1. Fasihi, P.; Kendall, O.; Abrahams, R.; Mutton, P.; Qiu, C.; Schläfer, T.; Yan, W. Tribological Properties of Laser Cladded Alloys for Repair of Rail Components. *Materials* **2022**, *15*, 7466. [CrossRef] [PubMed]
2. Zhong, W.; Hu, J.; Shen, P.; Wang, C.; Lius, Q. Experimental Investigation between Rolling Contact Fatigue and Wear of High-Speed and Heavy-Haul Railway and Selection of Rail Material. *Wear* **2011**, *271*, 2485–2493. [CrossRef]
3. Mädler, K.; Zoll, A.; Heyder, R.; Brehmer, M. *Rail Materials-Alternatives and Limits*; Deutsche Bahn AG, DB Systemtechnik: Brandenburg-Kirchmöser, Germany, 2008; pp. 18–22.
4. Fu, Z.; Ding, H.; Wang, W.; Liu, Q.; Guo, J.; Zhu, M. Investigation on Microstructure and Wear Characteristic of Laser Cladding Fe-Based Alloy on Wheel/Rail Materials. *Wear* **2015**, *330*, 592–599. [CrossRef]
5. Lazorenko, G.; Kasprzhitskii, A.; Nazdracheva, T. Anti-Corrosion Coatings for Protection of Steel Railway Structures Exposed to Atmospheric Environments: A Review. *Constr. Build. Mater.* **2021**, *288*, 123115. [CrossRef]
6. Wang, W.-J.; Fu, Z.-K.; Guo, J.; Zhang, Y.-Q.; Liu, Q.-Y.; Zhu, M.-H. Investigation on Wear Resistance and Fatigue Damage of Laser Cladding Coating on Wheel and Rail Materials under the Oil Lubrication Condition. *Tribol. Trans.* **2016**, *59*, 810–817. [CrossRef]
7. Kim, J.-D.; Peng, Y. Plunging Method for Nd: YAG Laser Cladding with Wire Feeding. *Opt. Lasers Eng.* **2000**, *33*, 299–309. [CrossRef]
8. Khamidullin, B.; Tsivilskiy, I.; Gorunov, A.; Gilmutdinov, A.K. Modeling of the Effect of Powder Parameters on Laser Cladding Using Coaxial Nozzle. *Surf. Coat. Technol.* **2019**, *364*, 430–443. [CrossRef]
9. Cui, C.; Guo, Z.; Liu, Y.; Xie, Q.; Wang, Z.; Hu, J.; Yao, Y. Characteristics of Cobalt-Based Alloy Coating on Tool Steel Prepared by Powder Feeding Laser Cladding. *Opt. Laser Technol.* **2007**, *39*, 1544–1550. [CrossRef]
10. Appiah, A.N.S.; Bialas, O.; Czupryński, A.; Adamiak, M. Powder Plasma Transferred Arc Welding of Ni-Si-B+ 60 Wt% WC and Ni-Cr-Si-B+ 45 Wt% WC for Surface Cladding of Structural Steel. *Materials* **2022**, *15*, 4956. [CrossRef]
11. Czupryński, A. Microstructure and Abrasive Wear Resistance of Metal Matrix Composite Coatings Deposited on Steel Grade AISI 4715 by Powder Plasma Transferred Arc Welding Part 1. Mechanical and Structural Properties of a Cobalt-Based Alloy Surface Layer Reinforced with Particles of Titanium Carbide and Synthetic Metal–Diamond Composite. *Materials* **2021**, *14*, 2382. [CrossRef]
12. Hou, Q.; He, Y.; Zhang, Q.; Gao, J. Influence of Molybdenum on the Microstructure and Wear Resistance of Nickel-Based Alloy Coating Obtained by Plasma Transferred Arc Process. *Mater. Des.* **2007**, *28*, 1982–1987. [CrossRef]
13. Rokanopoulou, A.; Skarvelis, P.; Papadimitriou, G. Microstructure and Wear Properties of the Surface of 2205 Duplex Stainless Steel Reinforced with Al2O3 Particles by the Plasma Transferred Arc Technique. *Surf. Coat. Technol.* **2014**, *254*, 376–381. [CrossRef]
14. Çelik, O.N. Microstructure and Wear Properties of WC Particle Reinforced Composite Coating on Ti6Al4V Alloy Produced by the Plasma Transferred Arc Method. *Appl. Surf. Sci.* **2013**, *274*, 334–340. [CrossRef]
15. Takatani, Y.; Tomita, T.; Tani, K.; Harada, Y. Corrosion Behavior of Aluminum Weld Overlay Alloys with Dispersed Niobium Carbide Particles in Sodium Chloride Solution. *J. Therm. Spray Technol.* **1996**, *5*, 154–160. [CrossRef]

16. Lai, C.-H.; Lin, S.-J.; Yeh, J.-W.; Chang, S.-Y. Preparation and Characterization of AlCrTaTiZr Multi-Element Nitride Coatings. *Surf. Coat. Technol.* **2006**, *201*, 3275–3280. [CrossRef]
17. Wang, L.M.; Chen, C.C.; Yeh, J.W.; Ke, S.T. The Microstructure and Strengthening Mechanism of Thermal Spray Coating NixCo0.6Fe0.2CrySizAlTi0.2 High-Entropy Alloys. *Mater. Chem. Phys.* **2011**, *126*, 880–885. [CrossRef]
18. Wadge, M.D.; Lowther, M.; Cooper, T.P.; Reynolds, W.J.; Speidel, A.; Carter, L.N.; Rabbitt, D.; Kudrynskyi, Z.R.; Felfel, R.M.; Ahmed, I.; et al. Tailoring Absorptivity of Highly Reflective Ag Powders by Pulsed-Direct Current Magnetron Sputtering for Additive Manufacturing Processes. *J. Mater. Process. Technol.* **2023**, *317*, 117985. [CrossRef]
19. Yang, J.; Bai, B.; Ke, H.; Cui, Z.; Liu, Z.; Zhou, Z.; Xu, H.; Xiao, J.; Liu, Q.; Li, H. Effect of Metallurgical Behavior on Microstructure and Properties of FeCrMoMn Coatings Prepared by High-Speed Laser Cladding. *Opt. Laser Technol.* **2021**, *144*, 107431. [CrossRef]
20. Sandhu, S.S.; Shahi, A.S. Metallurgical, Wear and Fatigue Performance of Inconel 625 Weld Claddings. *J. Mater. Process. Technol.* **2016**, *233*, 1–8. [CrossRef]
21. Kumar, V.; Rakshit, R.; Das, A.K. Mechanical and Tribological Performance of Fiber Laser Cladded H-BN + SS316 Composite on SS316 Surface. *J. Mater. Process. Technol.* **2020**, *278*, 116509. [CrossRef]
22. Ding, H.; Yang, T.; Wang, W.; Zhu, Y.; Lin, Q.; Guo, J.; Xiao, Q.; Gan, L.; Liu, Q. Optimization and Wear Behaviors of 316L Stainless Steel Laser Cladding on Rail Material. *24th Int. Conf. Wear Mater.* **2023**, *523*, 204830. [CrossRef]
23. Fu, K.; Zhong, C.; Zhang, L.; Wang, X.; Nie, B.; Xue, Y.; Gui, W.; Luan, B. Effect of Multiple Thermal Cycling on the Microstructure and Microhardness of Inconel 625 by High-Speed Laser Cladding. *J. Mater. Res. Technol.* **2023**, *24*, 1093–1107. [CrossRef]
24. Fu, Y.; Li, L.; Guo, X.; Li, M.; Pan, Z.; Wang, H.; Liu, C.; Zhao, L. Fe-Co-Based Coating with High Hardness and High Saturation Magnetization Deposited by Co-Axial Powder Feeding Plasma Transferred Arc Welding. *Mater. Lett.* **2022**, *315*, 131928. [CrossRef]
25. Farias, F.W.C.; da Cruz Payão Filho, J.; da Silva Júnior, D.A.; de Moura, R.N.; Rios, M.C.G. Microstructural Characterization of Ni-Based Superalloy 625 Clad Welded on a 9% Ni Steel Pipe by Plasma Powder Transferred Arc. *Surf. Coat. Technol.* **2019**, *374*, 1024–1037. [CrossRef]
26. Ulutan, M.; Kiliçay, K.; Çelik, O.N.; Er, Ü. Microstructure and Wear Behaviour of Plasma Transferred Arc (PTA)-Deposited FeCrC Composite Coatings on AISI 5115 Steel. *J. Mater. Process. Technol.* **2016**, *236*, 26–34. [CrossRef]
27. Wu, P.; Zhou, C.; Tang, X. Microstructural Characterization and Wear Behavior of Laser Cladded Nickel-Based and Tungsten Carbide Composite Coatings. *Surf. Coat. Technol.* **2003**, *166*, 84–88. [CrossRef]
28. Wu, T.; Shi, W.; Xie, L.; Gong, M.; Huang, J.; Xie, Y.; He, K. Study on the Effect of Ni60 Transition Coating on Microstructure and Mechanical Properties of Fe/WC Composite Coating by Laser Cladding. *Opt. Laser Technol.* **2023**, *163*, 109387. [CrossRef]
29. Yamaguchi, T.; Tanaka, K.; Hagino, H. Porosity Reduction in WC-12Co Laser Cladding by Aluminum Addition. *Int. J. Refract. Met. Hard Mater.* **2023**, *110*, 106020. [CrossRef]
30. Appiah, A.N.S.; Bialas, O.; Żuk, M.; Czupryński, A.; Sasu, D.K.; Adamiak, M. Hardfacing of Mild Steel with Wear-Resistant Ni-Based Powders Containing Tungsten Carbide Particles Using Powder Plasma Transferred Arc Welding Technology. *Mater. Sci.-Pol.* **2022**, *40*, 42–63. [CrossRef]
31. Czupryński, A.; Żuk, M. Matrix Composite Coatings Deposited on AISI 4715 Steel by Powder Plasma-Transferred Arc Welding. Part 3. Comparison of the Brittle Fracture Resistance of Wear-Resistant Composite Layers Surfaced Using the PPTAW Method. *Materials* **2021**, *14*, 6066. [CrossRef]
32. Kim, J.-S.; Lee, H.-W. Effect of Welding Heat Input on Microstructure and Texture of Inconel 625 Weld Overlay Studied Using the Electron Backscatter Diffraction Method. *Metall. Mater. Trans. A* **2016**, *47*, 6109–6120. [CrossRef]
33. Yoon, B.H.; Ahn, Y.S.; Lee, C.H. The Effect of Dilution on HAZ Liquation Cracking in PTAW Ni-Base Superalloys Overlay Deposit. *ISIJ Int.* **2002**, *42*, 178–183. [CrossRef]
34. DuPont, J. Solidification of an Alloy 625 Weld Overlay. *Metall. Mater. Trans. A* **1996**, *27*, 3612–3620. [CrossRef]
35. Gittos, M.; Gooch, T. Effect of Iron Dilution on Corrosion Resistance of Ni-Cr-Mo Alloy Cladding. *Br. Corros. J.* **1996**, *31*, 309–314. [CrossRef]
36. Örnek, C.; Mansoor, M.; Larsson, A.; Zhang, F.; Harlow, G.S.; Kroll, R.; Carlà, F.; Hussain, H.; Derin, B.; Kivisäkk, U.; et al. The Causation of Hydrogen Embrittlement of Duplex Stainless Steel: Phase Instability of the Austenite Phase and Ductile-to-Brittle Transition of the Ferrite Phase–Synergy between Experiments and Modelling. *Corros. Sci.* **2023**, *217*, 111140. [CrossRef]
37. Dong, L.; Peng, Q.; Xue, H.; Han, E.-H.; Ke, W.; Wang, L. Correlation of Microstructure and Stress Corrosion Cracking Initiation Behaviour of the Fusion Boundary Region in a SA508 Cl. 3-Alloy 52M Dissimilar Weld Joint in Primary Pressurized Water Reactor Environment. *Corros. Sci.* **2018**, *132*, 9–20. [CrossRef]
38. Wu, Y.; Cai, Y.; Sun, D.; Zhu, J.; Wu, Y. Microstructure and Properties of High-Power Laser Welding of SUS304 to SA553 for Cryogenic Applications. *J. Mater. Process. Technol.* **2015**, *225*, 56–66. [CrossRef]
39. Farias, F.W.C.; da Cruz Payão Filho, J.; de Azevedo, L.M.B. Microstructural and Mechanical Characterization of the Transition Zone of 9% Ni Steel Cladded with Ni-Based Superalloy 625 by GTAW-HW. *Metals* **2018**, *8*, 1007. [CrossRef]
40. Pan, C.; Ma, K.; Chen, X.; Feng, S.; Fang, C.; Yang, H.; He, P. Corrosion Behavior of Ni-Based Coating with Ni-Coated TiC-Reinforced Particles by Induction Cladding in Molten Brass. *Mater. Chem. Phys.* **2023**, *296*, 127257. [CrossRef]
41. Li, M.; Yang, J.; Han, B.; Song, L.; Li, P.; Dong, W.; Xue, X. Comparative Investigation on Microstructures and Properties of WC/Cr3C2 Reinforced Laser Cladding Ni-Based Composite Coatings Subjected to Ultrasonic Impact Treatment. *Mater. Today Commun.* **2023**, *34*, 105219. [CrossRef]

42. Wu, W.; Zhang, C.; Wang, R.; Zhang, Y.; Lu, X. A High Temperature Wear-Resistant Ni-Based Alloy Coating for Coppery Blast Furnace Tuyere Application. *Surf. Coat. Technol.* **2023**, *464*, 129550. [CrossRef]
43. *ISO 14175*; Welding Consumables—Gases and Gas Mixtures for Fusion Welding and Allied Processes. American Welding Society (ISO/TC 44/SC 3): Miami, FL, USA, 2008.
44. *PN-EN ISO 3452*; Non-Destructive Testing—Penetrant Testing. South African Bureau of Standards (ISO/TC 135/SC 2): Pretoria, South Africa, 2021.
45. ArunKumar, V.; Prakash, N.; Deenadayalan, K. Comparison between the Spherical and Angular Type WC Particles Reinforced in Nickel-Based Matrix Deposited Using Plasma Transferred Arc Welding Process. *Mater. Today Proc.* 2023, *in press*. [CrossRef]
46. Li, W.; Di, R.; Yuan, R.; Song, H.; Lei, J. Microstructure, Wear Resistance and Electrochemical Properties of Spherical/Non-Spherical WC Reinforced Inconel 625 Superalloy by Laser Melting Deposition. *J. Manuf. Process.* **2022**, *74*, 413–422. [CrossRef]
47. Phelan, D.; Dippenaar, R. Widmanstätten Ferrite Plate Formation in Low-Carbon Steels. *Metall. Mater. Trans. A* **2004**, *35*, 3701–3706. [CrossRef]
48. Igwemezie, V.; Dirisu, P.; Mehmanparast, A. Critical Assessment of the Fatigue Crack Growth Rate Sensitivity to Material Microstructure in Ferrite-Pearlite Steels in Air and Marine Environment. *Mater. Sci. Eng. A* **2019**, *754*, 750–765. [CrossRef]
49. Poloczek, T.; Lont, A.; Górka, J. The Structure and Properties of Laser-Cladded Inconel 625/TiC Composite Coatings. *Materials* **2023**, *16*, 1265. [CrossRef]
50. Czupryński, A. Microstructure and Abrasive Wear Resistance of Metal Matrix Composite Coatings Deposited on Steel Grade AISI 4715 by Powder Plasma Transferred Arc Welding Part 2. Mechanical and Structural Properties of a Nickel-Based Alloy Surface Layer Reinforced with Particles of Tungsten Carbide and Synthetic Metal–Diamond Composite. *Materials* **2021**, *14*, 2805. [CrossRef]
51. Shu, K.; Zhang, C.; Sun, D.; Wu, B.; Li, Z.; Wu, J.; Zheng, D.; Gu, L. Contact Cracking and Delamination of Thin Solid Lubricating Films on Deformable Substrates. *Tribol. Int.* **2021**, *163*, 107181. [CrossRef]
52. Kotarska, A.; Poloczek, T.; Janicki, D. Characterization of the Structure, Mechanical Properties and Erosive Resistance of the Laser Cladded Inconel 625-Based Coatings Reinforced by TiC Particles. *Materials* **2021**, *14*, 2225. [CrossRef]
53. Górka, J.; Czupryński, A.; Żuk, M.; Adamiak, M.; Kopyść, A. Properties and Structure of Deposited Nanocrystalline Coatings in Relation to Selected Construction Materials Resistant to Abrasive Wear. *Materials* **2018**, *11*, 1184. [CrossRef]
54. Mondal, J.; Das, K.; Das, S. An Investigation of Mechanical Property and Sliding Wear Behaviour of 400Hv Grade Martensitic Steels. *Wear* **2020**, *458*, 203436. [CrossRef]
55. Katinas, E.; Jankauskas, V.; Kazak, N.; Michailov, V. Improving Abrasive Wear Resistance for Steel Hardox 400 by Electro-Spark Deposition. *J. Frict. Wear* **2019**, *40*, 100–106. [CrossRef]
56. Białobrzeska, B. Comparative Analysis of Abrasive-Wear Resistance of Brinar and Hardox Steels. *Tribologia* **2017**, *272*, 7–16. [CrossRef]
57. Bansal, A.; Zafar, S.; Sharma, A.K. Microstructure and Abrasive Wear Performance of Ni-WC Composite Microwave Clad. *J. Mater. Eng. Perform.* **2015**, *24*, 3708–3716. [CrossRef]
58. Döscher, M.; Pauli, M.; Müller, J. A Study on WSi2 Thin Films, Formed by the Reaction of Tungsten with Solid or Liquid Silicon, by Rapid Thermal Annealing. *Thin Solid Films* **1994**, *239*, 251–258. [CrossRef]
59. Zygmuntowicz, J.; Miazga, A.; Konopka, K.; Edrysiak, K.J.; Kaszuwara, W. Alumina Matrix Ceramic-Nickel Composites Formed by Centrifugal Slip Casting. *Process. Appl. Ceram.* **2015**, *9*, 199–202. [CrossRef]
60. Poloczek, T.; Czupryński, A.; Żuk, M.; Chruściel, M. Structure and Tribological Properties of Wear-Resistant Layers Produced in Process of Plasma Powder Surfacing. *Weld. Technol. Rev.* **2019**, *91*, 35–41. [CrossRef]
61. Bianchi, L.; Leger, A.C.; Vardelle, M.; Vardelle, A.; Fauchais, P. Splat Formation and Cooling of Plasma-Sprayed Zirconia. *Thin Solid Films* **1997**, *305*, 35–47. [CrossRef]
62. Bhowmik, A.; Yang, Y.; Zhou, W.; Chew, Y.; Bi, G. On the Heterogeneous Cooling Rates in Laser-Clad Al-50Si Alloy. *Surf. Coat. Technol.* **2021**, *408*, 126780. [CrossRef]
63. Su, M.; Yuan, X.; Yue, C.; Zheng, W.; Kang, J. Effect of Liquid Film Constitutional Supercooling at the End of Solidification on the Hot Cracking Behavior of Al-4.0 Wt.%Cu Alloy. *Mater. Lett.* **2022**, *328*, 133116. [CrossRef]
64. Xibao, W.; Hua, L. Metal Powder Thermal Behaviour during the Plasma Transferred-Arc Surfacing Process. *Surf. Coat. Technol.* **1998**, *106*, 156–161. [CrossRef]
65. Nelson, T.W.; Lippold, J.; Mills, M. Investigation of Boundaries and Structures in Dissimilar Metal Welds. *Sci. Technol. Weld. Join.* **1998**, *3*, 249–255. [CrossRef]
66. Yang, Y.; Kou, S. Macrosegregation Mechanisms in Arc Welds Made with Dissimilar Filler Metals. *Sci. Technol. Weld. Join.* **2010**, *15*, 15–30. [CrossRef]
67. Soysal, T.; Kou, S.; Tat, D.; Pasang, T. Macrosegregation in Dissimilar-Metal Fusion Welding. *Acta Mater.* **2016**, *110*, 149–160. [CrossRef]
68. Lou, L.-Y.; Liu, K.-C.; Jia, Y.-J.; Ji, G.; Wang, W.; Li, C.-J.; Li, C.-X. Microstructure and Properties of Lightweight Al0.2CrNbTiV Refractory High Entropy Alloy Coating with Different Dilutions Deposited by High Speed Laser Cladding. *Surf. Coat. Technol.* **2022**, *447*, 128873. [CrossRef]

69. Rajeev, G.P.; Kamaraj, M.; Bakshi, S.R. Comparison of Microstructure, Dilution and Wear Behavior of Stellite 21 Hardfacing on H13 Steel Using Cold Metal Transfer and Plasma Transferred Arc Welding Processes. *Surf. Coat. Technol.* **2019**, *375*, 383–394. [CrossRef]
70. Barrick, E.J.; Jain, D.; DuPont, J.N.; Seidman, D.N. Effects of Heating and Cooling Rates on Phase Transformations in 10 Wt Pct Ni Steel and Their Application to Gas Tungsten Arc Welding. *Metall. Mater. Trans. A* **2017**, *48*, 5890–5910. [CrossRef]
71. Nippes, E.; Balaguer, J. A Study of the Weld Heat-Affected Zone Toughness of 9% Nickel Steel. *Weld J* **1986**, *65*, 237.
72. Jang, J.-I.; Ju, J.-B.; Lee, B.-W.; Kwon, D.; Kim, W.-S. Micromechanism of Local Brittle Zone Phenomenon in Weld Heat-Affected Zones of Advanced 9% Ni Steel. *J. Mater. Sci. Lett.* **2001**, *20*, 2149–2152. [CrossRef]

materials

MDPI

Article

Powder Plasma Transferred Arc Welding of Ni-Si-B+60 wt%WC and Ni-Cr-Si-B+45 wt%WC for Surface Cladding of Structural Steel

Augustine Nana Sekyi Appiah [1], Oktawian Bialas [2], Artur Czupryński [3] and Marcin Adamiak [1,*]

1 Materials Research Laboratory, Faculty of Mechanical Engineering, Silesian University of Technology, 18A Konarskiego Street, 44-100 Gliwice, Poland; augustine.appiah@polsl.pl
2 Department of Engineering Materials and Biomaterials, Faculty of Mechanical Engineering, Silesian University of Technology, 18A Konarskiego Street, 44-100 Gliwice, Poland; oktawian.bialas@polsl.pl
3 Welding Department, Faculty of Mechanical Engineering, Silesian University of Technology, 18A Konarskiego Street, 44-100 Gliwice, Poland; artur.czuprynski@polsl.pl
* Correspondence: marcin.adamiak@polsl.pl; Tel.: +48-502-155-968

Abstract: Increasing demand for sustainable approaches to mining and raw material extraction, has prompted the need to explore advanced methods of surface modification for structural steels used in the extractive industry. The technology of powder plasma transferred arc welding (PPTAW), was used in this study as a surface modification technique to improve upon the abrasive wear resistance of structural steel grade EN S355. PPTAW process parameters, namely, plasma transferred arc (PTA) current and plasma gas flow rate (PGFR), were varied, and the effects of the variation were studied and used as criteria for selecting optimum conditions for further studies and parametric reproducibility. Two metal matrix composite (MMC) powders were used in the process, having compositions of Ni-Si-B+60 wt%WC (PG) and Ni-Cr-Si-B+45 wt%WC (PE). Microstructural observation under a scanning electron microscope (SEM) revealed a dendritic, multi-directional microstructure consisting of partially dissolved primary tungsten carbide particles and secondary tungsten carbide precipitates within the MMC solid solution. The hardness of the surface layers was higher than that of a reference AR400 steel by more than 263 HV. Final surface layers obtained from the MMC powders had abrasive wear resistance up to 5.7 times that of abrasion-resistant reference AR400 steel. Alloying the MMC matrix with chromium increased the hardness by 29.4%. Under the same process conditions, MMC powder with 60 wt% WC reinforcement had better abrasive wear resistance by up to 45.8% more than the MMC powder with 45 wt% WC.

Keywords: PPTAW; microstructure; abrasive wear; surface layer; SEM; precipitation; hardness

check for updates

Citation: Appiah, A.N.S.; Bialas, O.; Czupryński, A.; Adamiak, M. Powder Plasma Transferred Arc Welding of Ni-Si-B+60 wt%WC and Ni-Cr-Si-B+45 wt%WC for Surface Cladding of Structural Steel. *Materials* **2022**, *15*, 4956. https://doi.org/10.3390/ma15144956

Academic Editors: Francisco J. G. Silva and Tomasz Trzepieciński

Received: 28 June 2022
Accepted: 15 July 2022
Published: 16 July 2022

1. Introduction

During the service lives of engineering materials, there is constant exposure to conditions, such as creep, wear, corrosion, etc., that have significant damaging impacts on the properties of the materials. With time, there is material failure, mostly accompanied by catastrophic consequences, workplace hazards, financial loss, inter alia. The use of structural steel is frequently seen in processes such as underground mining, or oil and gas extraction. These processes subject the steels to rapid wear and failure. A major contribution to this observation in the extractive industry is the complex nature of the layered rocks as well as the geological inhomogeneity of these structures in the extraction sites [1]. The surfaces of the tools and machines used for drilling activities in this sector form the basis for accelerated material wear. The surface properties are, therefore, of much influence on the overall performance of the material. In the extractive sector, handling and transporting raw materials present an inevitability of contact of tool parts with hard components, including sand and rocks. Whereas industries in the heavy machinery and power generation sector constantly develop novel approaches to addressing challenges of material wear, the fossil fuel sector is still plagued by this limitation, due to the complexity of operation and interaction of tools with different particles [2]. The major approach adopted by such industries to address this is keeping an inventory of spare parts of machines and tools that are frequently exposed to wear [3]. This, however, encourages the need for regular downtimes for maintenance, resulting in massive tolls taken on company finances and time management, originating from the cost of tool parts, which keeps increasing, as well as both scheduled and abrupt maintenance times. Another approach adopted by the mining industry is the purchase of tools and machine parts with high quality, that can be in service for longer periods, reducing the need for regular downtimes [4]. The downside to this strategy is the high cost of such tools and machine parts, which would also need to be replaced at some point anyway. The conditions to which tools are exposed vary geologically, making it difficult to predict the service lives of the same tool used in different locations. The type of environment and abrasive wear encountered by the tools, therefore, have direct effects on their service lives.

This challenge in the extractive industry has been highlighted by research [5–8], stressing the need for reducing the abrasive wear of tools and machine parts. As some researchers [9–11] focus on designing new tools, by optimizing parameters to extend the service lives of such tools, surface modification approaches have been acclaimed as being more sustainable in addressing abrasive wear challenges in the extractive sector. By using surface modification strategies, techniques and materials, engineering materials, structural steel inclusive, have experienced enhanced performance and properties, such as corrosion resistance, hardness, wear resistance, etc. Surface modification, due to its flexibility of use and wide availability of material choices and techniques, is considered an economical and highly effective way of addressing challenges of engineering materials failure. Surface modification can completely alter the structure of the material's surface, giving rise to different material properties, evidenced by the structure—property relationship of materials. Amongst the several surface modification techniques currently in use, such as physical vapor deposition (PVD), chemical vapor deposition (CVD), thermal spray, etc., the most effective strategies that are used for wear resistance applications are those involving laser and plasma technologies, which are high energy density processes [12]. Plasma transferred arc welding (PTAW) has been used for such applications, because it has advantages such as better adhesion and relatively lower dilution, compared to other techniques. The earliest use of the PTAW technique was reported in the early 1960s, for the production of overlays [13]. The PTAW techniques use a torch cathode, inside of which lies an electrode, e.g., tungsten, and a plasma arc is built around this electrode. The anode is the substrate material onto the surface of which the coating material, e.g., powders, are deposited, with shielding from an inert gas. The shield gas is used to limit oxidation during the melting of the coating material. This technique makes it less challenging to achieve desired conditions regarding dilution and penetration, enabled by the high ionization of the PTAW process [14]. PTAW can be considered a sustainable method for reasons such as low cost of implementation, high quality final surface layers, better adhesion between the substrate and surface layers, highly stable energy flux and high efficiency of melting and deposition [15]. Depending on the type of filler material, the process is called either plasma arc welding (PAW) or powder plasma transferred arc welding (PPTAW), when the filler material is a wire or powder, respectively. PPTAW generally proceeds with lower melting heat demands. The resulting property of the final product has contributions from the type of powders used, and the PPTAW process parameters, such as plasma gas flow rate (PGFR) and plasma arc current.

The flexibility of PPTAW to be used with different powders has made it a popular new surface modification technique for enhanced strength, hardness, corrosion, and abrasive wear resistance applications. Table 1 presents some metal matrix composite (MMC) powders used for PPTAW surface modification, as reported by other studies. The use of MMC with carbide reinforcements, such as WC, TiC, etc., fosters the combination of the properties of the metal matrix, such as plasticity, corrosion resistance and wear resistance, with that of the hard carbide reinforcement [16]. In application, such surface layers enable the material to withstand high impact loads and to perform better against abrasive wear. These properties are relatively hard to achieve when the surface layer is made up of only metal or only ceramic powders. This has motivated the use of MMC powders for the surface layers to address abrasive wear resistance challenges in tools and machine parts in the extractive industry. An advantage MMCs have over monolithic alloys is that the MMC could transfer higher tensile and compressive stresses than the monolithic alloy. This is enabled by the high degree of dispersion of the carbide phases in the matrix, where the existential bonds between them help transfer applied load from the matrix phase to the reinforcement [17].

Similar studies in this area of research [2,18,19] have focused on the mechanical properties and wear behavior of PPTAW hard-facing of structural steel with MMC powders of different compositions. The microstructural behavior of the applied surface layers has direct translation into the resulting properties of the layers, and it is, therefore, critical to study how the microstructure is modified as the process conditions are varied. To the best of our knowledge, at the time of conducting this study, there have not been results reported from the literature which focus on the effects of the PPTAW process parameters, and the use of different alloying elements in Ni-based MMC powders reinforced with ceramic WC particles, on the microstructure, hardness, and wear properties of the obtained surface layers on a structural steel substrate for applications in the extractive industry. Findings from this research will therefore contribute to providing more insight into this state.

Table 1. Chemical Composition of MMC powders used in PPTAW processes for abrasive wear resistance applications, as reported from literature.

Substrate Material	Metal Matrix Composite (MMC) Powders		References
	Matrix	Reinforcement	
Aluminum	Al-Ni	SiC, TiC	[17]
Stainless steel	Co-Mo-Si-Cr-Ni-Fe	WC-Cr, Al_2O_3	[18]
Stainless steel	Co-Cr-W-C	WC, TiC, NbC	[19]
Titanium	Ti	NbC	[20]
Structural steel	Co-Cr-W-C	Cr_3C_2	[21]
Structural steel	Co-Cr-W-C	W-TiC+PD	[2]
Structural steel	Fe-C-B-Mn-Si	B_4C	[22]
Structural steel	Fe-Cr-C-Ni	Cr_3C_2, Cr_7C_3, $Cr_{23}C_6$	[23]
Structural steel	Ni-Cr-B-Si	WC/Co	[24]

2. Materials and Methods

2.1. Sample Preparation

The substrate material used in this study was structural steel grade EN S355. Steel plate specimens were prepared according to the specifications indicated in Figure 1a. The powder plasma transferred arc welding (PPTAW) technique was used to deposit the MMC powders onto the surface of the substrate material. The process was carried out using PPTAW system EuTronic Gap 3511 DC synergic (Castolin Eutectic, Gliwice, Poland), shown in Figure 2. The cathode was made up of a plasma torch consisting of a central tungsten electrode around which a plasma arc was built. The powders were fed through the torch. The anode was the structural steel plate substrate material. Argon gas was used as the shield to protect the padding weld from possible oxidation resulting from interaction with the environment during melting. To obtain the optimal process parameters for the PPTAW surface modification technology, process parameters were varied. The parameters varied were the plasma gas flow rate (PGFR) and the plasma transferred arc (PTA) current. Varied values of the process parameters are given in Table 2. Varying the process parameters provided additional control over the technique for easier reproducibility. Optimum process parameters are those that proceed with high deposition rate, low dilution, less distortion, and the least surface imperfections. In Figure 3, representative images of the specimen after the PPTAW process are presented for each powder used. Under the same process conditions, the final surface clad obtained from the various powders were not significantly distinguishable from each other. Ni-based MMC powders were obtained commercially from Castolin Eutectic® in Gliwice Poland. Two types of MMC powders were acquired with varying concentrations of WC reinforcement. Powder designated by PG has chemical composition Ni-Si-B+60 wt%WC and powder designated by PE has chemical composition Ni-Cr-Si-B+45 wt%WC. The concentration of WC was varied for both powders to study the effects of the reinforcement concentration on the mechanical properties of the final surface clad. This would also help in understanding the mechanism of abrasive wear better. SEM morphological images of the powders were generated using Supra 35 (Zeiss, Oberkochen, Germany). MMC powder particle sizes were analyzed using laser particle sizer ANALYSETTE 22 (Fritsch, East Windsor, NJ, USA).

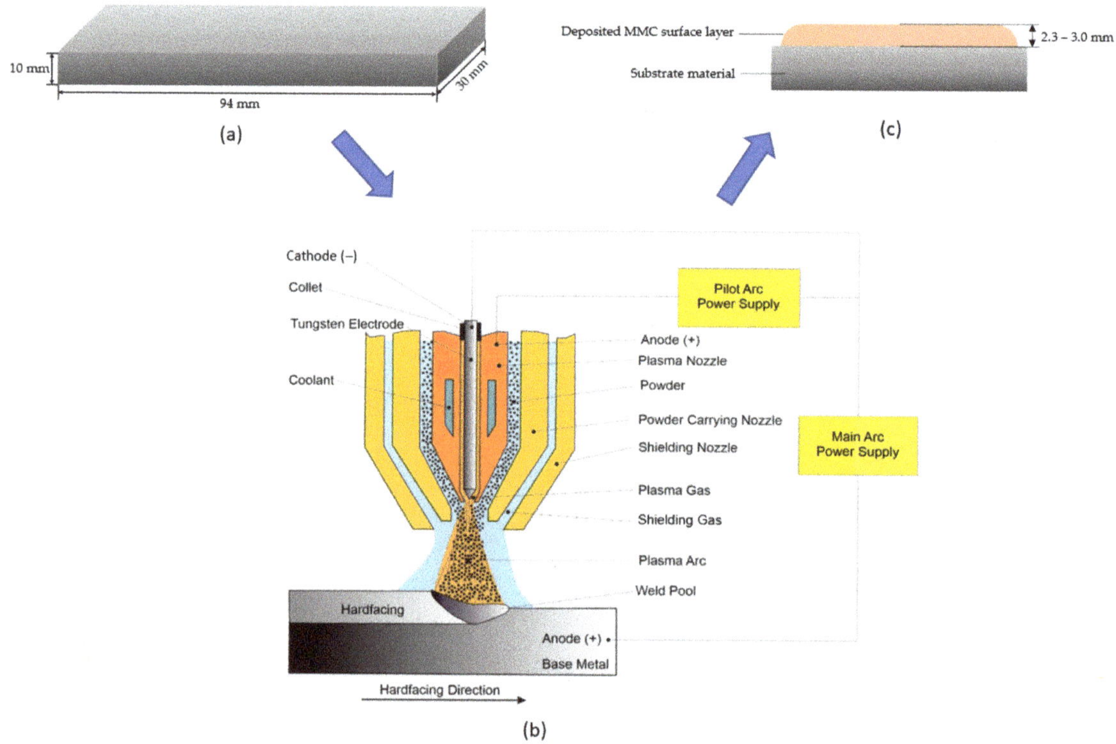

Figure 1. Process flow diagram for the surface cladding process (**a**) Image of the prepared structural steel plate specimen used for the study, including its dimensions (**b**) Schematic diagram of the PPTAW process (**c**) Schematic image of the cross-section of the final product after surface coating.

2.2. *Characterizaion and Testing Methods*

2.2.1. Microscopic Observations

The digital microscope, Leica DVM6 (Leica Microsystems, Heerbrugg, Switzerland) was used to obtain digital images of the surface layers, and to examine the depth of wear on the coated surfaces after the abrasive wear resistance tests. Light microscopy was used to generate micrographs of the deposited MMC after the PPTAW process. The light microscope used was AxioVision (ZEISS, Jena, Germany).

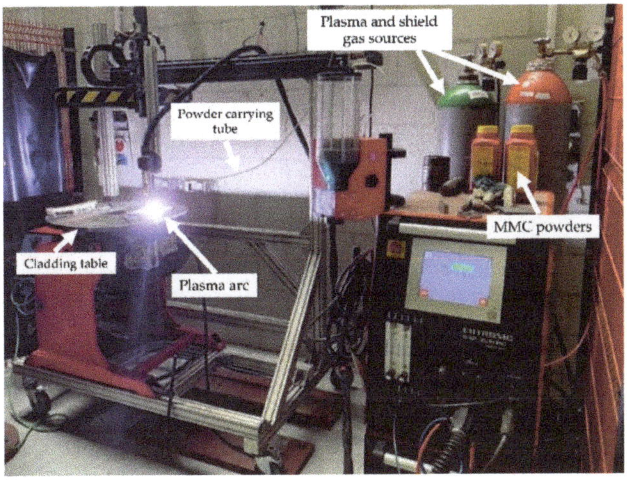

Figure 2. EuTronic© Gap 3511 DC synergic system used for PPTAW process.

Table 2. PPTAW process parameter variation for specimen preparation, Travel Speed, V = 1.3 mm/s.

Specimen	MMC Powder	Plasma Arc Current (A)	Plasma Gas Flow Rate (L/min)
1-PG	PG *	110	1.0
2-PG	PG	110	1.2
3-PG	PG	150	1.2
4-PG	PG	110	1.5
1-PE	PE **	110	1.0
2-PE	PE	110	1.2
3-PE	PE	150	1.2
4-PE	PE	110	1.5

* PG—MMC with composition Ni-Si-B+60 wt%WC; ** PE—MMC with composition Ni-Cr-Si-B+45 wt%WC.

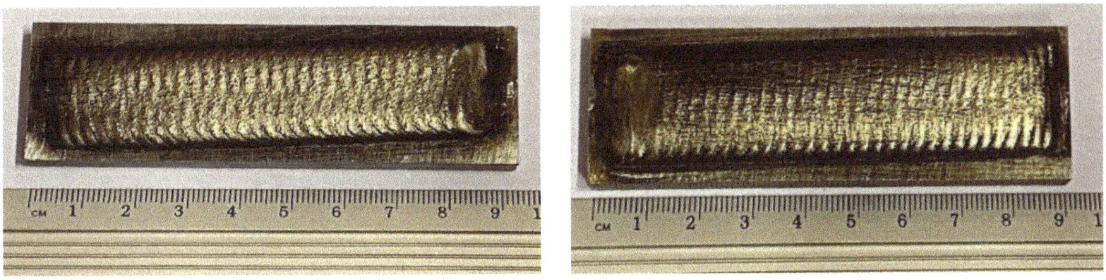

(**a**) (**b**)

Figure 3. Images of prepared samples showing the as-deposited MMC surface coatings on the substrate material (**a**) specimen 2-PG prepared with PG MMC (**b**) specimen 2-PE prepared with PE MMC.

2.2.2. Hardness Tests

The hardness of the as-deposited surface layers was measured using the Rockwell hardness testing equipment SHR-15E (GSTI Co. Ltd., Guiyang, China), and a load of 150 kg. The test was carried out on the extreme surface of each tested sample after thorough surface cleaning with compressed air. In total, 8 measurements were taken at an even space of

10 mm between the points. The average hardness was then computed and recorded. The hardness tester FM-ARS 9000 (Future Tech Corporation, Tokyo, Japan) and a load of 9.81 N was used to measure the Vickers hardness of the surface layers. These microhardness measurements were carried out along the cross-section of the clad after metallographic polishing, with 8 measurements taken at random points in the matrix as well as the reinforcing WC. The average was then computed and recorded.

2.3. Relative Abrasive Wear Resistance Measurement

2.3.1. Abrasive Wear Test

Abrasive wear resistance tests were performed on the as-deposited MMC surface layers as well as on a reference material (abrasive wear resistant AR400 steel). The test followed the guidelines of standard ASTM G65 which is often referred to as the "rubber wheel" method. This method sandwiches abrasive materials between the surface under study and a rubber lined wheel, as shown in Figure 4. The rubber lined wheel is set to revolving motion while the surface under study is set in a stationary position over a period. The process parameters used in this test are outlined in Table 3.

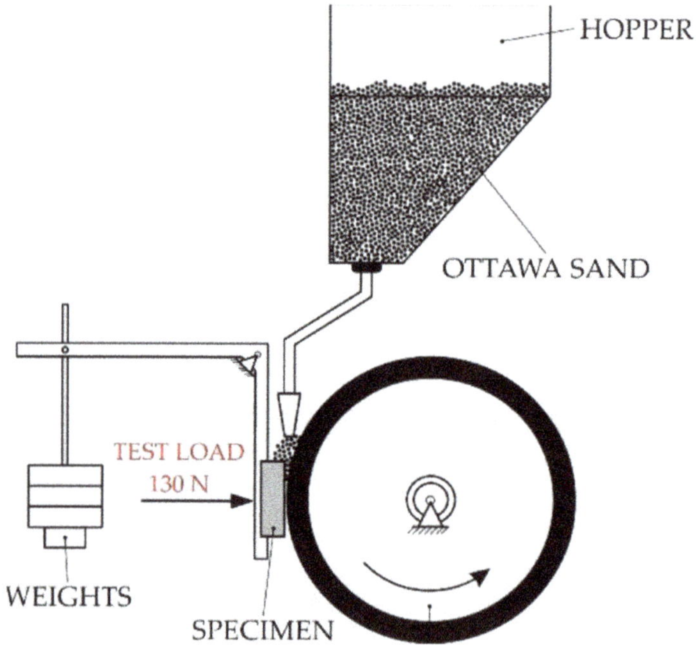

Figure 4. Abrasive wear resistance testing schematic diagram showing the abrasive quartz sand particles sandwiched between the rotating rubber-lined wheel and the stationary surface of the MMC hard-faced layer.

Table 3. Abrasive wear resistance testing conditions.

Parameter	Value	Unit
Abrasive particle grain size	210–297	μm
Feed rate	335	g/min
Pressure	130	Pa
Rubber wheel turns	6000	turns
Test time	30	min

2.3.2. Calculation of Relative Abrasive Wear Resistance

To determine the abrasive wear resistance of the prepared surface layers relative to that of the reference material, firstly, the weights of the tested specimen were taken before and after the test. The mass loss was then determined using Equation (1). The volume loss was estimated using Equation (2). Then, the relative abrasive wear resistance of the specimen under study was computed with Equation (3).

$$\text{Mass Loss, } M_L\ [g] = M_B\ [g] - M_A\ [g] \tag{1}$$

$$\text{Volume Loss, } V_L\ [mm^3] = \frac{M_L[g]}{\rho[\frac{g}{cm^3}]} \times 1000 \tag{2}$$

$$\text{Relative Abrasive Wear Resistance} = \frac{V_{LR}[mm^3]}{V_{LS}[mm^3]} \tag{3}$$

where M_B is the mass of the specimen before abrasive wear test; M_A is the mass of the specimen after abrasive wear test; ρ is the density of the material; V_{LR} is the average volume loss of the reference material; V_{LS} is the average volume loss of the specimen.

3. Results and Discussion

3.1. MMC Powder Morphology and Particle Size Distribution

SEM images of the powders PG and PE are shown in Figures 5a and 6a, respectively. The morphology of the MMC powders under the SEM revealed two distinct structures. These structures were subjected to microanalysis (Figures 5 and 6), coupled with the chemical compositions in Tables 4 and 5, revealed that the matrix particles appeared to be more spherical, whereas the reinforcing carbides had sharp-edged morphology. Spherical powder particles are more desirable in applications where high toughness is required, and sharp-edged or angular shaped powder particles are more desirable in applications where wear resistance is required. Nonetheless, MMC powder particles having a mixture of both spherical and angular morphologies results in more mechanical stability, suitable for applications requiring high toughness, wear resistance and high hardness [25]. Figure 7 shows the results of laser-assisted particle size analysis. Statistical deductions from the plots show that for both powders, analyzed with a specific quantity of particles, denoted by n, 10% of these particles were smaller than d_{10}, 50% of these particles were smaller than d_{50}, and 90% of these particles were smaller than d_{90}. These values could be used to determine the degree of uniformity, or span of the powder particles, a dimensionless value given by the equation

$$Span = \frac{d_{90} - d_{10}}{d_{50}} \tag{4}$$

When span is equal to 0, it signifies complete particle uniformity, and a higher value of span conversely shows a higher degree of particle non-uniformity. The PG powder had a span of 1.3 and, as shown in Figure 7b, the size distribution presented as a group, even though there were two distinct morphologies. The PE powder had a higher span value of 4.3 and, as shown in Figure 7a, the size distribution presented as two different groups, to make up for the higher disparity in the particle sizes. The PG powder particles can, therefore, be said to be relatively larger than the PE powder particles. This also accounts for the observation of larger particle sizes of PG powders (Figure 5a) than the PE powders (Figure 6a) at the same magnification under the scanning electron microscope.

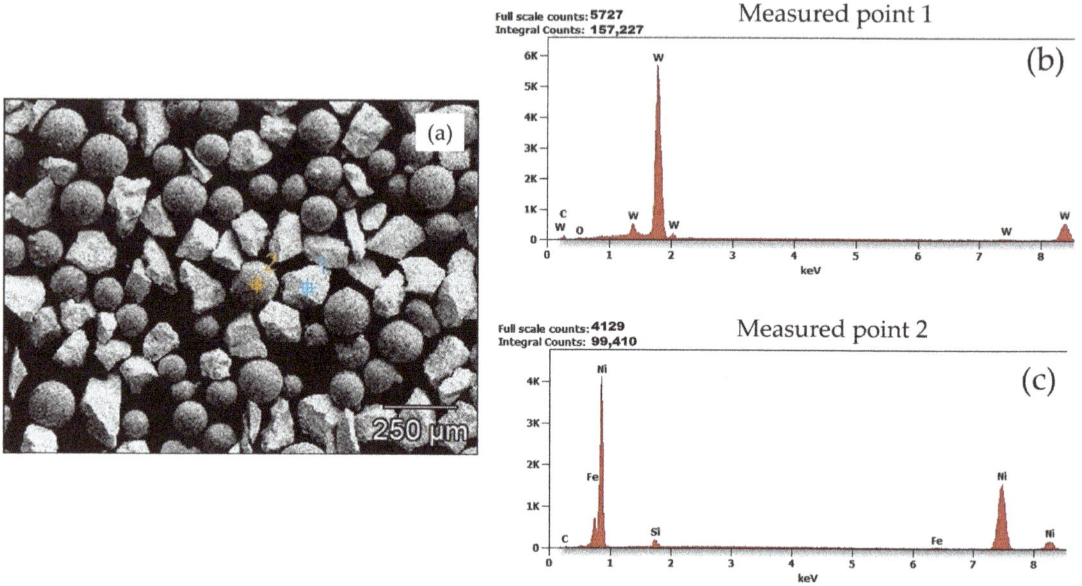

Figure 5. Microanalysis of powder particles (**a**) acquired SEM image of PG powder showing its morphology and measured points under analysis (**b**,**c**) Energy-dispersive X-ray spectrometry (EDS) diagrams of measured points 1 and 2.

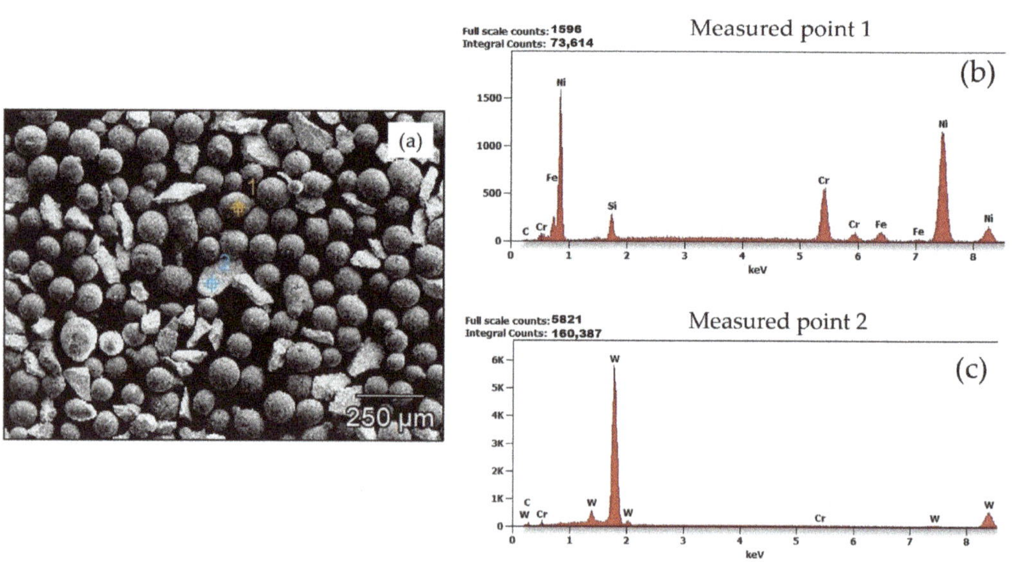

Figure 6. Microanalysis of powder particles (**a**) acquired SEM image of PE powder showing its morphology and measured points under analysis (**b**,**c**) Energy-dispersive X-ray spectrometry (EDS) diagrams of measured points 1 and 2.

Table 4. Chemical composition in weight % and atom % of measured points of PG powder under analysis.

		C	O	Si	Fe	Ni	W
Measured point 1	Weight %	3.9	0.9	-	-	-	95.2
	Atom %	36.2	6.4	-	-	-	57.5
Measured point 2	Weight %	1.5	-	2.4	0.7	95.4	-
	Atom %	6.9	-	4.7	0.6	87.8	-

Table 5. Chemical composition in weight % and atom % of measured points of PE powder under analysis.

		C	O	Si	Cr	Fe	Ni	W
Measured point 1	Weight %	0.7	-	3.5	15.3	3.5	76.9	-
	Atom %	3.9	-	6.7	15.8	3.4	70.3	-
Measured point 2	Weight %	3.8	3.7	-	0.5	-	-	92.0
	Atom %	29.8	22.0	-	0.9	-	-	47.3

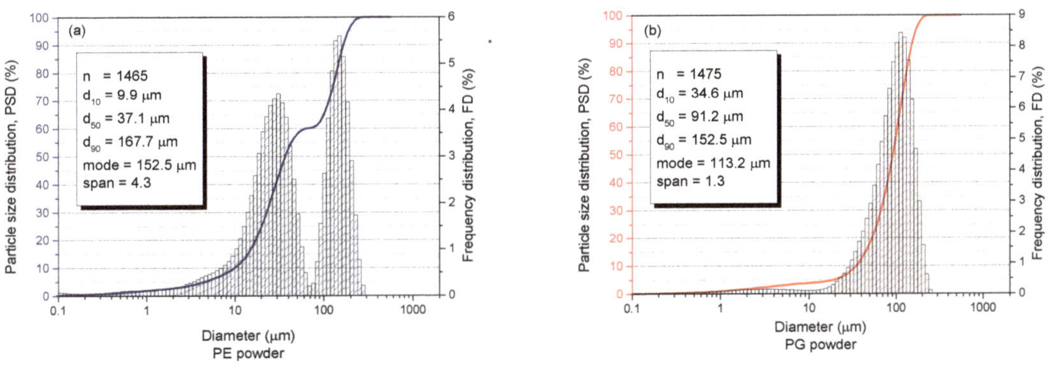

Figure 7. Particle size distribution (PSD) plots for the MMC powders (**a**) PE powder PSD plot with particle statistics (**b**) PG powder PSD plot with particle statistics.

3.2. Metallographic Analysis and Precipitation

Microstructural observation of the coated layer revealed the degree of dispersion of the WC reinforcement and its precipitates in the Ni-based matrix throughout the coating. The observable zones along the cross-section of the surface layer are shown in Figure 8.

An exemplary observation of the middle zone of sample 4-PE, prepared under conditions of 110 A PTA current and 1.5 L/min PGFR, under the SEM, is shown in Figure 9. Figure 9a shows partially dissolved primary carbides and precipitates of secondary carbides throughout the matrix. They formed a dendritic, multi-directional microstructure. The chemical composition of the area of the middle zone shown in Figure 9a was analyzed and the scattered X-ray energy diagram in Figure 9b shows the chemical elements present in this zone. The elemental maps in Figure 9c–j show the presence and surface distribution of the chemical elements present in the middle zone. The bright grey regions in Figure 9a are the primary carbides and precipitates of secondary carbide particles, evidenced by the high concentrations of carbon and tungsten in Figure 9d,i. The black regions in Figure 9a are the Ni-based matrix sites, evidenced by Figure 9i. Ceramic reinforcement particles in MMCs usually agglomerate when solidified, due to the difference in density of the carbides from the matrix [26]. The PPTAW technique has been reported to result in surface clads that have agglomerated ceramic reinforcements, weak wettability, and imperfections on the surface of the clads [27]. In this work, however, the ceramic carbides, owing to the effectiveness of the method of mixing the powder particles, were seen to be dispersed in the matrix without agglomeration. The dendritic structure of the surface layer, formed by the precipitation of complex secondary carbides on the boundary of the matrix and carbide was made possible by the partial melting of the WC particles (T_m = 2785 °C) and the complete melting of the Ni-based matrix powders (T_m = 1555 °C) during the deposition [28]. Poloczek et al. (2019) [29] have stipulated that the stability of the deposited clad is possible by the bonding of the primary carbides to the matrix by the secondary carbides through accelerated diffusion. This SEM microstructural observation was like all samples prepared by each of the PG and PE MMC powders used. The precipitates of the secondary carbides in the matrix were subjected to microanalysis and the results of this microanalysis are presented in Figure 10. It was observed from this analysis that these precipitates were composed of W, C, Ni, Fe and Si. The proportions of these compositions are given in Table 6. The dissolution of the substrate material into the surface coating during deposition accounted for the presence of Fe in the secondary carbide precipitates.

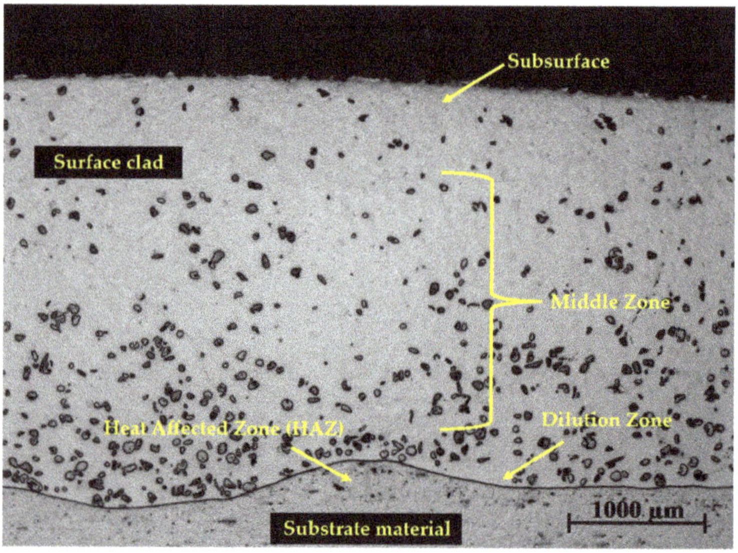

Figure 8. Cross-section of deposited surface layer of sample 1-PE showing the subsurface of the layer, middle zone, dilution zone, and HAZ.

At the coating–substrate interface, shown in Figure 11, two major cross-sectional zones were observed, i.e., the dilution zone and the heat affected zone (HAZ). Microanalytical determination of the concentrations of chemical elements across the interface, in the direction from the coating to the substrate is presented in Figure 11. The concentrations of W and Ni were seen to be higher in the coating and in the dilution zone. However, they gradually dropped to their minima within the HAZ and maintained the least concentrations within the substrate material. A contrasting observation is seen regarding Fe. Its concentration was the least at the dilution zone and rose gradually within the HAZ and reached its maximum in the substrate material. Carbon maintained a constant concentration across the interface. These observations at the interface suggested a strong adhesion between the coating and the substrate material, by dissolution of chemical components across the interface from the substrate material into the coating and vice versa, this being metallurgical bonding.

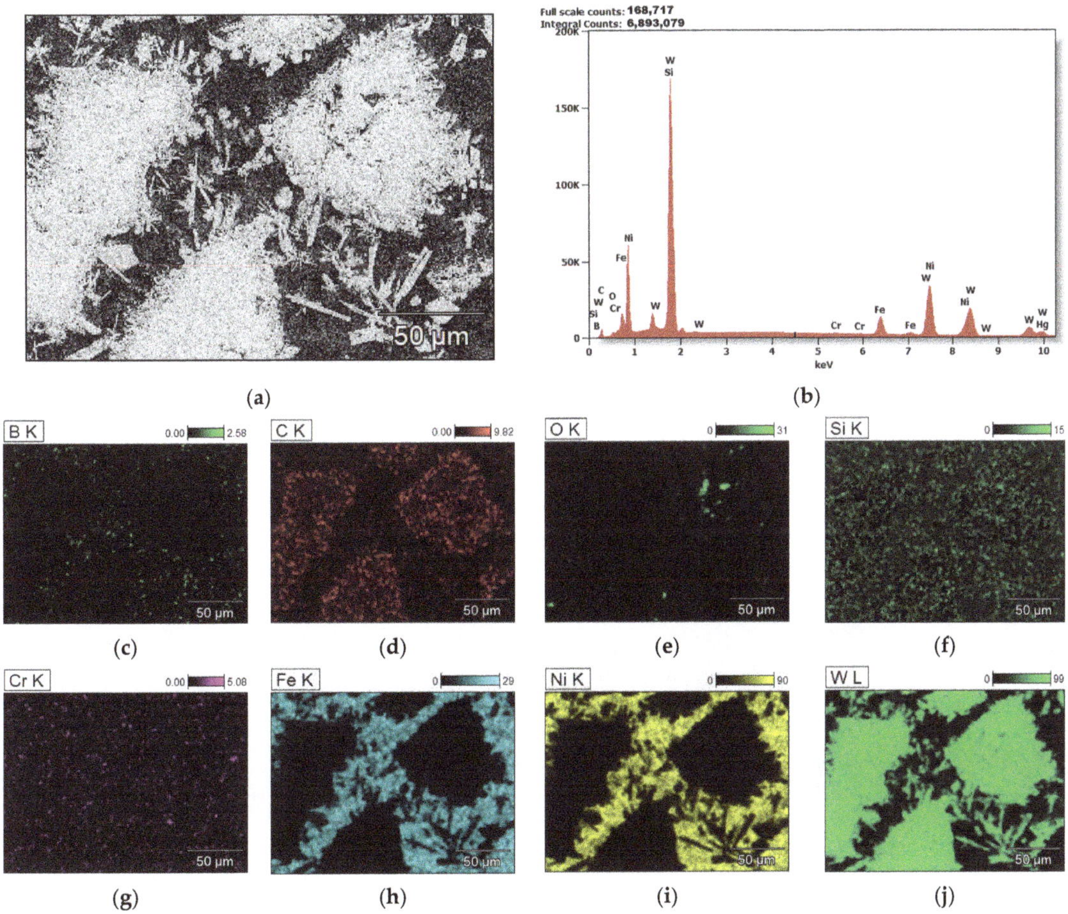

Figure 9. Elemental analysis of the middle zone of the surface layer of sample 4-PE (**a**) area under observation (**b**) X-radiation energy diagram of area under observation (**c**–**j**) elemental maps showing the position and amounts of elements present in the area under observation.

3.3. Hardness Tests Results

Microhardness measurement results of the Ni-based matrix and WC reinforcement for surface layers prepared by each MMC powder, are presented in Table 7, in a comparative observation regarding the microhardness of the reference abrasion-resistant steel AR400. The measured average microhardness of the surface layers was observed to be significantly higher than that of the reference material by more than 263 HV. The observed high value of standard deviation was accounted for by the fact that the measurements were taken at random points throughout the clad for both the matrix and the WC phases. The non-uniform dispersion of the WC particles in the clad influenced this measurement approach. It was consequently observed that the microhardness measurements increased along the clad towards the coating–substrate interface. This could be attributed to the weld dilution at this interface, and the effects of the HAZ. Such measurement difficulties have been reported by research works [28,30], which studied the wear behavior of MMC cladded surfaces. Surface hardness of the prepared coatings are also presented in Table 8. Results from microhardness measurements on the surface were influenced by the distribution of stress from the matrix to the reinforcing WC particles. This was because the gradient of the surface of the samples comprised of MMC solid solution had dispersed primary WC particles, as well as precipitates of secondary carbides. The observed increase in hardness could be explained by the partial saturation of the matrix caused by the partial dissolution of the primary tungsten carbides in the matrix, as well as the alloying of the Ni-based matrix with Cr, in the case of samples prepared with PE MMC powder. The use of Cr as an alloying element influences the mechanical properties by increasing the hardness of the alloy [31]. Cr can form different carbides in the presence of carbon. Due to this property, it improves the hardenability of steel during production [32]. In a study conducted by Lin et al. (2018b) [31] to investigate the effects of Cr on the microstructure and properties of TiC-steel composites, it was reported that an increase in the Cr content caused an increase in the hardness, with a corresponding decrease in the transverse rapture strength (TRS). Similar results were obtained in this work: the average microhardness of the matrix of the PE MMC powder was about 29.4% higher than that of the matrix of the PG MMC powder. This was significantly attributed to the use of Cr as an alloying element in the PE MMC powder.

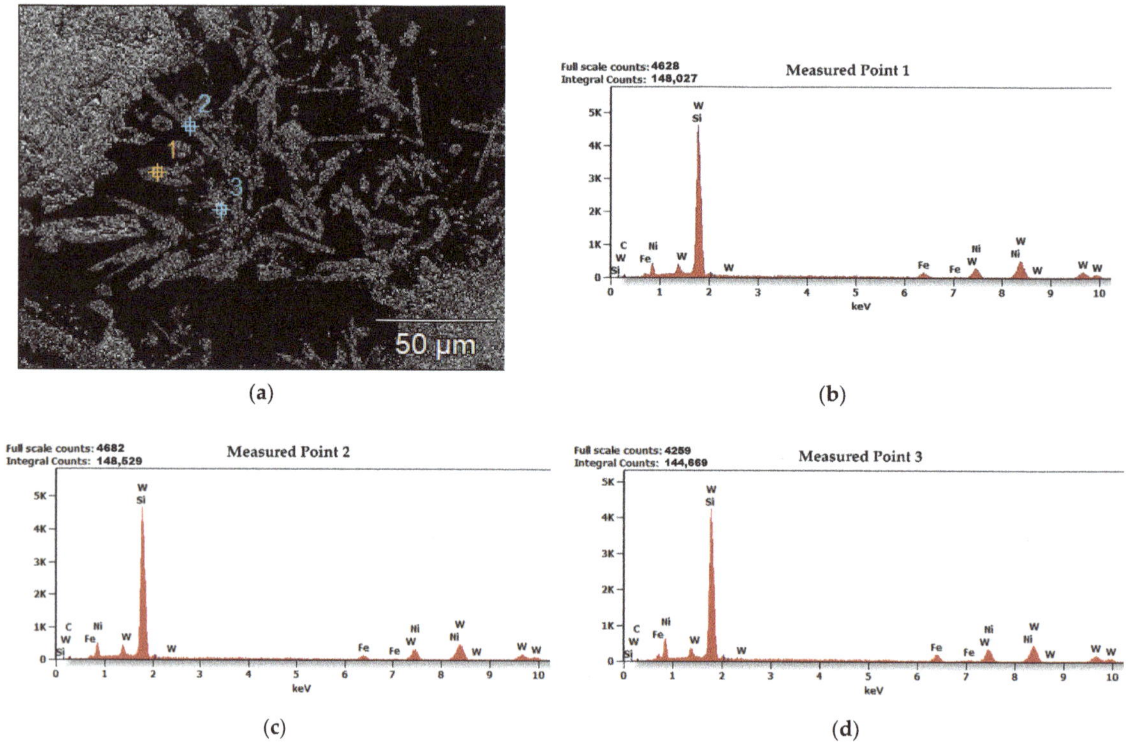

Figure 10. Results of microanalysis of precipitates of the WC reinforcement within the surface layer of sample 4-PE prepared under conditions of 110 A PTA current and 1.5L/min PGFR (**a**) Image of microanalytical area showing points of interests for analysis (**b–d**) Energy-dispersive X-ray spectrometry (EDS) diagrams of measured points 1 through 3.

Table 6. Chemical composition in Weight % and Atom % of measured points of precipitation subjected to microanalysis of sample 4-PE prepared under condition of 110 A PTA current and 1.5 L/min PGFR.

		C	Si	Fe	Ni	W
Measured Point 1	Weight %	3.0	5.0	3.6	10.0	78.4
	Atom %	22.6	16.5	6.0	15.7	39.2
Measured Point 2	Weight %	3.1	3.9	2.8	11.0	79.3
	Atom %	24.2	13.0	4.7	17.6	40.6
Measured Point 3	Weight %	2.6	5.1	4.8	12.4	75.1
	Atom %	19.4	16.5	7.9	19.2	37.1

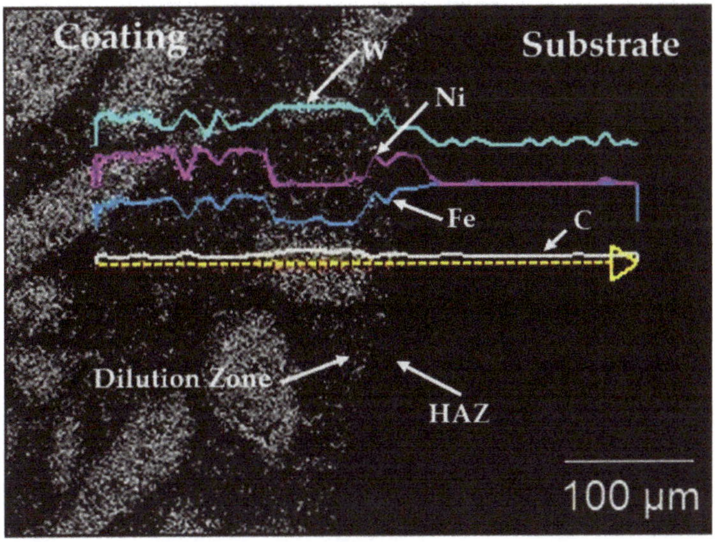

Figure 11. SEM image of the coating-substrate interface of sample 3-PE, showing the dilution zone, HAZ and the diffusion gradients of the chemical components involved in the interface adhesion between the coating and the surface.

Table 7. Measured microhardness of the matrix and reinforcing carbides through the cross-section of the deposited surface layers for each MMC powder used, compared to the microhardness of abrasive wear-resistant steel, AR400.

Specimen	**Microhardness of Matrix (HV1)**		**Microhardness of Reinforcement (HV1)**	
	Mean	Standard Deviation	Mean	Standard Deviation
1-PG	590.7	5.2	2413.0	62.9
2-PG	573.3	10.2	2128.7	33.3
3-PG	687.0	2.4	2162.7	76.1
4-PG	673.0	18.5	2275.0	49.5
1-PE	888.7	18.0	2349.3	38.7
2-PE	844.7	23.2	2436.3	24.1
3-PE	888.7	23.8	2343.3	61.6
4-PE	893.0	16.1	2391.3	80.5
		Reference Material		
AR400	424.6	6.2	-	-

Table 8. Measured Rockwell hardness on the surface of the surface layers for samples prepared by each MMC powder.

Specimen	**Surface Hardness**		
	Mean, HV1	Mean, HRC	Standard Deviation
1-PG	462	46.3	0.5
2-PG	475	47.3	2.6
3-PG	480	47.7	2.5
4-PG	488	48.3	1.2
1-PE	660	58.3	3.7
2-PE	556	52.7	3.3
3-PE	602	55.3	2.9
4-PE	610	55.7	1.2

3.4. *Abrasive Wear Resistance Test Results*

The abrasive wear resistance of the surface layers prepared by each MMC powder, relative to the wear resistance of reference material AR400, was carried out using the methods described in Section 2.3. The results were computed using Equations (1)–(3), and presented in Table 9. The marginal mass loss for each sample set the right measurement precision to three decimal places. However, results are presented in four decimal places to make up for measurement anomalies and external environmental influences, which, in the end, did not have any significant influence on the reported relative abrasive wear resistance values. It was observed that the abrasive wear resistance of the surface layers was much higher than the reference material in the cases of both MMC powders used. It was again observed in this test that, on average, under the same process conditions, the MMC powder PG, having 60 wt% WC, had better abrasive wear performance than the PE MMC powder having 45 wt% WC, by up to 45.8%. The higher volume of WC present in the PG powder explained the observation of higher abrasive wear resistance to that of the PE powder MMC. Figure 12 shows the surface structure of the MMC surface layers prepared by each powder under the same conditions of 110 A PTA current and 1.0 L/min PGFR, after undergoing the abrasive wear test. The mechanism of abrasive wear resistance observed in this work was studied and can be described as occurring in two major stages. At the initial stage, as the rubber wheel was set in motion with the quartz sand in contact with the surface of the layer, the surface began to wear off. This continued until the quartz sand removed enough material from the surface and came into closer proximity with wear debris rich in tungsten carbide. Contact with these WC particles initially caused smearing of the WC particle. As the contact time increased, WC particles caused reduction in the friction from the quartz sand and the revolution of the rubber wheel, significantly reducing wear. The image of the surface layer after the abrasion test in Figure 13a reveals exposed WC particles at the surface when further abrasion was hindered. The depth map in Figure 13b shows that the region on the surface which lacked WC particles were abraded deeper than the regions rich in WC particles. The wear mechanism observed in the reference material, which was WC free, presented in Figure 12c and Figure 16a,e,i, shows that during the dent formation, the exposed metal surface was ploughed from the micro scale, enhancing the transfer of abrasive material onto the dent surface. This resulted in deeper grooves and displacement of metal to the sides of the abrasive material as contact time increased. This observed mechanism has also been reported in a similar study by Chung et al. (2021) [33] who studied the erosion–corrosion wear behavior of AR400 steel and found micro ploughing to be the main wear mechanism in this WC free steel. It was deduced from the studied mechanisms that MMC powders rich in WC have better abrasive wear resistance, than those with relatively lower volumes of WC. This mechanism also supported the observation of better abrasive wear resistance in samples prepared with PG powder MMC than in those prepared with PE powder MMC, which had relatively lower volume of WC reinforcement.

Table 9. Abrasive wear resistance tests results.

Sample ID	Mass before Test, g	Mass after Test, g	Average Mass Loss, g	Material Density, g/cm^3	Average Volume Loss, mm^3	Relative Abrasive Wear Resistance *
			Surface layers prepared with PG powder, NiSiB+60%WC			
1-PG	195.6418	195.3264	0.3154	11.1935	28.1771	4.7
2-PG	209.0038	208.7471	0.2567	11.1935	22.9329	5.7
3-PG	196.0594	195.6905	0.3689	11.1935	32.9566	4.0
4-PG	227.8358	227.5179	0.3179	11.1935	28.4004	4.7
			Surface layers prepared with PE powder, NiCrSiB+45%WC			
1-PE	226.4951	226.1412	0.3539	9.8274	36.0116	3.7
2-PE	228.6697	228.3604	0.3093	9.8274	31.4732	4.2
3-PE	231.6575	230.8754	0.7821	9.8274	79.5836	1.7
4-PE	221.7090	221.3348	0.3742	9.8274	38.0772	3.5
			Reference Marterial AR400 steel			
H1	104.6219	103.4971	1.0318	7.7836	132.5607	1.0
H2	111.7377	110.7989				

* Abrasive wear resistance relative to reference material AR400.

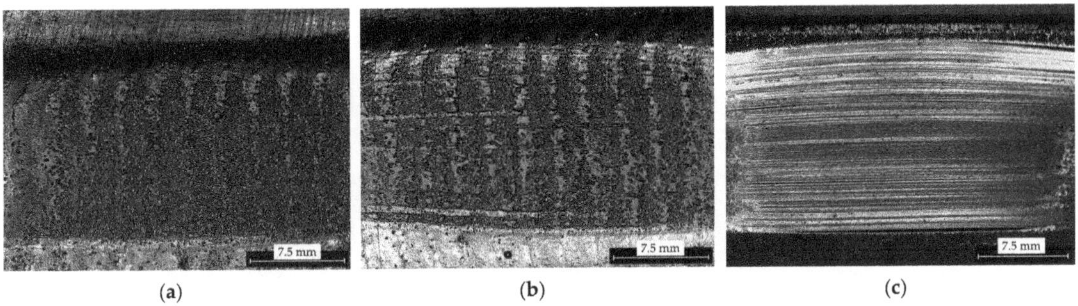

Figure 12. Surface of coating layers after abrasive wear resistance tests (**a**) surface of layer of sample 1-PG prepared using PG powder (**b**) surface of layer of sample 1-PE prepared using PE powder (**c**) surface of reference material AR400.

3.5. Parameter Optimization

Identifying optimum process parameters for high energy density processes like PPTAW is of much importance. This is because it significantly increases productivity by optimizing the quality and quantity of deposition [34]. PPTAW process parameters, plasma transferred arc current and plasma gas flow rate, were varied to investigate their effects on the resulting properties of the surface layers, to aid in establishing the criteria for selecting optimum values for parametric reproducibility.

3.5.1. Plasma Transferred Arc (PTA) Current

The PTA current was varied between 110 A and 150 A, while keeping all other parameters constant, to prepare samples from the MMC powders. An observation of the cross-section of the surface layers revealed that increasing the PTA current caused more dissolution of the WC reinforcements into the matrix. To illustrate this, cross-sections of the surface clad of samples 2-PG and 3-PG, prepared under the same process conditions but at two different PTA current values (110 A and 150 A), are shown in Figure 14. At PTA current of 110 A, the WC particles were seen to be more evenly distributed throughout the cross-section of the coating (Figure 14a). After the PTA current was increased to 150 A, the WC particles were seen to be less clustered, having a relatively higher degree of dispersion within the matrix (Figure 14b). It was particularly observed that at higher PTA current, the subsurface of the layer in the cross-sectional view was mostly free of WC particles. This significantly reduced the abrasive wear performance of the layer, as shown in the comparative plots in Figure 15c,d.

(**a**)

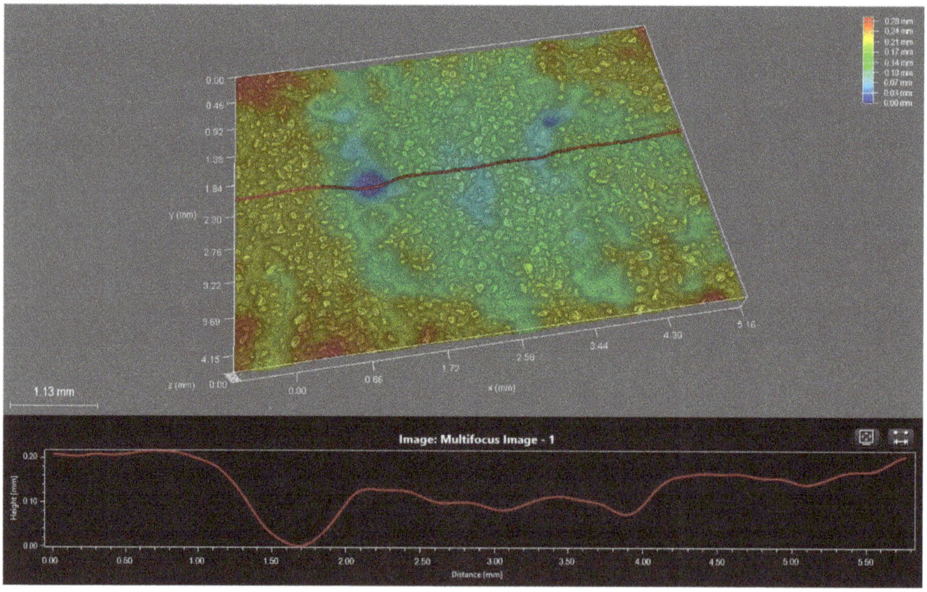

(**b**)

Figure 13. Images of surface layer of sample 4-PG after abrasive wear resistance test (**a**) post-abrasive wear test surface showing dispersed WC particles and the matrix phase (**b**) depth map showing the depth of abrasion on the surface of the layer.

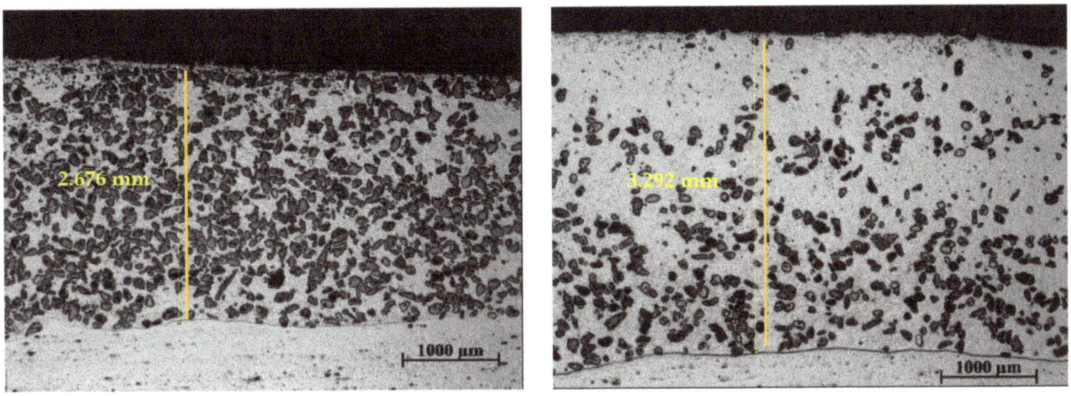

(**a**) (**b**)

Figure 14. Cross-section of surface layers prepared with PG powder MMC at different PTA current values showing the measured thickness of surface layers (**a**) surface layer of 2-PG prepared at PTA current of 110 A (**b**) surface layer of 3-PG prepared at PTA current of 150 A.

Figure 14 also shows the measured average thickness of the coating along its cross-section. At PTA current of 110 A, the thickness of the weld bead averaged around 2.7 mm. At a higher PTA current, the average of the weld bead was around 3.3 mm. An increase in the PTA current resulted in higher overall thickness of the deposited layers. This was caused by a rise in temperature upon an increase in PTA current. This led to an increase in supplied energy, causing more dissolution of the reinforcing particles. Consequently, adhesion was also improved, due to the higher dissolution of the base material into the surface layer at the HAZ [35]. The dissolution of the WC particles in the matrix increased complex carbon compound formation with the alloying elements, such as Cr, and this slightly increased the overall hardness of the coating matrix [36]. This is also supported by the plots in Figure 15a,b.

3.5.2. Plasma Gas Flow Rate (PGFR)

The mechanical effects of the PGFR on the abrasive wear resistance and hardness of the prepared surface coatings were evaluated by keeping all process conditions constant and varying the PGFR values as 1.0 L/min, 1.2 L/min and 1.5 L/min. Representative images of pre- and post-abrasive wear tests are illustrated in Figure 16, with samples 1-PE, 2-PE and 4-PE having PGFRs of 1.0, 1.2 and 1.5 L/min, respectively. As can be seen from the plots in Figure 17a,b, the relative abrasive wear resistance of the coatings, for each MMC powder used, increased from 1.0 L/min to 1.2 L/min, which could be considered as a critical value, then reduced as the PGFR was further increased to 1.5 L/min. As gas flow rate is the measure of the volume of gas passing a particular point over a period, it appears that an increase in the PGFR increased thermal activity, causing increased velocity of powder particles along the axis of the torch. This increased the compressive forces between the powder particles, and at the coating-substrate interface. As a result, there was greater adhesion between dissolved particles, reducing the overall porosity of the clad. This formed the basis for the observation of a general increase in the microhardness of the matrix of the coatings as the PGFR increased from 1.0 L/min to 1.5 L/min in Figure 17c,d, and the increase in abrasive wear resistance as the PGFR increased from 1.0 L/min to 1.2 L/min. However, as the PGFR increased, turbulent flow of the powder particles resulted in increased energy of the process, similar to increasing the PTA current, resulting in less WC particles present at the coating subsurface. From Figure 16f–h, it can be seen that at PGFR of 1.0 L/min, the grooves on the surface after contact with abrasive quartz sand were seen to be less deep, compared to the grooves on the surfaces of the samples having PGFR of 1.2 L/min and 1.5 L/min. The lower presence of the WC particles resulting from higher dissolution of the WC particles at the subsurface was the reason for the deeper grooves caused by the quartz sand at higher PGFR values and the observation of a decline in the abrasive wear resistance at a PGFR of 1.5 L/min in Figure 17a,b. These observations were consistent with other research works [37,38] that investigated the effects of gas flow rate on the mechanical and wear behavior of surface coatings using TIG and plasma spray approaches, respectively.

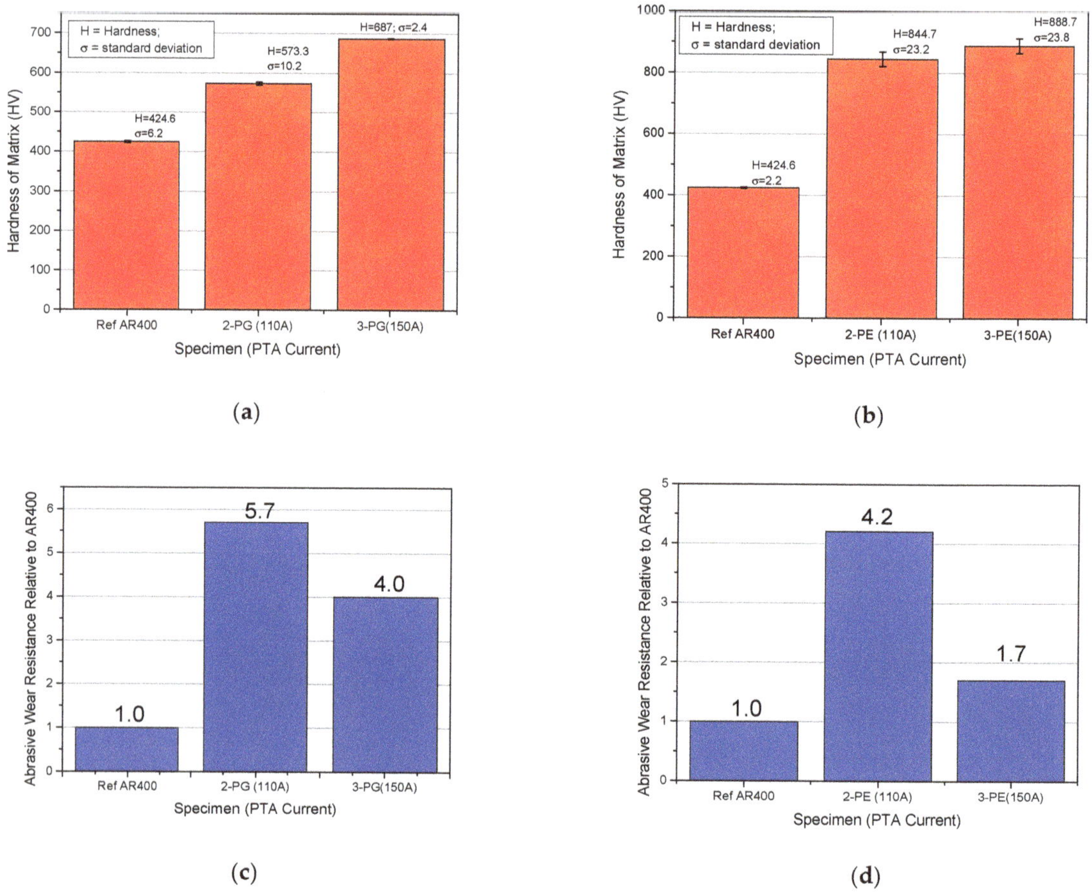

Figure 15. Comparative plots on the effects of PTA current on hardness and abrasive wear performance of surface coating (**a**,**b**) effects of PTA current on hardness of matrix of samples prepared with PG and PE powder MMCs respectively (**c**,**d**) effects of PTA current on the abrasive wear resistance of samples prepared with PG and PE powder MMCs respectively.

The optimum process parameters from those listed in Table 2, coupled with PPTAW device working parameters are summarized in Table 10. These parameters were used extensively for the study, and they formed the basis for the presented results.

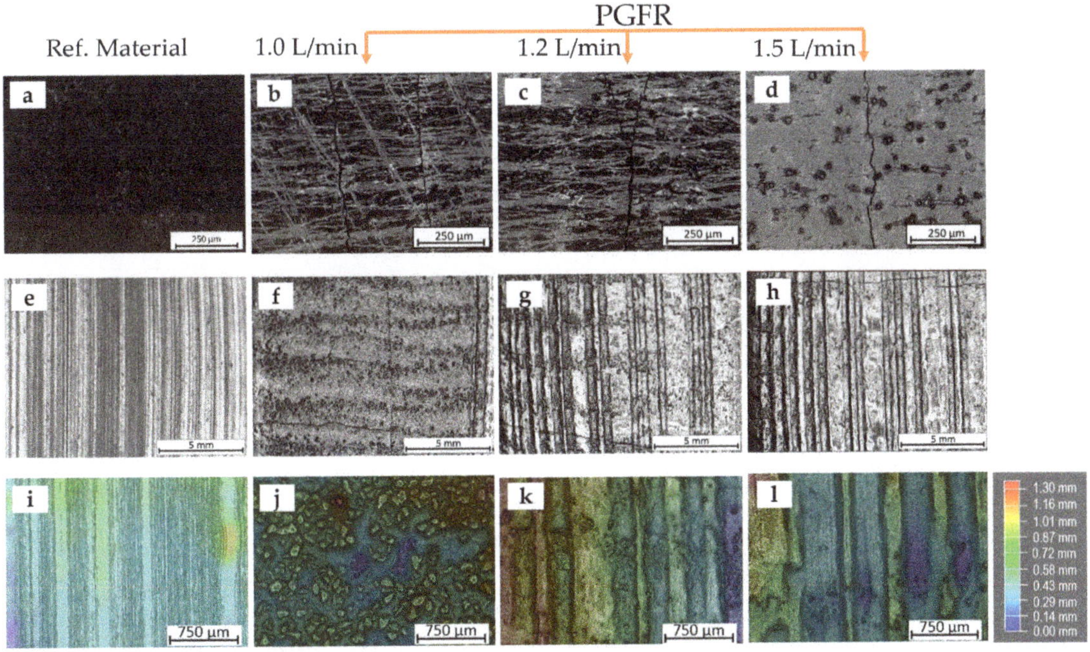

Figure 16. Images of the surfaces of specimens before and after abrasive wear test (**a**,**e**,**i**), reference material AR400 steel before abrasive wear test, after abrasive wear test and wear depth after abrasive wear, respectively (**b**,**f**,**j**) specimen 1-PE with PGFR of 1.0 L/min and PTA current of 110 A before abrasive wear test, after abrasive wear test and wear depth after abrasive wear, respectively (**c**,**g**,**k**) specimen 2-PE with PGFR of 1.2 L/min and PTA current of 110 A before abrasive wear test, after abrasive wear test and wear depth after abrasive wear, respectively (**d**,**h**,**l**) specimen 4-PE with PGFR of 1.5 L/min and PTA current of 110 A before abrasive wear test, after abrasive wear test and wear depth after abrasive wear, respectively.

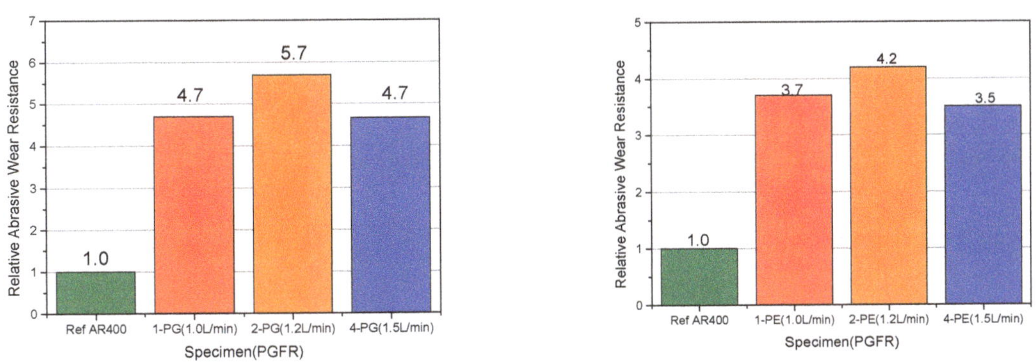

(**a**) (**b**)

Figure 17. *Cont.*

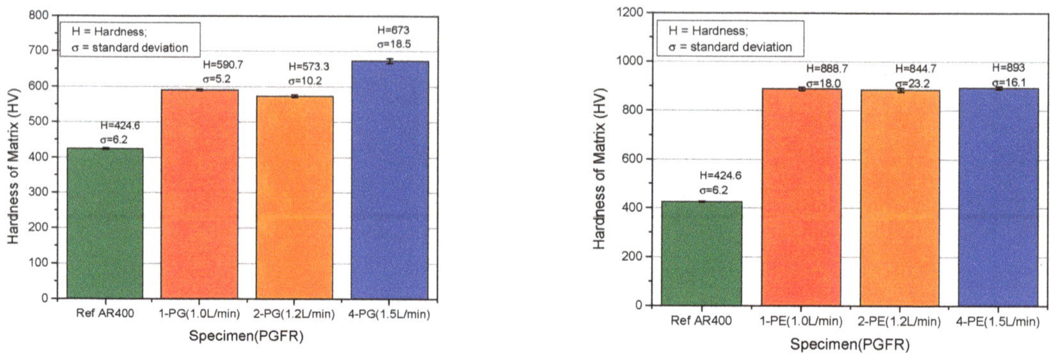

(c) (d)

Figure 17. Comparative plots on the effects of PGFR on abrasive wear performance and hardness of surface coatings (**a**,**b**) effects of PGFR on the abrasive wear resistance of samples prepared with PG and PE powder MMCs, respectively (**c**,**d**) effects of PGFR on hardness of matrix of samples prepared with PG and PE powder MMCs, respectively.

Table 10. Optimum process parameters used for sample preparation, applicable to the two MMC powders used for this study.

Process Parameter	Value	Unit
Plasma arc current	110	A
Pilot arc current	30	A
Travel speed	1.3	mm/s
Plasma gas flow rate	1.2	L/min
Open circuit voltage	95	V DC
Power flow (Cos phi)	0.99	-

4. Conclusions

- The technology of powder plasma transferred arc welding (PPTAW) was used in this study as a surface modification technique to improve upon the abrasive wear resistance of structural steel grade EN S355 for applications in the mining and raw materials extraction industry.
- Two metal matrix composite (MMC) powders were used in the process, having compositions of Ni-Si-B+60 wt%WC (PG) and Ni-Cr-Si-B+45 wt%WC (PE).
- PPTAW process parameters, namely, plasma arc (PTA) current and plasma gas flow rate (PGFR), were varied, and the effects of the variation were studied and used as criteria for selecting optimum conditions for further studies and parametric reproducibility.
- PTA current had influence on the degree of dissolution of the primary carbides in the matrix, the thickness of the surface clad, the surface hardness and the abrasive resistance of the surface layer.
- PGFR was observed to have influence on the hardness and abrasive wear resistance of the surface coatings as it increased from 1.0 L/min to 1.2 L/min and then to 1.5 L/min.
- Microstructural observation revealed a dendritic, multi-directional microstructure, consisting of partially dissolved primary tungsten carbide particles and secondary tungsten carbide precipitates, within the Ni-based matrix.
- The hardness of the surface layers was higher than that of a reference material AR400 by more than 263 HV. Final surface layers obtained from the MMC powders had abrasive wear resistance up to 5.7 times that of abrasion-resistant reference Hardox 400 (AR400) steel. Alloying the MMC matrix with chromium increased the surface

hardness by 29.4%. Under the same process conditions, MMC powder with 60 wt% WC reinforcement had better abrasive wear resistance by up to 45.8% more than the MMC powder with 45 wt% WC.

Author Contributions: Conceptualization and methodology, M.A. and A.C.; formal analysis and investigation, A.N.S.A., O.B., A.C. and M.A.; hardness, microscopy and metallography, A.N.S.A., O.B. and M.A.; Wear Resistance and NDT, A.C. and A.N.S.A.; writing—original draft preparation, A.N.S.A. and O.B.; writing—review and editing, A.C., M.A., O.B. and A.N.S.A. All authors have read and agreed to the published version of the manuscript.

Funding: This research was supported in part by the Polish Budget Funds for Scientific Research in 2022 as a core funding for R&D activities in the Silesian University of Technology—Grant No. 10/100/BK22/0024, Grant No. 10/100/BKM22/0036 and Grant No. 10/100/BKM22/1110.

Institutional Review Board Statement: Not applicable.

Informed Consent Statement: Not applicable.

Data Availability Statement: Not applicable.

Acknowledgments: Co-financed by the European Union through the European Social Fund (grant POWR.03.05.00-00-Z305). The Authors would like to express gratitude to Castolin Eutectic Sp. z o.o., Leonarda da Vinci 5 Str., 44-109 Gliwice, Poland for technical support in the preparation of test specimen.

Conflicts of Interest: The authors declare no conflict of interest.

References

1. Bremerstein, T.; Potthoff, A.; Michaelis, A.; Schmiedel, C.; Uhlmann, E.; Blug, B.; Amann, T. Wear of Abrasive Media and Its Effect on Abrasive Flow Machining Results. *Wear* **2015**, *342*, 44–51. [CrossRef]
2. Czupryński, A. Microstructure and Abrasive Wear Resistance of Metal Matrix Composite Coatings Deposited on Steel Grade AISI 4715 by Powder Plasma Transferred Arc Welding Part 1. Mechanical and Structural Properties of a Cobalt-Based Alloy Surface Layer Reinforced with Particles of Titanium Carbide and Synthetic Metal–Diamond Composite. *Materials* **2021**, *14*, 2382. [CrossRef] [PubMed]
3. Gooneratne, C.P.; Magana-Mora, A.; Otalvora, W.C.; Affleck, M.; Singh, P.; Zhan, G.D.; Moellendick, T.E. Drilling in the Fourth Industrial Revolution—Vision and Challenges. *IEEE Eng. Manag. Rev.* **2020**, *48*, 144–159. [CrossRef]
4. Temizel, C.; Canbaz, C.H.; Palabiyik, Y.; Putra, D.; Asena, A.; Ranjith, R.; Jongkittinarukorn, K. A Comprehensive Review of Smart/Intelligent Oilfield Technologies and Applications in the Oil and Gas Industry. In Proceedings of the SPE Middle East Oil and Gas Show and Conference, Manama, Bahrain, 18–21 March 2019.
5. Gharahbagh, E.A.; Qiu, T.; Rostami, J. Evaluation of Granular Soil Abrasivity for Wear on Cutting Tools in Excavation and Tunneling Equipment. *J. Geotech. Geoenviron. Eng.* **2013**, *139*, 1718–1726. [CrossRef]
6. Dewangan, S.; Chattopadhyaya, S.; Hloch, S. Wear Assessment of Conical Pick Used in Coal Cutting Operation. *Rock Mech. Rock Eng.* **2015**, *48*, 2129–2139. [CrossRef]
7. Nahak, S.; Dewangan, S.; Chattopadhyaya, S.; Królczyk, G.; Hloch, S. Discussion on Importance of Tungsten Carbide–Cobalt (WC-Co) Cemented Carbide and Its Critical Characterization for Wear Mechanisms Based on Mining Applications. *Arch. Min. Sci.* **2018**, *63*. [CrossRef]
8. Dewangan, S.; Chattopadhyaya, S. Characterization of Wear Mechanisms in Distorted Conical Picks after Coal Cutting. *Rock Mech. Rock Eng.* **2016**, *49*, 225–242. [CrossRef]
9. Kučerová, L.; Zetková, I.; Jeníček, Š.; Burdová, K. Production of Hybrid Joints by Selective Laser Melting of Maraging Tool Steel 1.2709 on Conventionally Produced Parts of the Same Steel. *Materials* **2021**, *14*, 2105. [CrossRef]
10. Ali, Y.; Henckell, P.; Hildebrand, J.; Reimann, J.; Bergmann, J.; Barnikol-Oettler, S. Wire Arc Additive Manufacturing of Hot Work Tool Steel with CMT Process. *J. Mater. Processing Technol.* **2019**, *269*, 109–116. [CrossRef]
11. Ackermann, M.; Šafka, J.; Voleský, L.; Bobek, J.; Kondapally, J.R. Impact Testing of H13 Tool Steel Processed with Use of Selective Laser Melting Technology. *Mater. Sci. Forum* **2018**, *919*, 43–51. [CrossRef]
12. Drake, R.P. Introduction to High-Energy-Density Physics. In *High-Energy-Density Physics;* Springer: Berlin/Heidelberg, Germany, 2018; pp. 1–20.
13. Veinthal, R.; Sergejev, F.; Zikin, A.; Tarbe, R.; Hornung, J. Abrasive Impact Wear and Surface Fatigue Wear Behaviour of Fe–Cr–C PTA Overlays. *Wear* **2013**, *301*, 102–108. [CrossRef]
14. Rohan, P.; Boxanova, M.; Zhang, L.; Kramar, T.; Lukac, F. High Speed Steel Deposited by Pulsed PTA-Frequency Influence. In Proceedings of the 25th Anniversary International Conference on Metallurgy and Materials, Brno, Czech Republic, 25–37 May 2016.

15. Skowrońska, B.; Sokołowski, W.; Rostamian, R. Structural Investigation of the Plasma Transferred Arc Hardfaced Glass Mold after Operation. *Weld. Technol. Rev.* **2020**, *92*, 55–65. [CrossRef]
16. Mele, C.; Lionetto, F.; Bozzini, B. An Erosion-Corrosion Investigation of Coated Steel for Applications in the Oil and Gas Field, Based on Bipolar Electrochemistry. *Coatings* **2020**, *10*, 92. [CrossRef]
17. Deuis, R.; Subramanian, C.; Yellup, J. Abrasive Wear of Composite Coatings in a Saline Sand Slurry Environment. *Wear* **1997**, *203*, 119–128. [CrossRef]
18. Durejko, T.; Łazińska, M.; Dworecka-Wójcik, J.; Lipiński, S.; Varin, R.A.; Czujko, T. The Tribaloy T-800 Coatings Deposited by Laser Engineered Net Shaping (LENSTM). *Materials* **2019**, *12*, 1366. [CrossRef] [PubMed]
19. Wu, J.B.; Redman, J. Hardfacing with Cobalt and Nickel Alloys. *Weld. J. Miami United States* **1994**, *73*, 9.
20. Hung, F.-Y.; Yan, Z.-Y.; Chen, L.-H.; Lui, T.-S. Microstructural Characteristics of PTA-Overlayed NbC on Pure Ti. *Surf. Coat. Technol.* **2006**, *200*, 6881–6887. [CrossRef]
21. Aoh, J.-N.; Jeng, Y.-R.; Chu, E.-L.; Wu, L.-T. On the Wear Behavior of Surface Clad Layers under High Temperature. *Wear* **1999**, *225*, 1114–1122. [CrossRef]
22. Xibao, W. The Metallurgical Behavior of B4C in the Iron-Based Surfacing Alloy during PTA Powder Surfacing. *Appl. Surf. Sci.* **2005**, *252*, 2021–2028. [CrossRef]
23. Liu, Y.-F.; Han, J.-M.; Li, R.-H.; Li, W.-J.; Xu, X.-Y.; Wang, J.-H.; Yang, S.-Z. Microstructure and Dry-Sliding Wear Resistance of PTA Clad (Cr, Fe)7C3/γ-Fe Ceramal Composite Coating. *Appl. Surf. Sci.* **2006**, *252*, 7539–7544. [CrossRef]
24. Smirnov, A.N.; Kozlov, E.V.; Radchenko, M.V.; Knyaz'kov, K.V.; Knyaz'kov, V.L. State of Ni–Cr–B–Si–Fe/WC Coatings after Plasma Powder Surfacing with Nanopowder Modifier. *Steel Transl.* **2016**, *46*, 251–255. [CrossRef]
25. Qi, C.; Zhan, X.; Gao, Q.; Liu, L.; Song, Y.; Li, Y. The Influence of the Pre-Placed Powder Layers on the Morphology, Microscopic Characteristics and Microhardness of Ti-6Al-4V/WC MMC Coatings during Laser Cladding. *Opt. Laser Technol.* **2019**, *119*, 105572. [CrossRef]
26. Zygmuntowicz, J.; Miazga, A.; Konopka, K.; Edrysiak, K.J.; Kaszuwara, W. Alumina Matrix Ceramic-Nickel Composites Formed by Centrifugal Slip Casting. *Processing Appl. Ceram.* **2015**, *9*, 199–202. [CrossRef]
27. Bober, M. Composite Coatings Deposited by the Plasma Transferred Arc–Characterization and Coating Formation. Weld. *Technol. Rev* **2011**, *83*, 43–47.
28. Czupryński, A. Microstructure and Abrasive Wear Resistance of Metal Matrix Composite Coatings Deposited on Steel Grade AISI 4715 by Powder Plasma Transferred Arc Welding Part 2. Mechanical and Structural Properties of a Nickel-Based Alloy Surface Layer Reinforced with Particles of Tungsten Carbide and Synthetic Metal–Diamond Composite. *Materials* **2021**, *14*, 2805. [CrossRef] [PubMed]
29. Poloczek, T.; Czupryński, A.; Żuk, M.; Chruściel, M. Structure and Tribological Properties of Wear-Resistant Layers Produced in Process of Plasma Powder Surfacing. *Weld. Technol. Rev.* **2019**, *91*. [CrossRef]
30. Yahiaoui, M.; Paris, J.-Y.; Delbé, K.; Denape, J.; Gerbaud, L.; Colin, C.; Ther, O.; Dourfaye, A. Quality and Wear Behavior of Graded Polycrystalline Diamond Compact Cutters. *Int. J. Refract. Met. Hard Mater.* **2016**, *56*, 87–95. [CrossRef]
31. Lin, T.; Guo, Y.; Wang, Z.; Shao, H.; Lu, H.; Li, F.; He, X. Effects of Chromium and Carbon Content on Microstructure and Properties of TiC-Steel Composites. *Int. J. Refract. Met. Hard Mater.* **2018**, *72*, 228–235. [CrossRef]
32. Pariser, H.; Backeberg, N.; Masson, O.; Bedder, J. Changing Nickel and Chromium Stainless Steel Markets-a Review. *J. S. Afr. Inst. Min. Metall.* **2018**, *118*, 563–568. [CrossRef]
33. Chung, R.J.; Jiang, J.; Pang, C.; Yu, B.; Eadie, R.; Li, D.Y. Erosion-Corrosion Behaviour of Steels Used in Slurry Pipelines. *Wear* **2021**, *477*, 203771. [CrossRef]
34. Nikam, S.H.; Jain, N.K.; Sawant, M.S. Optimization of Parameters of Micro-Plasma Transferred Arc Additive Manufacturing Process Using Real Coded Genetic Algorithm. *Int. J. Adv. Manuf. Technol.* **2020**, *106*, 1239–1252. [CrossRef]
35. Liu, T.; Chang, M.; Cheng, X.; Zeng, X.; Shao, H.; Liu, F. Characteristics of WC Reinforced Ni-Based Alloy Coatings Prepared by PTA + PMI Method. *Surf. Coat. Technol.* **2020**, *383*, 125232. [CrossRef]
36. Xiong, Y.; Lin, D.; Zheng, Z.; Li, J.; Deng, T. The Effect of Different Arc Currents on the Microstructure and Tribological Behaviors of Cu Particle Composite Coating Synthesized on GCr15 Steel by PTA Surface Alloying. *Metals* **2018**, *8*, 984. [CrossRef]
37. Singh, J.; Thakur, L.; Angra, S. Effect of Argon Flow Rate and Standoff Distance on the Microstructure and Wear Behaviour of WC-CoCr TIG Cladding. *J. Phys. Conf. Ser.* **2019**, *1240*, 012162. [CrossRef]
38. Goudarzi, M.; Ghaziasadi, H. The Effect of Gas Flow Rate on Structural, Mechanical and Antibacterial Properties of Atmospheric Plasma Sprayed Cu Coatings. *Physica Scripta* **2021**, *96*, 075601. [CrossRef]

materials

MDPI

Article

Comparison of Wear Resistance of Overlay Welded Layers and Thermal Sprayed Coatings in Real Conditions

Michał Szymura [1], Grzegorz Gąsiorek [1], Artur Czupryński [2,*], Waldemar Kwaśny [2] and Viktor Kvasnytskyi [3]

1 FPM S.A., Wyzwolenia Street 12C, 43-190 Mikołów, Poland; mich.szymura@gmail.com (M.S.); grzegorz.gasiorek@fpmsa.com (G.G.)
2 Department of Welding, Silesian University of Technology, Konarskiego Street 18A, 44-100 Gliwice, Poland; waldemar.kwasny@polsl.pl
3 Igor Sikorsky Kyiv Polytechnic Institute, 37, Peremohy Ave., 03056 Kyiv, Ukraine; kvas69@ukr.net
* Correspondence: artur.czuprynski@polsl.pl

Abstract: Tribological tests in real conditions enable obtaining full data on the life of interacting machine parts. This article presents the results of operational tests on the elements of the support ring guidance system in a vertical ball-race mill. The guide and active armour operate under abrasive wear conditions with moderate-impact loads. The wear resistance of elements with overlay welding layers deposited with flux cored wire with a structure of high-alloy chrome cast iron and with a coating flame-sprayed with nickel-based powder was compared. The wear intensity of the overlay weld deposits was much lower than that of the sprayed coatings. The scope of this study also included the analysis of the chemical and phase composition, macro- and microscopic metallographic examinations, and the measurement of the hardness of the deposited layers and coatings.

Keywords: operational tests; wear; flux cored arc surfacing; thermal flame spraying; chrome cast iron; nickel alloy; vertical ball-race mill

Citation: Szymura, M.; Gąsiorek, G.; Czupryński, A.; Kwaśny, W.; Kvasnytskyi, V. Comparison of Wear Resistance of Overlay Welded Layers and Thermal Sprayed Coatings in Real Conditions. *Materials* **2023**, *16*, 7215. https://doi.org/10.3390/ma16227215

Academic Editor: Chih-Chun Hsieh

Received: 19 October 2023
Revised: 12 November 2023
Accepted: 16 November 2023
Published: 17 November 2023

1. Introduction

Coal dust used in power plants is obtained from crumble coal grains in a mill. Vertical roller mills and vertical ball-race mills are most often used for grinding hard coal. In a ball-race mill, the milling process takes place by crushing and grinding the milling material. This process takes place between the crushing ring and the grinding balls, which are pressed by the pressure ring and the stop ring located on it. The guidance system consisting of guide and armour (active and passive) is responsible for maintaining the pressure ring in the mill axis through the stop ring. The dominant wear process of the guide and active armour is abrasion occurring at moderate-impact loads. The result of wear is an increase in the gap between the surfaces of the guide and the passive armour. The lack of systematic adjustment of the gap length leads to an improper pressure of the balls on the crushing ring, increasing the vibration level of the mill and then leading to its failure as a result of damage caused to the spring pressure system (spring fracture) [1,2].

Welding technologies enable the application of abrasion-resistant layers and coatings. This resistance depends on the type of additional material used and the surfacing technology [3–5]. During the surfacing with arc welding process, dilution occurs, which may cause differences in the structure of the surfacing layer, and thus changes in its abrasion resistance depending on the degree of wear [6]. Thermal spray processes are characterized by the absence of melting of the base material. According to Dallaire [7], the submerged-arc surfacing Fe-Cr-C layer showed a much smaller (an over seven-fold) volume loss in the ASTM G-65 abrasion resistance test compared to an arc-sprayed coating consisting of a nickel-based alloy containing fine borides and carbides.

Laboratory tests of wear resistance enable the simulation of the wear of elements for certain specific real conditions; therefore, they are comparative tests [7–10]. Generally, abrasion resistance testing devices can be divided into three categories according to the type of abrasion: (two-body and three-body), contact kinematics (sliding, rolling, and reciprocating, impact), and contact medium (dry, wet, and slurry) [11]. Most abrasion testing equipment is designed in accordance with the prescribed standards (e.g., ISO and ASTM). Changing the design of testing devices or test parameters may lead to differences in the results obtained [11–13]. The study [9] on the comparison of abrasion resistance of layers deposited by flux cored arc surfacing (including Fe-Cr-Nb-C alloy) in various abrasion tests (dry-sand rubber-wheel tester and impeller–tumbler tests at low and high energy) shows that the Fe-Cr-Nb-C alloy does not wear evenly under high loads because the plastic deformation of the matrix occurs and the chromium carbides crack. The study [6] on the influence of the type of abrasive material on the abrasive wear resistance of wear plates with a high-alloy chromium cast iron overlay made by self-shielded flux cored arc surfacing in an abrasion test based on ASTM G-65 showed that the type of abrasive material (analysed materials: rounded quartz grain sand, clinker sand, and blast furnace slag) has a decisive influence on the abrasive wear resistance of surface layers. In the study [14] comparing the wear intensity of flame-sprayed NiCrBSi coatings and cast iron samples in the ball on disc wear test, it was found that the ratio of their abrasion resistance is not the same when their sliding speed and their load was changed. Abrasive wear is a complex process and, in relation to the operation of parts of machines and devices in industrial conditions, it may be impossible to develop a model on a laboratory scale due to the inability to precisely determine the wear conditions (forces, unit pressures, friction speed, type of abrasive, etc.), among other factors.

The available literature presents the results of laboratory tests of abrasion resistance of both overlay welding layers with high-alloy chromium cast iron [6,8,15–19] and flame-sprayed coatings with nickel-based powder [20–24]. However, there was no quantitative data enabling the comparison of the wear intensity of layers and coatings applied to guides and active armour in real conditions. The data obtained in this way can be directly used in industrial practice. Therefore, an attempt was made to compare the resistance to abrasive wear with moderate-impact loads of the mentioned kinematic pair with overlay welded wear-resistant layers and with thermal sprayed wear-resistant coatings.

This article presents the results of operational tests of the elements of the support ring guidance system in a vertical ball-race mill. The analysed guides and active armour were protected with a layer deposited by flux cored arc surfacing with the structure of high-alloy chrome cast iron and a flame-sprayed coating with nickel-based powder. The scope of this study also included the analysis of the chemical and phase composition, macro- and microscopic metallographic examinations, and the measurement of the hardness of the deposited layers and coatings.

2. Materials and Research Methods

Materials and Their Preparation

For operational tests, two sets (guide and active armour) of the EM-70 mill thrust ring guidance system manufactured by FPM were made. The guides are made of cast steel for general engineering uses, GS240 according to EN 10293 [25], and the active armour is made of non-alloy structural steel, S235JR according to EN 10025-2 [26]. The chemical composition and mechanical properties of these material are provided in Tables 1 and 2. The working surfaces of the elements of the first set were surfaced with self-shielded flux cored wire SK A43-O produced by voestalpine Böhler Welding (Voestalpine AG, Linz, Austria), ensuring an overlay weld deposit of high-alloy chrome cast iron, Fe15 according to EN 14700 [27]. This wire is recommended for protecting surfaces exposed to intense abrasive wear with moderate- and high-impact load. The chemical composition and hardness of the weld deposit of the filler material are provided in Table 3. Metco 15E powder from Oerlikon Metco (OC Oerlikon Corporation AG, Pfäffikon, Schwyz, Switzerland) was used to flame spray the surface of the guide and the active armour of the second set. According to the manufacturer, coatings that are thermally sprayed with Metco 15E powder are resistant to tribological wear and corrosion. The chemical composition and hardness of the coating sprayed with this powder are presented in Table 4. Compared to the layers deposited by arc welding with self-shielded flux cored wire, coatings flame-sprayed with self-fluxing powders after melting are characterized by a much greater surface smoothness, which may contribute to an increase in the durability of the friction pair [20]. The technological parameters of deposited layers and coatings are presented in Tables 5 and 6. Photos of elements surface-coated by welding and thermal spraying for operational tests are presented in Figure 1. For laboratory tests, elements cut from S235JR non-alloy steel sheet with dimensions of 250 × 100 × 20 mm were used as the base material.

Table 1. Chemical composition of materials for guides and active armour according to the manufacturer (Huta Stalowa Wola S.A., Stalowa Wola, Poland).

Element	Material	Chemical Composition, wt.%										
		C	Si	Mn	P	S	Cr	Mo	Ni	V	Cu	N
Active Armor	S235JR	0.15	0.24	1.15	0.022	0.007	0.01	-	0.01	0.001	0.01	0.003
Guide	GS240	0.19	0.43	1.12	0.015	0.012	0.09	0.01	0.04	0.001	0.06	- [1]

Note: [1] lack of data.

Table 2. Mechanical properties of materials for guides and active armour according to the manufacturer (Huta Stalowa Wola S.A., Stalowa Wola, Poland).

Element	Material	Yield Strength, MPa	Tensile Strength, MPa	Elongation, %
Active Armor	S235JR	285	468	31
Guide	GS240	307	491	23

Table 3. Chemical composition and hardness of the weld deposit of the self-shielded flux cored wire SK A43-O according to the manufacturer (Voestalpine AG, Linz, Austria).

Chemical Composition, wt.%						Hardness, HRC
C	Cr	Nb	Si	Mn	Fe	
5.6	20.2	6.7	1.3	0.2	bal.	64

Table 4. Chemical composition and hardness of the powder Oerlikon Metco 15E according to the manufacturer (OC Oerlikon Corporation AG, Pfäffikon, Schwyz, Switzerland).

Chemical Composition, wt.%						Hardness, HRC
Cr	B	Si	C	Fe	Ni	
17.0	3.5	4.0	1.0	4.0	bal.	60

Table 5. Parameters of surfacing layers with SK A43-O self-shielded flux cored wire.

Wire Diameter, mm	Welding Current, A	Arc Voltage, V	Electrode Extension, mm	Travel Speed, cm/min
2.4	280	28	35	50

Note: the overlay welds were surfacing automatically, with stringer beads parallel to the long side of the guide. Two layers were deposited for operational tests.

Table 6. Parameters of flame spraying of coatings with Metco 15E powder.

Oxygen Pressure, Bar	Acetylene Pressure, Bar	Air Pressure, Bar	Number of the Orifice for the Powder
4.0	0.7	1	4

Note: standard modular nozzles regulating the flame outlet (SSM 20) were used. Slightly carburizing flame was used. The spraying distance was 200 mm. The surface of the base material for spraying was prepared with electrocorundum abrasive blasting. Before spraying, the surface of the substrate was heated to 280 °C. After heating, a thin powder coating was applied to protect the surface against oxidation, and then, a coating with a thickness of approx. 1.0 mm was applied. The coating was melted using a normal flame.

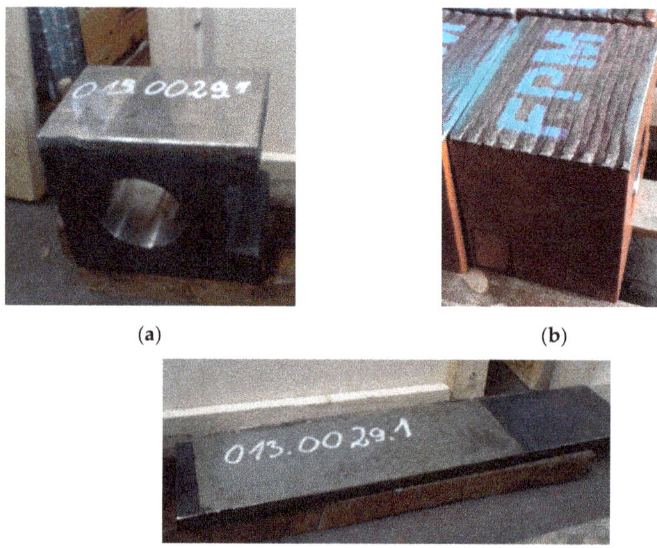

Figure 1. Images of (**a**) guide with a thermal sprayed coating; (**b**) guide with a flux cored arc surfaced layer; and (**c**) active armour with a thermal sprayed coating.

3. Characteristics of the Tests

3.1. Operational Tests

Operational tests of two sets of support ring guidance systems were carried out in the EM-70 ball-race mill located in a combined heat and power plant powered by hard coal. The tests were performed in real conditions during the standard operation of the heat and power plant. The wear resistance of the made layers and coatings was defined as the operating time (in hours) per unit of wear (corresponding to the size of the structural gap between the guide and the passive armour). The larger value of the pair of gap width measurements in the upper and lower areas was considered as the gap size (Figure 2).

The correct installation of the guidance system and its appropriate adjustment during the mill's operation period requires maintaining the mentioned construction gap in the range of 2 ± 1 mm. The gap width measurements were performed during scheduled mill inspections, i.e., approximately every 700–800 h of operation. In accordance with the technical and operational documentation, in order to assemble the guide set, the guidance chamber was opened, and the previous guide and active armour were demounted. Then, the guiding set with the sprayed coating was installed, Figure 3a, and the construction gap was established, Figure 3b. Before closing, the conduction chamber was sealed. The same procedure was followed for the set with the overlay welded layer.

Figure 2. Location of a pair of measurements of the size of the gap between the guide and the passive armour.

(a) (b)

Figure 3. Installation of the guide and active armour with a sprayed coating in the guide chamber: (**a**) view of the installed elements and (**b**) adjustment of the gap between the guide and the passive armour.

3.2. Chemical Composition Analysis

The analysis of the chemical composition in the micro-areas of the samples was performed using a Hitach S-3400N scanning electron microscope (Hitachi, Ltd., Marunouchi, Chiyoda, Tokyo, Japan) equipped with a Noran SYSTEM SIX EDS spectrometer (Thermo Fisher Scientific Inc., Waltham, MA, USA).

3.3. X-ray Qualitative Phase Analysis

X-ray phase analysis of the samples was performed using a JDX-7S diffractometer from JEOL (JEOL Ltd. Akishima, Tokyo, Japan). Filtered radiation from a copper X-ray tube was used. Measurements were performed in the angular range from 30° to 90° of angle with a step of 0.05° 2θ and a counting time of 5 s. Phase identification was performed based on the International Centre for Diffraction Data PDF-4+ database, year 2023.

3.4. Macro- and Microscopic Metallographic Examinations

In order to determine the quality of layers and coatings, macro- and microscopic metallographic examinations were performed. Macro- and microscopic metallographic examinations were carried out on transverse metallographic sections for each sample using SZX9 and GX71 light microscopes from Olympus (Olympus Corporation, Shinjuku, Tokyo, Japan). Microscopic observations at higher magnifications were performed on a Hitach S-3400N scanning electron microscope (Hitachi, Ltd., Marunouchi, Chiyoda, Tokyo, Japan). The porosity of the thermal sprayed coating was determined based on the analysis of ten microstructure images at ×200 magnification using the Special AnalySIS pro program, version JMP9 (JMP Statistical Discovery LLC, Cary, NC, USA).

3.5. Hardness Measurement

The hardness measurements of the overlay welded layer and the sprayed coating were carried out using the Vickers method at a load of 2.942 N, in accordance with the ISO 6507-1 [28] standard using the Future-Tech FM-ARS 9000 hardness tester with an automatic measurement line and an image analysis system (Future-Tech Corp., Kawasaki, Japan). Measurements were performed on the cross-sectional surface according to Figure 4.

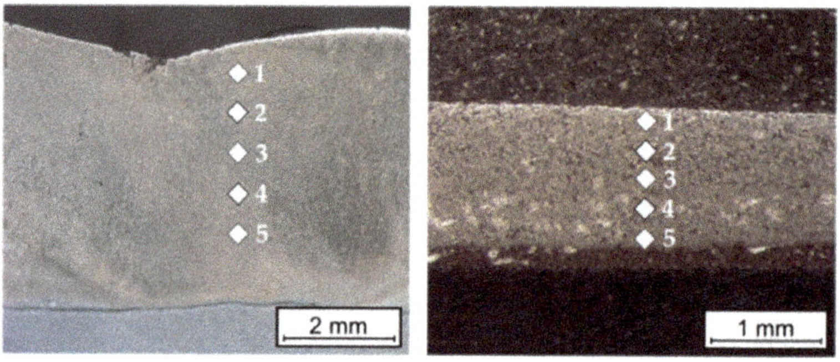

(**a**) (**b**)

Figure 4. Pictorial diagram of the location of hardness measurement points on the cross-sectional surface of (**a**) the overlay welded layer and (**b**) sprayed coating.

4. Results and Discussion

4.1. Operational Tests

The aim of the study was to compare the wear resistance of layers and coatings applied to guides and active armour in real conditions. The results of measurements of the gap width in relation to the operating time of the analysed guide sets and the determined wear resistance are presented in Table 7. A view of the guidance system with sprayed coatings after 1578 h of operation is shown in Figure 5, and the view of the sprayed active armour after disassembly (after 1578 h of operation) is shown in Figure 6.

The guidance system with overlay welded layers was operated for 4540 h, i.e., for the entire assumed test period in real conditions. During this time, in accordance with the technical and operational documentation of the mill manufacturer, five (every 700–800 h of operation) adjustments were made to the width of the construction gap between the guide and the passive armour. The measured gap widths in the upper part of the set were in the range of 2.6–3.8 mm and in the lower part were in the range of 2.4–3.6 mm. For the group of measurements carried out in the upper and lower parts of the set after a given period of operation, the smallest difference was 0.1 mm and the largest was 0.2 mm. There was no hole present in the overlay welded layer of base materials on the guide and active armour. The analysed layers wore out evenly. No further propagation of existing cracks or spalling of the deposited layers was observed. The determined wear resistance of the set was 416–1190 h/mm. The coefficient of variation of wear resistance in the time intervals between subsequent gap measurements was 26.59%, while for five periods (excluding the first period), it was 10.64%. The calculated values of the coefficients of variation may indicate the most common course of the friction pair wear process. In the first period, running-in took place (with a decrease in the wear intensity), resulting from the topography of the surface of the layer with straight beads. As the running-in process progressed, the surface elevations (irregularities) decreased, and thus, the actual contact area increased. As a result, the average unit pressure decreased. After the running-in process, there was a period of stabilized wear with constant wear intensity (wear resistance: 883–1190 h/mm). Most likely, during the operation time of 4540 h, the third wear period (i.e., accelerated wear) did not begin [29].

Table 7. Results of wear resistance tests of overlay welded and thermal sprayed guides and active armour in operational conditions.

Working Time of Guide Sets (to Their Mounting), h	Width of the Gap between the Guide and Passive Armor, mm				Remarks		Wear Resistance, h/mm [1]	
	Set of Overlay Welded Parts		Set of Thermal Sprayed Parts					
	Dimension "a" (Upper Part of the Set), mm	Dimension "b" (Bottom Part of the Set), mm	Dimension "a" (Upper Part of the Set), mm	Dimension "b" (Bottom Part of the Set), mm	Set of Overlay Welded Parts	Set of Thermal Sprayed Parts	Set of Overlay Welded Parts	Set of Thermal Sprayed Parts
0	2.0	2.0	2.0	2.0	Assembly	Assembly	-	-
748	3.8	3.6	6.7	2.8	Gap adjustment	Gap adjustment	416	159
1578	2.9	2.7	11.0	2.5	Gap adjustment	Removal	922	92
2313	2.7	2.6	-	-	Gap adjustment	-	1050	-
3120	2.8	2.7	-	-	Gap adjustment	-	1009	-
3834	2.6	2.4	-	-	Gap adjustment	-	1190	-
4540	2.8	2.7	-	-	Removal	-	883	-

Note: [1] wear resistance is defined as the operating time (in hours) per unit of wear (corresponding to the size of the structural gap between the guide and the passive armour).

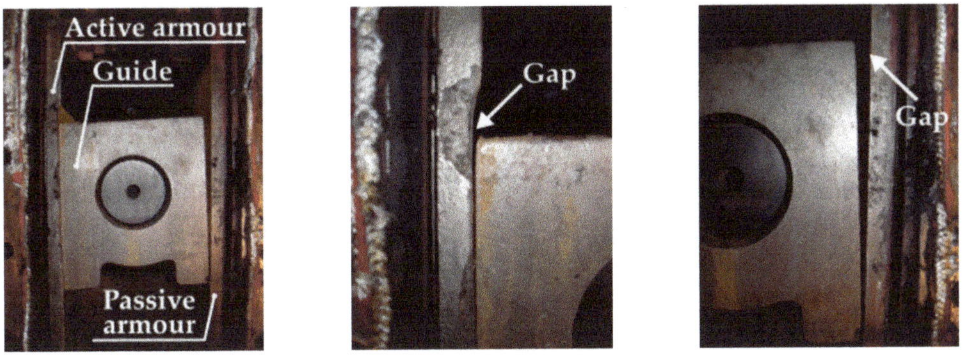

(**a**) (**b**) (**c**)

Figure 5. Guidance system with thermal sprayed coatings after 1578 h of operation: (**a**) view of the entire system, (**b**) view of the gap between the guide and the active armour, and (**c**) view of the gap between the guide and the passive armour.

The set with thermal sprayed coatings was used for 1578 h and was dismantled because after that period the gap between the guide and the passive armour was 11.0 mm, and this could result in an improper operation of the mill. At 748 h after installation, the gap was adjusted, which was 6.7 mm in the upper area and 2.8 mm in the lower area. In both periods of operation (after 748 h and after 1578 h), the sprayed coating wore out unevenly. The sprayed coating was abraded on the analysed elements (Figure 6), which resulted in a significant increase in the intensity of wear. After the overwearing of local coating (holes formed in the base material), the guide was further localized into the active armour, and as a result, the construction gap reached a value of 11.0 mm. Small areas of adhering ground carbon were visible on the surface of the sprayed coatings. The determined wear resistance of the set with sprayed coatings was 159 h/mm for the first period (748 h into the operation from installation) and 92 h/mm for the second period (another 830 h after installation). In both analogous periods, the wear resistance of the sprayed guidance system was much lower than that of the overlay welded system. In the study [7], thermal sprayed coatings with the composition of a nickel alloy with fine borides and carbides also showed a lower abrasion resistance than the hardfacing layer with the structure of high-alloy chrome cast iron.

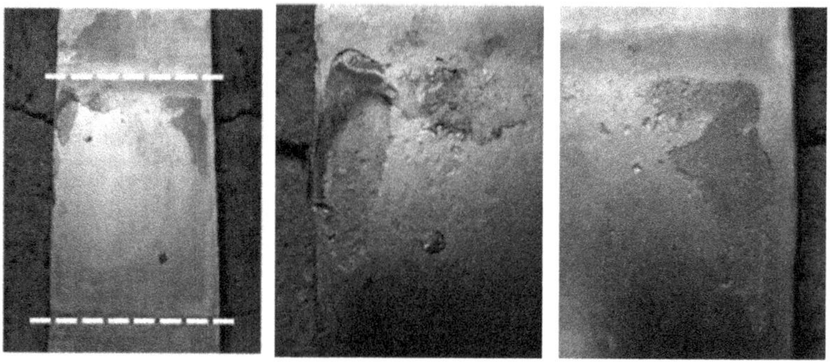

(**a**) (**b**) (**c**)

Figure 6. Active armour with a flame-sprayed coating after 1578 h of operation: (**a**) view of the coating with the working area of the friction pair marked (white-dashed lines) and (**b**,**c**) view of the coating in the areas with the greatest wear.

4.2. Macro- and Microstructure Studies

The metallographic tests carried out did not reveal any defects in the applied layers and coatings, Figure 7.

(**a**) (**b**)

Figure 7. Macrostructure of (**a**) self-shielded flux cored arc surfacing layer and (**b**) flame-sprayed coating.

The cracks were found in the overlay welded layer (Figure 8), which ran perpendicular to the direction of deposited beads and did not penetrate the base material. These types of cracks (relief-check cracks) enable the relaxation of stresses arising from the solidification and cooling of the layer during the overlay welding process. In surfaced layers with chromium cast iron alloys intended for operation in abrasive wear conditions with low-impact loads, cracks are permissible if they do not penetrate the base material, are not oriented parallel to the fusion surface, and do not lead to the detachment of fragments of the welded layer [30–32]. In tribological tests, where abrasion occurs under high pressures or significant impact loads, the uneven wear of the weld deposits could occur as a result of fragments being chipped off [9,30]. No cracks were observed in the sprayed coatings. Both overlay welded layers and thermal sprayed coatings are properly bonded to the substrate material. The sprayed coating was completely melted, which proves that the melting process was carried out correctly [22,33]. There was no melting of the surface of the base material of samples flame-sprayed with the powder. The coating had a homogenous and non-lamellar structure, which is characteristic of sprayed coatings [34].

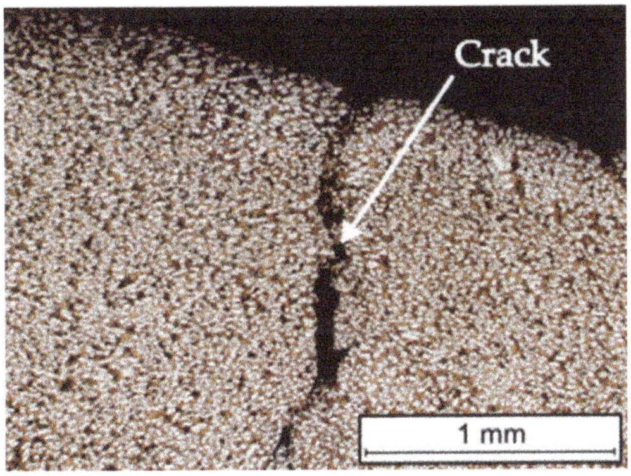

Figure 8. View of a crack in a layer produced by self-shielded flux cored arc surfacing.

Based on the analysis of the chemical composition (Figure 9) and the identification of the phase composition (Figure 10), as well as microscopic metallographic examinations (Figure 11), it was possible to conclude that the overlay weld had a structure composed of larger Cr_7C_3 carbides and smaller NbC carbides in an austenite matrix. The sprayed coating consisted of CrB borides and Cr_7C_3 carbides in a matrix of γ-Ni solid solution and Ni_3B. According to Kim et al. [35], melting the sprayed Ni-Cr-B-Si-C coating results in obtaining CrB and Cr_7C_3. The analysis of the chemical composition in micro-areas (Figure 9) indicated the occurrence of Si mainly in the matrix of the γ-Ni solid solution.

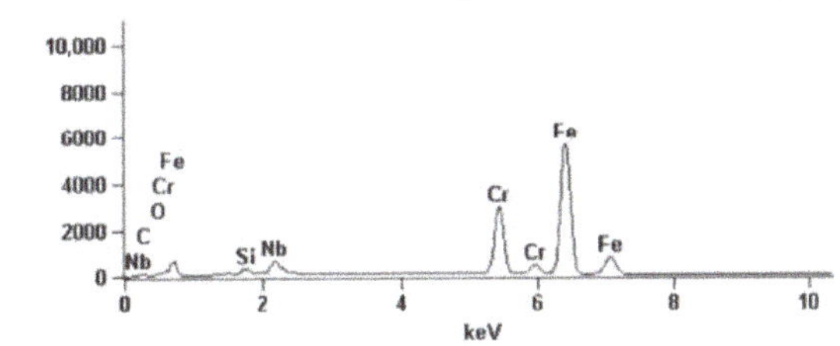

Mass concentration, wt.%			
Si	Cr	Fe	Nb
1.6	22.3	69.9	6.4

(**a**)

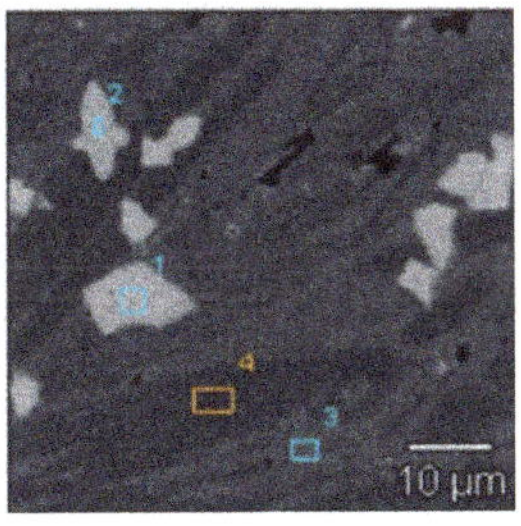

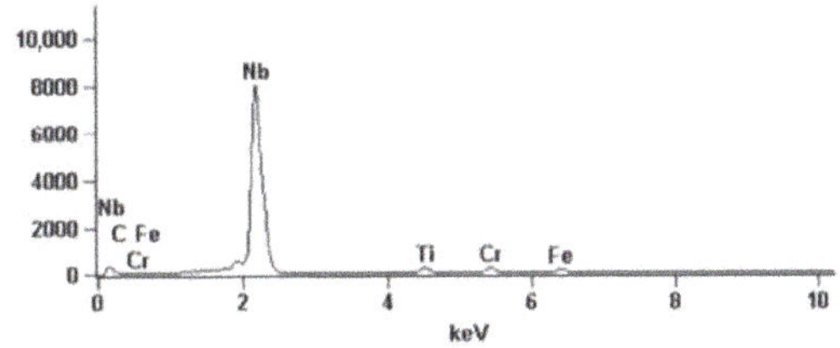

Figure 9. *Cont.*

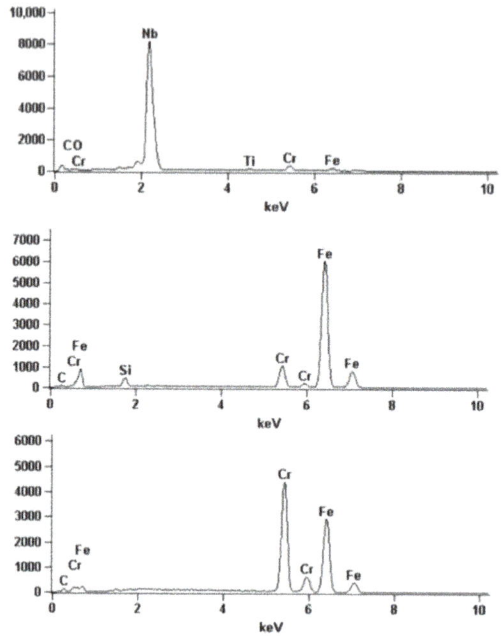

Micro-area	Mass concentration, wt.%				
	Si	Ti	Cr	Fe	Nb
1	-	1.5	2.2	2.6	93.7
2	-	1.4	4.1	2.9	91.6
3	2.9	-	8.8	88.3	-
4	-	-	48.0	52.0	-

(**b**)

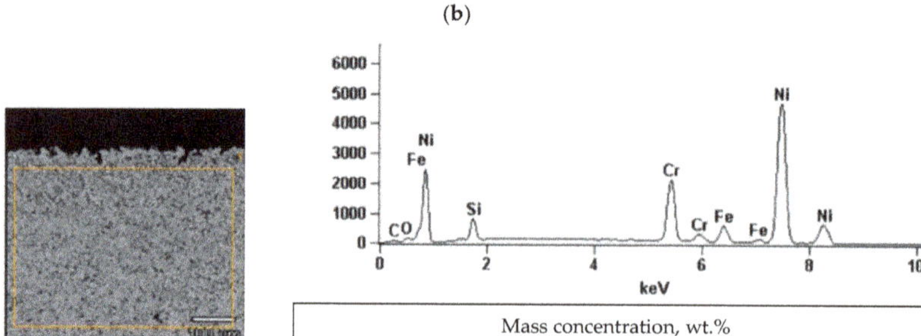

Mass concentration, wt.%			
Si	Cr	Fe	Ni
5.0	16.2	5.4	73.4

(c)

Figure 9. *Cont.*

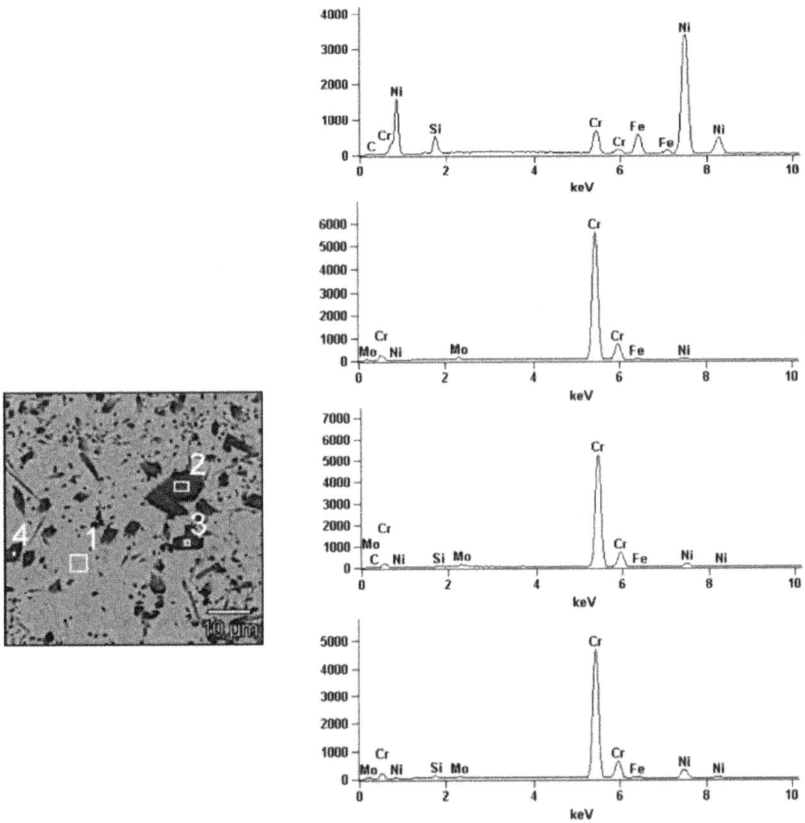

<table>
<tr><td rowspan="2">Micro-area</td><td colspan="5">Mass concentration, wt.%</td></tr>
<tr><td>Si</td><td>Cr</td><td>Fe</td><td>Ni</td><td>Mo</td></tr>
<tr><td>1</td><td>4.9</td><td>7.5</td><td>7.2</td><td>80.4</td><td>-</td></tr>
<tr><td>2</td><td>-</td><td>94.7</td><td>0.9</td><td>2.8</td><td>1.6</td></tr>
<tr><td>3</td><td>0.5</td><td>90.9</td><td>1.2</td><td>6.1</td><td>1.2</td></tr>
<tr><td>4</td><td>0.9</td><td>85.3</td><td>1.5</td><td>11.4</td><td>0.9</td></tr>
</table>

(**d**)

Figure 9. The examined micro-area along with the results of the analysis of the chemical composition of EDS: (**a**,**b**) self-shielded flux cored arc surfacing layer and (**c**,**d**) flame-sprayed coating.

The porosity of the thermal sprayed coating, determined as the arithmetic mean of ten measurements based on the analysis of ten microstructure images, is 1.08%. Self-fluxing coatings after melting are characterized by a lower porosity [35]; hence, the very low value of the determined porosity most likely results from the melting of the coating after spraying.

4.3. Hardness Measurements

The hardness of the overlay welding layer measured on the cross-sectional surface was within the range of 837–1061 HV0.3, and that of the sprayed coating was within the range of 579–714 HV0.3 (Table 8). The hardness measurements showed a very high average hardness of the overlay welding layer (932 HV0.3) and a relatively high average hardness of the sprayed coating (637 HV0.3), respectively. The range of hardness measurements for the sprayed coating was smaller (135 HV0.3) than that of the overlay welded layer (230 HV0.3).

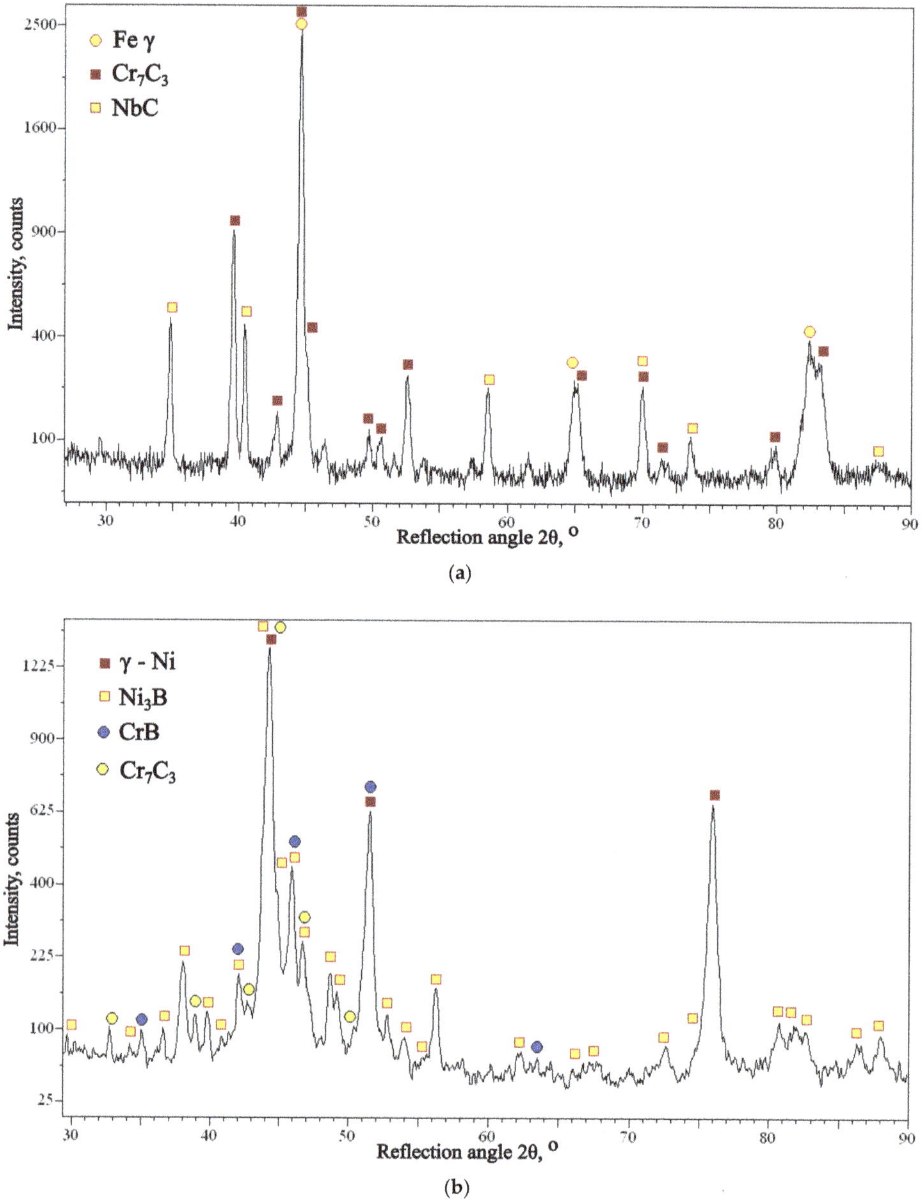

Figure 10. X-ray diffractogram of (**a**) self-shielded flux cored arc surfacing layer and (**b**) flame-sprayed coating.

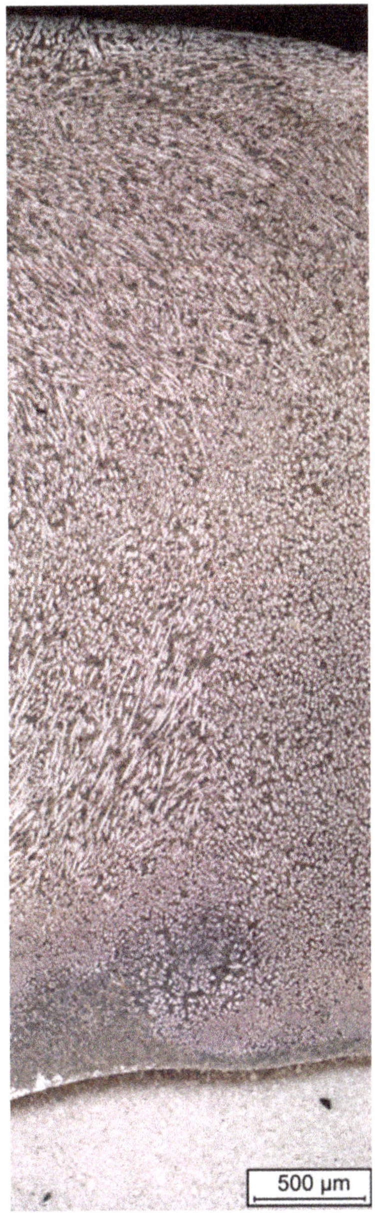

(a)

Figure 11. *Cont.*

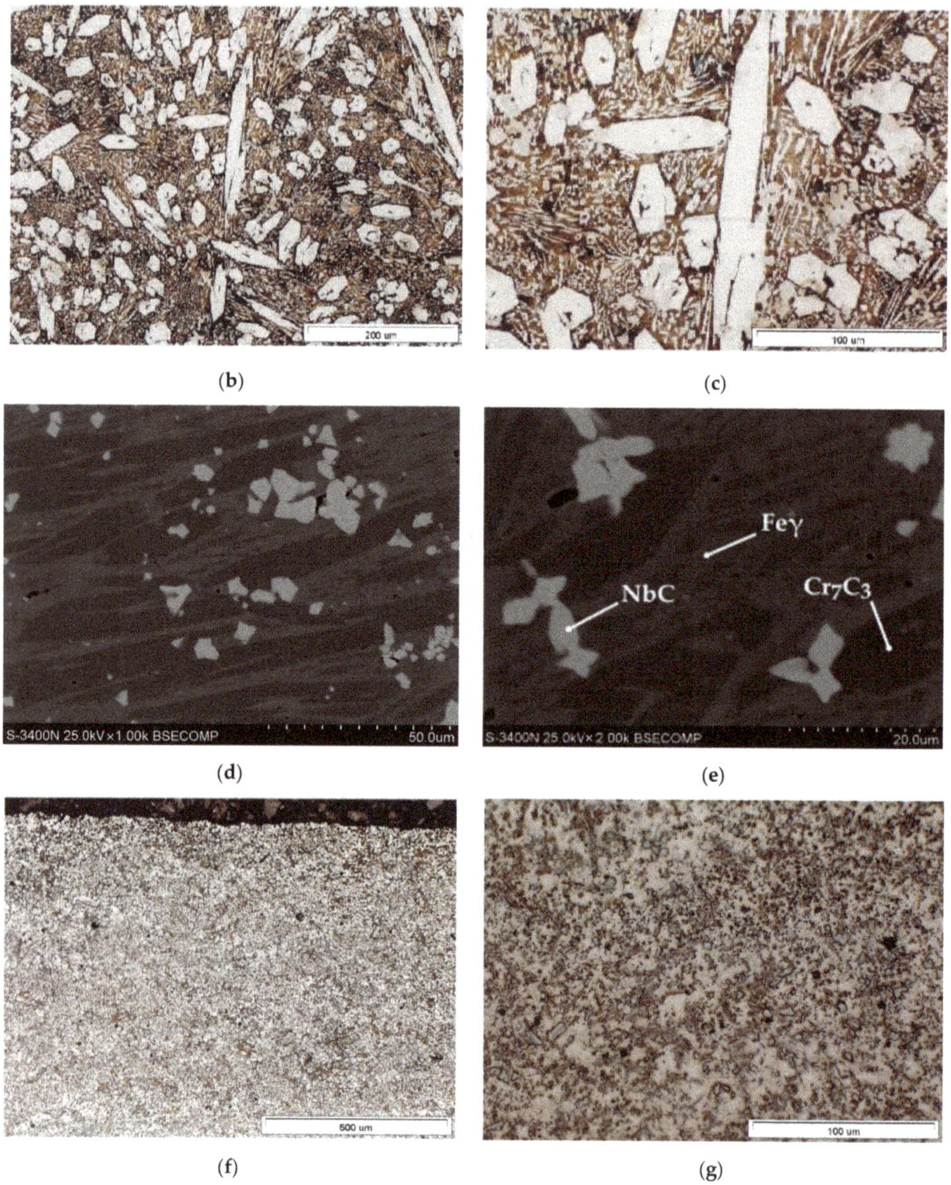

Figure 11. *Cont.*

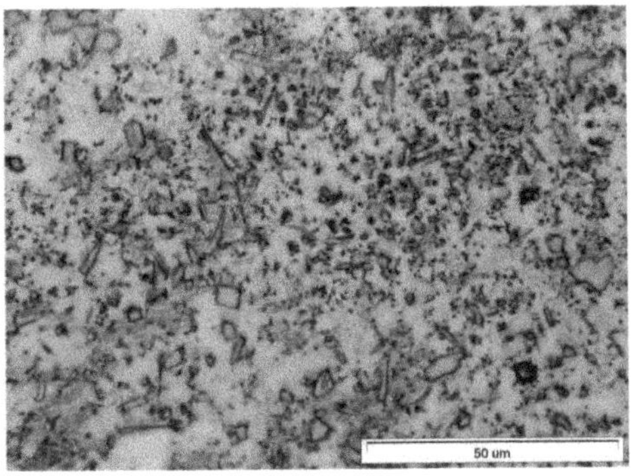

(**h**)

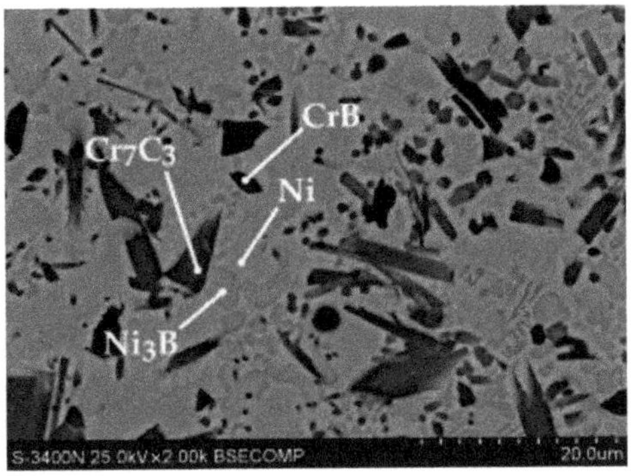

(**i**)

Figure 11. Microstructure of (**a–e**) self-shielded flux cored arc surfacing layer and (**f–i**) flame-sprayed coating.

Table 8. Results of HV0.3 hardness measurements on the cross-sectional surface of the overlay welded layer and the sprayed coating.

Method	Filler Metal	Hardness Measurement Point HV0.3 According to Figure 4				
		1	2	3	4	5
Self-shielded flux cored arc surfacing	SK A43-O	928	965	1067	864	837
Flame spraying	Metco 15E	634	616	579	640	714

5. Conclusions

Based on the study conducted for the comparison of overlay welded layers and flame-sprayed coatings in terms of the elements of the retaining ring guidance system in a ball-race mill for coal grinding, the conclusions that can be drawn are as follows:

1. In operational tests, the wear resistance of self-shielded flux cored arc surfacing layers with the structure of high-alloy chromium cast iron is much greater than that of coatings flame-sprayed with nickel-based powder. For comparable periods of use, this resistance is 416 and 922 h/mm for overlay welded layers and 159 and 92 h/mm for thermal sprayed coatings.
2. The layers deposited by flux cored arc surfacing wore out evenly, and there was no propagation of existing cracks or spalling of the layer. No hole in the surfacing layer of the base material was found after 4540 h of use. The thermal sprayed coating wore out locally (the exposure of the base material) in up to 748 h. This resulted in an increased level of wear.

Author Contributions: Conceptualization, M.S., G.G. and A.C.; methodology, M.S., G.G., A.C., W.K. and V.K.; formal analysis, M.S., G.G., A.C., W.K. and V.K.; investigation, M.S., G.G., A.C., W.K. and V.K.; resources, G.G., A.C., W.K. and V.K.; writing—original draft preparation, M.S., G.G. and A.C.; writing—review and editing, M.S., G.G., A.C., W.K. and V.K.; visualization, G.G., A.C. and W.K.; supervision, M.S. and A.C.; project administration, M.S., A.C. and V.K.; funding acquisition, A.C. and W.K. All authors have read and agreed to the published version of the manuscript.

Funding: This research received no external funding.

Institutional Review Board Statement: Not applicable.

Informed Consent Statement: Not applicable.

Data Availability Statement: Research data cannot be made available.

Conflicts of Interest: The authors declare that they have no financial or other affiliation with other persons or organisations that may unduly influence the work of the authors.

References

1. Woodruff, E.B.; Lammers, H.B.; Lammers, T.F. Chapter 3: Design and Construction of Boilers. In *Steam Plant Operation*, 10th ed.; McGraw-Hill Education: New York, NY, USA, 2017.
2. Ganapathy, V. Industrial Boilers and Heat Recovery Steam Generators: Design, Applications, and Calculations. In *Physics, Engineering, Environmental Science*; CRC Press: Boca Raton, FL, USA, 2002.
3. Bober, M.; Senkara, J.; Li, H. Comparative Analysis of the Phase Interaction in Plasma Surfaced NiBSi Overlays with IVB and VIB Transition Metal Carbides. *Materials* **2021**, *14*, 6617. [CrossRef] [PubMed]
4. Szymura, M.; Różański, M. The influence of positioning deposited beads towards direction of abrasive movement, to resistance on grind wear plates' abrasive wear. *Weld. Technol. Rev.* **2019**, *91*, 11–16. [CrossRef]
5. Adamiak, M.; Tomiczek, B.; Górka, J.; Czupryński, A. Joining of the AMC Composites Reinforced with Ti3Al Intermetallic Particles by Resistance Butt Welding. *Arch. Metall. Mater.* **2016**, *61*, 847–852. [CrossRef]
6. Klimpel, A.; Górka, J.; Czupryński, A. Comparison of chromium cast iron deposits of wear plates. *J. Achiev. Mater. Manuf. Eng.* **2006**, *8*, 387–390.
7. Dallaire, S. Hard arc-sprayed coating with enhanced erosion and abrasion wear resistance. *J. Therm. Spray Technol.* **2001**, *10*, 511–519. [CrossRef]
8. Neale, M.; Gee, M. *A Guide to Wear Problems and Testing for Industry*; William Andrew: Norwich, NY, USA, 2001.
9. Kirchgaßner, M.; Badisch, E.; Franek, F. Behaviour of iron-based hardfacing alloys under abrasion and impact. *Wear* **2008**, *265*, 772–779. [CrossRef]
10. Bialobrzeska, B.; Kostencki, P. Abrasive wear characteristics of selected low alloy boron steels as measured in both field experiments and laboratory tests. *Wear* **2015**, *328–329*, 149–159. [CrossRef]
11. Pondicherry, K.; Sukumaran, J.; De Baets, P.; Keresztes, R. Abrasion test facilities—A review. *Mech. Eng. Lett. Szent István Univ.* **2015**, *12*, 19–29.
12. Adachi, K.; Hutchings, I.M. Wear-mode mapping for the micro-scale abrasion test. *Wear* **2003**, *255*, 23–29. [CrossRef]
13. Klimpel, A.; Janicki, D.; Bryś, M.; Klimpel, A. Influence of the type of an abrasive on wear resistance of the abrasive panel made of chromium cast iron. *Weld. Technol. Rev.* **2006**, *78*, 3–6.
14. Güney, B.; Mutlu, İ.; Küçüksariyildiz, H. The effect of flame spray coating on the tribological properties of brake disc. *J. Polytech.* **2020**, *23*, 755–761. [CrossRef]
15. Zum Gahr, K.H. Wear by hard particles. *Tribol. Int.* **1998**, *10*, 587–596. [CrossRef]
16. Szymura, M.; Czupryński, A.; Różański, M. Research on the properties of high chromium cast iron overlay welds deposited by tubular electrodes. *Weld. Technol. Rev.* **2018**, *90*, 26–30. [CrossRef]
17. Adamiak, M.; Górka, J.; Kik, T. Comparison of abrasion resistance of selected constructional materials. *J. Achiev. Mater. Manuf. Eng.* **2009**, *37*, 375–380.
18. Kotarska, A.; Poloczek, T.; Janicki, D. Characterization of the Structure, Mechanical Properties and Erosive Resistance of the Laser Cladded Inconel 625-Based Coatings Reinforced by TiC Particles. *Materials* **2021**, *14*, 2225. [CrossRef] [PubMed]
19. Czupryński, A. Microstructure and Abrasive Wear Resistance of Metal Matrix Composite Coatings Deposited on Steel Grade AISI 4715 by Powder Plasma Transferred Arc Welding Part 2. Mechanical and Structural Properties of a Nickel-Based Alloy Surface Layer Reinforced with Particles of Tungsten Carbide and Synthetic Metal–Diamond Composite. *Materials* **2021**, *14*, 2805. [CrossRef] [PubMed]
20. Abderrahmane, A.; Gaceb, M.; Cheikh, M.; Le Roux, S. Wear Behavior and Microstructure of Thermally Sprayed NiCrBSiFeC and Composite NiCrBSiFeC-WC (Co) Coatings. *Mater. Sci. (Medžiag.)* **2012**, *27*, 175–183. [CrossRef]
21. Kazamer, N.; Muntean, R.; Vălean, P.C.; Pascal, D.T.; Mărginean, G.; Şerban, V.-A. Comparison of Ni-Based Self-Fluxing Remelted Coatings for Wear and Corrosion Applications. *Materials* **2021**, *14*, 3293. [CrossRef]
22. Adamiec, J. High temperature corrosion of power boiler components cladded with nickel alloys. *Mater. Charact.* **2009**, *60*, 1093–1099. [CrossRef]
23. Navas, C.; Colaço, R.; De Damborenea, J.; Vilar, R. Abrasive wear behaviour of laser clad and flame sprayed-melted NiCrBSi coatings. *Surf. Coat. Technol.* **2006**, *200*, 6854–6862. [CrossRef]
24. Azzoug, R.; Mebdoua, Y.; Hellal, F.; Marra, F. Analysis of microstructure, mechanical indentation and corrosive behavior of a thermally sprayed NiFeCrBSi-WC composite coating. *J. Alloy. Compd.* **2022**, *900*, 163505. [CrossRef]
25. *EN 10293*; Steel Castings—Steel Castings for General Engineering Uses. CEN: Brussels, Belgium, 2015.
26. *EN 10025-2*; Hot Rolled Products of Structural Steels—Part 2: Technical Delivery Conditions for Non-Alloy Structural Steels. CEN: Brussels, Belgium, 2019.
27. *EN 14700*; Welding Consumables. Welding Consumables for Hard-Facing. CEN: Brussels, Belgium, 2014.
28. *ISO 6507*; Metallic Materials—Vickers Hardness Test—Part 1: Test Method. ISO: Geneva, Switzerland, 2018.
29. Nosal, S. Tribological aspect of creating abrasion resistance of machine elements. *J. Res. Appl. Agric. Eng.* **2018**, *63*, 35–39.
30. Riabcew, I.A.; Rosert, R.; Senczenkow, I.K.; Turyk, E. Welding Imperfections in Surfaced Layers. *Bull. Inst. Weld.* **2017**, *3*, 26–35. [CrossRef] [PubMed]

31. Menon, R.; Wallin, J. Specialty Cored Wires for Wear and Corrosion Applications. *Weld. J.* **2008**, *87*, 31–37.
32. Lisiecki, A. Study of laser cladding of composite coatings. *Int. J. Mod. Manuf. Technol.* **2022**, *14*, 121–127. [CrossRef]
33. Bergant, Z.; Grum, J. Quality improvement of flame sprayed, heat treated, and remelted NiCrBSi coatings. *J. Therm. Spray Technol.* **2009**, *18*, 380–391. [CrossRef]
34. Rachidi, R.; El Kihel, B.; Delaunois, F. Microstructure and mechanical characterization of NiCrBSi alloy and NiCrBSi-WC composite coatings produced by flame spraying. *Mater. Sci. Eng. B* **2019**, *241*, 13–21. [CrossRef]
35. Kim, H.J.; Hwang, S.Y.; Lee, C.H.; Juvanon, P. Assessment of wear performance of flame sprayed and fused Ni-based coatings. *Surf. Coat. Technol.* **2003**, *172*, 262–269. [CrossRef]

materials

Article

High-Power Diode Laser Surface Transformation Hardening of Ferrous Alloys

Artur Czupryński [1,*], Damian Janicki [1], Jacek Górka [1], Andrzej Grabowski [2], Bernard Wyględacz [1], Krzysztof Matus [3] and Wojciech Karski [1]

1 Department of Welding, Silesian University of Technology, Konarskiego 18A, 44-100 Gliwice, Poland; damian.janicki@polsl.pl (D.J.); jacek.gorka@polsl.pl (J.G.); bernard.wygledacz@polsl.pl (B.W.); wojciech.karski@polsl.pl (W.K.)
2 Institute of Physics—CSE, Silesian University of Technology, Krasińskiego 8, 40-019 Katowice, Poland; andrzej.grabowski@polsl.pl
3 Materials Research Laboratory, Silesian University of Technology, Konarskiego 18A, 44-100 Gliwice, Poland; krzysztof.matus@polsl.pl
* Correspondence: artur.czuprynski@polsl.pl

Abstract: A high-power direct diode laser (HPDDL) having a rectangular beam with a top-hat intensity distribution was used to produce surface-hardened layers on a ferrous alloy. The thermal conditions in the hardened zone were estimated by using numerical simulations and infrared (IR) thermography and then referred to the thickness and microstructure of the hardened layers. The microstructural characteristics of the hardened layers were investigated using optical, scanning electron and transmission electron microscopy together with X-ray diffraction. It was found that the major factor that controls the thickness of the hardened layer is laser power density, which determines the optimal range of the traverse speed, and in consequence the temperature distribution in the hardened zone. The increase in the cooling rate led to the suppression of the martensitic transformation and a decrease in the hardened layer hardness. The precipitation of the nanometric plate-like and spherical cementite was observed throughout the hardened layer.

Keywords: laser surface transformation hardening; high-power diode laser; ferrous alloys; thermography; cooling rate

check for updates

Citation: Czupryński, A.; Janicki, D.; Górka, J.; Grabowski, A.; Wyględacz, B.; Matus, K.; Karski, W. High-Power Diode Laser Surface Transformation Hardening of Ferrous Alloys. *Materials* **2022**, *15*, 1915. https://doi.org/10.3390/ma15051915

Academic Editor: Antonio Riveiro

Received: 8 February 2022
Accepted: 26 February 2022
Published: 4 March 2022

1. Introduction

Laser surface treatment methods have been intensively employed for the past three decades for the enhancement of surface properties of different cast iron grades [1–13]. Much research has been devoted to improving the tribological properties of cast irons via a laser surface transformation hardening (LSTH) process [10–12,14–18]. In contrast to conventional surface hardening processes such as induction and flame hardening, LSTH ensures excellent surface finish and very low distortion of the workpiece. In many cases, this eliminates any need for post-treatment finishing [19]. The precise control of the heat input together with the ability to heat localized areas make LSTH a unique technology for the fabrication of locally selected wear-resistant hard surfaces on complex-shaped components [20]. In recent years, there has been significant interest in the application of LSTH to increase the tribological performance of ductile cast irons (DCI) in both the as-cast condition and after austempering heat treatment [14–17].

Many researchers have reported on the significant increase in the wear properties of DCI after the LSTH process [14,16]. The tribological properties of the laser hardened layers on DCIs depend upon the layer hardness, which, in turn, is affected by the phase composition of the layer [16]. On the other hand, the microstructural characteristics of the laser hardened layer and also the layer thickness are directly determined by the thermal conditions in the hardened zone [17]. For successful laser hardening, it is necessary to form a relatively homogeneous austenite structure in the hardened zone during a very short period of the laser-induced thermal cycle. This requires notably higher temperatures on the hardened surfaces than those during conventional hardening to achieve a sufficiently high diffusion rate for the homogeneous austenitic structure [18]. As a result, to obtain the desired layer thickness and to avoid partial melting, precise control of the surface temperature is needed. However, scant data are available on the effect of LSTH parameters on the thermal conditions during the processing of DCIs and the resulting microstructure of the hardened layers.

The main objective of this work was to assess the effect of the LSTH parameters on the thermal conditions in the hardened zone and the resulting change in the thickness and the microstructure of the surface-hardened layer. The LSTH process was performed on a pearlitic DCI using a high-power direct diode laser having a rectangular beam with a top-hat intensity distribution. The thermal conditions in the hardened zone were determined by using infrared (IR) thermography and numerical simulations. The surface-hardened layers were characterized in terms of their surface finish, microstructure and hardness.

2. Materials and Methods

The chemical composition of the DCI (EN-GJS-700-2) used in this study is listed in Table 1. The DCI composition was measured using both a LECO GDS 500A optical emission spectrometer (LECO Corporation, St. Joseph, MI, USA) and a LECO CS125 Carbon Analyzer (LECO Corporation, St. Joseph, MI, USA). The microstructure of the as-received DCI was composed of a pearlitic matrix (ferrite content <15%) and graphite nodules with an average diameter of about 30 μm (Figure 1a). The average thickness of the cementite lamellae was 160 nm (Figure 1b). The hardness of the DCI in the as-received condition was 290 ± 15 HV0.5. The DCI specimens in the shape of a rectangular plate (40 × 30 × 10 mm) were ground to an average roughness (R_a) of approximately0.05 μm. The surface roughness measurements were conducted using a Surftest SJ-210 portable surface roughness tester (Mitutoyo Corporation, Kawasaki, Kanagawa, Japan). Prior to obtaining the reflection measurements and the laser hardening, the test plates were cleaned in acetone.

Table 1. Chemical composition of the investigated DCI substrate (wt%).

C	Si	Cu	Mn	Cr	Ni	S	P	Fe
3.52	2.62	0.80	0.24	0.02	0.04	0.008	0.016	balance

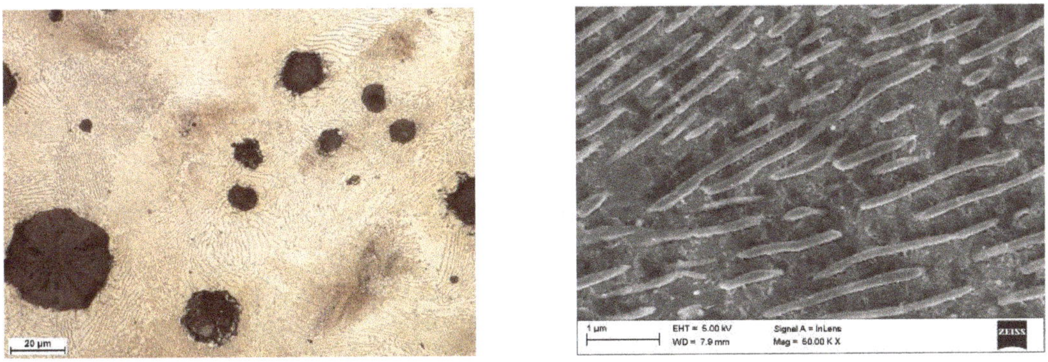

(a) (b)

Figure 1. (**a**) Optical image showing the microstructure of the as-received DCI grade EN-GJS-700-2; (**b**) In-lens SEM image showing the perlite matrix.

The reflectivity of the DCI specimens at room temperature was measured using a setup comprising an Ocean Optics PC2000-ISA-PC Plug in Fiber Optics Spectrometer (wavelength range 500–1000 nm; resolution: ±0.5 nm), an ISP-REF Integrating Sphere and a WS-1 diffuse reflectance standard (Ocean Optics Inc., Dunedin, FL, USA). Reflectance measurements were made using both diffuse and specularly reflected light.

The thermographic measurements were carried out with the use of a FLIR A655SC thermal camera (Teledyne FLIR LLC., Wilsonville, OR, USA). The camera works in the 7.5–14.0 μm wavelength range and a 16-bit dynamic range at a 50 Hz polling rate and has an accuracy of ±2% of reading.

The hardening trials were carried out with a high-power direct diode laser (HPDDL) with a rectangular beam with a top-hat intensity distribution in the slow-axis direction (HPDDL, Rofin-Sinar Laser GmbH, Hamburg, Germany). The dominant wavelength of HPDDL used was 808 nm. The HPDDL beam spot size was 1.5 × 6.6 mm (length × width). The focal plane of the laser optics was located at the DCI specimen and the slow HPDDL beam axis was set to be perpendicular to the scanning direction. Argon (at a flow rate of 10 L/min) was used during all hardening trials to prevent oxidation of the surface. Additional details of the HPDD laser source have been previously given [21]. The selected hardening conditions are listed in Table 2. The selection of the optimal range of processing conditions was aimed at avoiding the finish grinding operation in potential industrial applications. For this reason, the occurrence of partial surface melting was not accepted.

Table 2. Selected processing conditions of laser surface transformation hardening.

Processing Condition No./SHB No.	Laser Power (W)	Laser Power Density (W/mm^2)	Traverse Speed (m/min)	Heat Input [1] (J/mm)	Surface Hardness HV0.5	Layer Thickness [2] (μm)	Remarks/ Quality [3]
H1	400	40.4	0.1	240	867 ± 60	458 ± 14	PM
H2			0.2	120	755 ± 32	61 ± 11	NM
H3			0.3	80	290 ± 15	-	NH
H4	600	60.6	0.2	180	841 ± 54	437 ± 15	PM
H5			0.4	90	724 ± 29	192 ± 11	NM
H6			0.6	60	689 ± 31	78 ± 10	NM
H7	800	80.8	0.4	120	796 ± 58	393 ± 16	PM
H8			0.6	80	690 ± 28	196 ± 12	NM
H9			0.8	60	684 ± 30	92 ± 11	NM

[1] defined by the ratio of the laser power and the traverse speed; [2] measured in the center of SHB; [3] assessed in terms of the surface condition: PM—partial melting, NM—no surface melting, NH—no hardening.

Single-pass hardened beads (SHBs) were examined by optical microscopy (Eclipse MA100, Nikon Instech Co., Ltd., Tokyo, Japan), scanning electron microscopy (SEM, ZEISS SUPRA 35, Carl Zeiss Microscopy GmbH, Jena, Germany), transmission electron microscopy (FEI Titan 80/300, Scientific and Technical Instruments, Hillsboro, OR, USA) and X-ray diffraction (XRD). SEM investigations were conducted using both secondary electrons and in-lens imaging modes. For both the SEM and optical investigations, the specimens were etched with 4% Nital. The thickness of the hardened layer was measured using Nikon image processing software on optical and SEM images taken from the SHB cross-sections. The TEM investigations were performed using the conventional TEM and scanning transmission electron microscopy (STEM) modes. STEM observations were carried out using a high-angle annular dark-field (HAADF) detector. TEM samples were prepared using an FIB technique using an SEM/FIB DualBeam microscope (FEI Helios NanoLab 600i FIB, Scientific and Technical Instruments, Hillsboro, OR, USA).

The phase composition of the hardened layers was examined by using a PANalytical X'Pert PRO MPD X-ray diffractometer (Malvern Panalytical, Almelo, The Netherlands). XRD patterns were obtained using Co-Kα radiation (λ = 0.179 nm).

The hardness measurements were performed with a Wilson Wolpert 401 MVD microhardness intender (Wilson Wolpert Instruments, Aachen, Germany). The measurements on the hardened surface and the polished SHB cross-sections were conducted using 500 g and 100 g loads, respectively.

The laser hardening process was simulated by means of a nonlinear 3D transient thermal simulation in Visual Weld 16 (Sysweld core-ESI Group). The material database consisted of specific heat and thermal conductivity over the range of 20 °C to the melting point. The simulations take into consideration solid-state transformation with the Johnson–Mehl–Avrami–Kolmogorov equation for diffusive transformations and the Koistinen–Marburger equation for non-diffusive transformation. The equation parameters were chosen to fit the simulated TTT diagram to the experimental TTT diagram. A 3D Model with 691,620 elements and 187,640 nodes was developed (Figure 2). The dimensions of the developed 3D model were consistent with those of the specimens used in the experimental stage. The model was prepared in such a manner that where the simulated laser processing took place, hexa elements with a 0.05 × 0.05 × 0.05 mm size were used, and the rest of the model was composed of tetra elements with element size increasing with the distance from the processed area. Only half of the planar symmetrical sample was simulated to decrease computational complexity. This did not affect the simulation accuracy. The thermal boundary conditions were set to free air with a temperature of 20 °C exchange on the outer surfaces of the model excluding the symmetry plane. The process was simulated with the usage of a modified load 3D double ellipsoidal heat source proposed by Kik [22]. The double ellipsoidal heat source dimensions were calibrated to achieve the 800 °C isotherm dimensions equal to the dimensions of the hardening zone. The heat source parameters were as follows: length of 1.6 mm, depth of 0.1 mm and load zone width of 6.6 mm. The results of reflection measurements (Figure 3) were taken into account during the selection of energy input. The final heat source efficiency was in the 0.5–0.6 range.

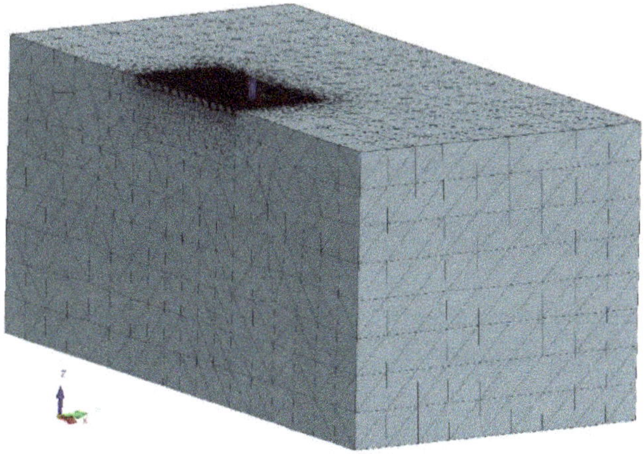

Figure 2. The developed 3D model and meshing arrangement.

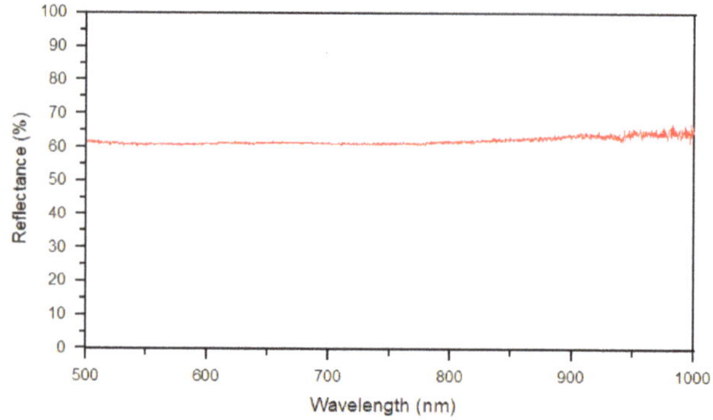

Figure 3. Spectral reflectance curve for a test plate of the DCI having an average roughness (R_a) of approximately 0.05 μm.

3. Results

3.1. Surface Condition Analysis

Figures 4 and 5 show the appearance of the laser hardened surface of the selected DCI in different processing conditions. The results of the hardness measurements on the hardened surfaces are listed in Table 2. The laser power level, determining the laser power density, has a direct impact on the optimal range of the traverse speed. With an increasing laser power density, the lower limit of the traverse speed, which ensures avoidance of partial melting, is notably increased (Table 2). The partial melting occurred in the regions directly adjacent to the graphite nodules and led to the formation of ledeburite eutectic structures on the processed surface (Figures 4a and 5a). The presence of ledeburite eutectic regions increased the overall surface hardness of the processed layers (Table 2). Note that, to avoid finish grinding operations, surface melting is undesirable in transformation hardening. The surface hardness of the SHB fabricated in the optimal range of parameters (providing no surface melting) was in the range of 755 to 684 HV0.5. It should be noted that with increasing traverse speed, the maximum surface hardness decreased. To understand the effect of the processing conditions on the microstructure and the hardness of the laser surface-hardened layers on DCI, the next sections present detailed investigations on the microstructure and hardness of the SHBs produced under processing condition no. H2 and H8, as well as the thermal conditions in the hardened zone under those two processing conditions (Table 2).

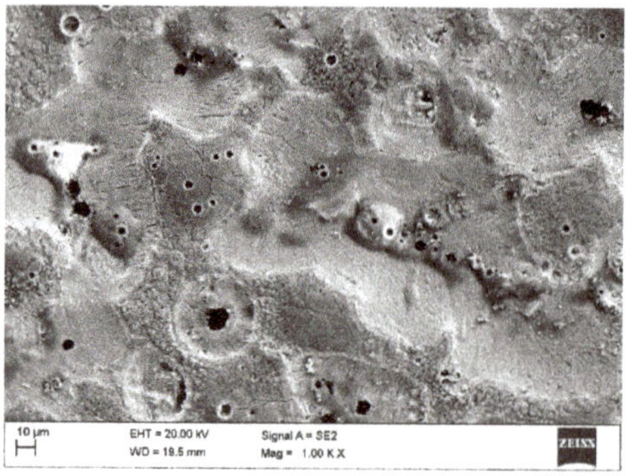

(**a**)

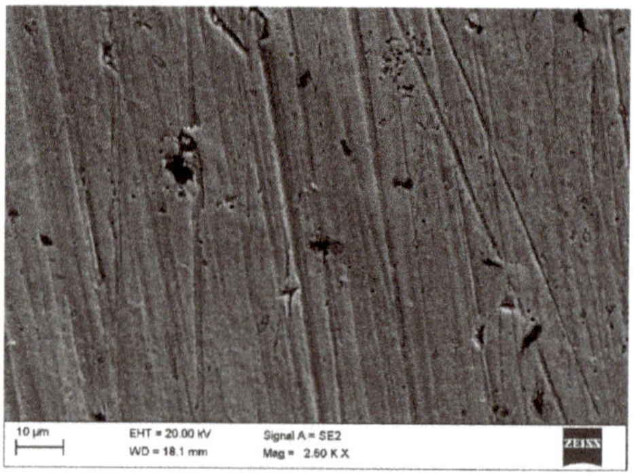

(**b**)

Figure 4. SEM images of the hardened surface at laser power of 400 W and traverse speed of: (**a**) 0.1 m/min; (**b**) 0.2 m/min.

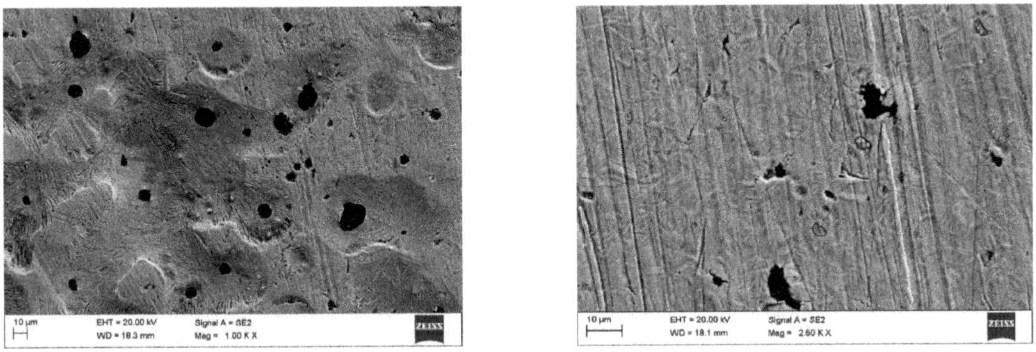

(a) (b)

Figure 5. SEM images of the hardened surface at a laser power of 800 W and a traverse speed of: (**a**) 0.4 m/min; (**b**) 0.6 m/min.

3.2. Macro and Microstructural Analysis

Figure 6a,b show the cross-sectional macrographs of the SHBs produced under processing conditions no. H2 and H8, respectively (Table 2). The corresponding low magnification optical micrographs of the central area of the SHB cross-sections are presented in Figure 7. Optical microscope images taken from the undersurface area of the above-mentioned SHBs are shown in Figure 8. XRD patterns for the investigated DCI in the as-received condition and after the LSTH under processing conditions no. H2 and H8 are presented in Figure 9a–c, respectively.

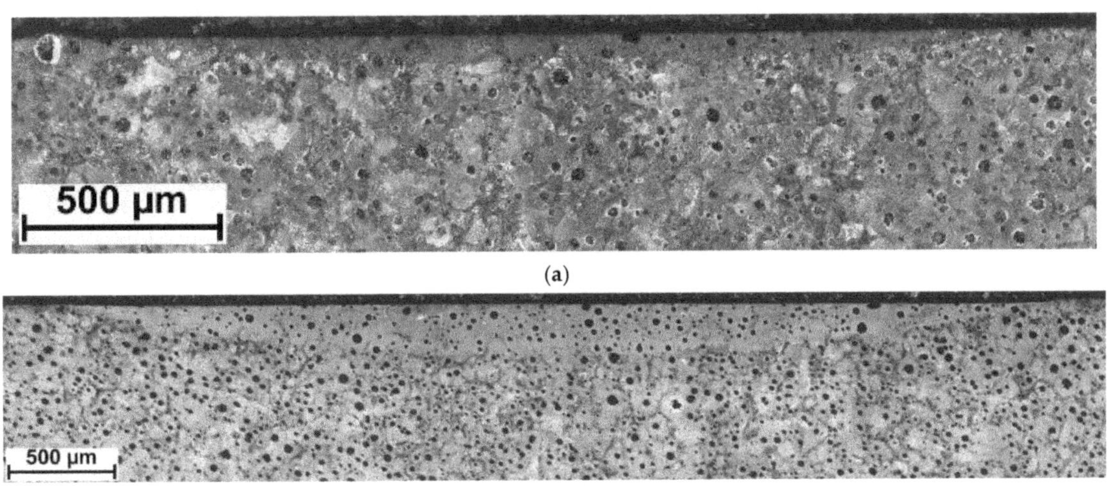

(a)

(b)

Figure 6. Optical macrographs of the SHBs no. (**a**) H2; (**b**) H8 (Table 2).

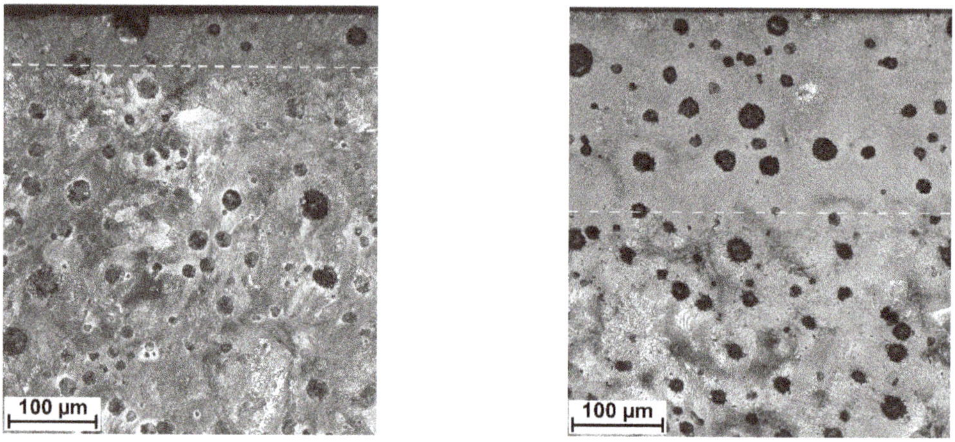

(**a**) (**b**)

Figure 7. Low magnification optical micrographs of the SHB no. (**a**) H2; (**b**) H8 (Table 2). Dashed lines indicate an approximate boundary between the hardened layer and the core material.

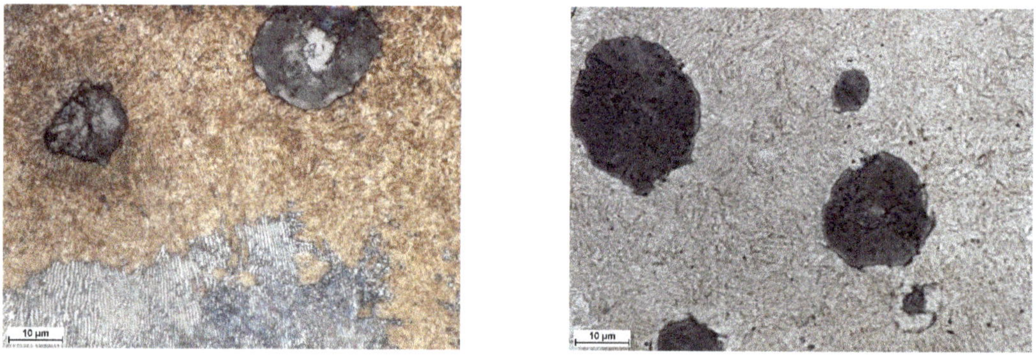

(**a**) (**b**)

Figure 8. Optical micrographs showing the undersurface area of the SHB (in the center of the bead) no. (**a**) H2 (hardened layer and transition boundary) and (**b**) H8 (hardened zone) (Table 2).

The micrographs in Figure 7 suggest that the thickness of the hardened layer was approximately 60 and 200 µm for conditions no. H2 and H8, respectively. Based on the optical micrographs presented in Figure 8, one can observe that the microstructure of the hardened layer is composed of graphite nodules and the martensitic matrix. The XRD analysis indicated the additional presence of the retained austenite phase in the matrix of the hardened layer produced under processing condition no. H8 (Figure 9c). Note that the retained austenite fraction in the hardened layer produced under processing condition no. H2 was negligible (Figure 9b). Moreover, XRD patterns from all hardened layers gave peaks that can be identified as belonging to the cementite phase.

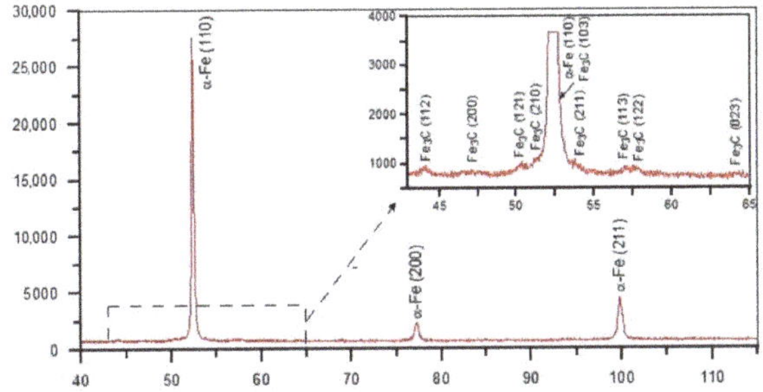

(a)

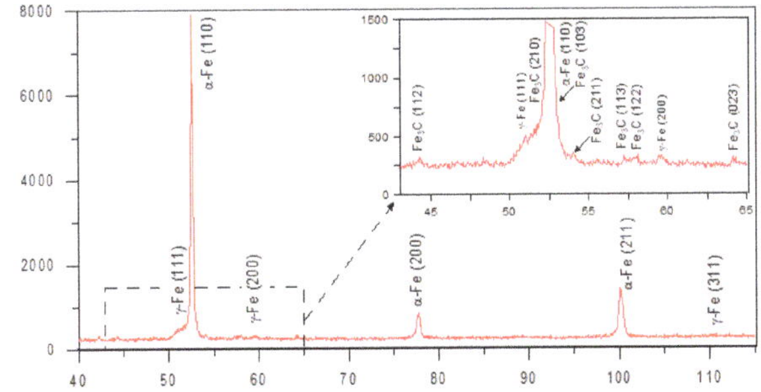

(b)

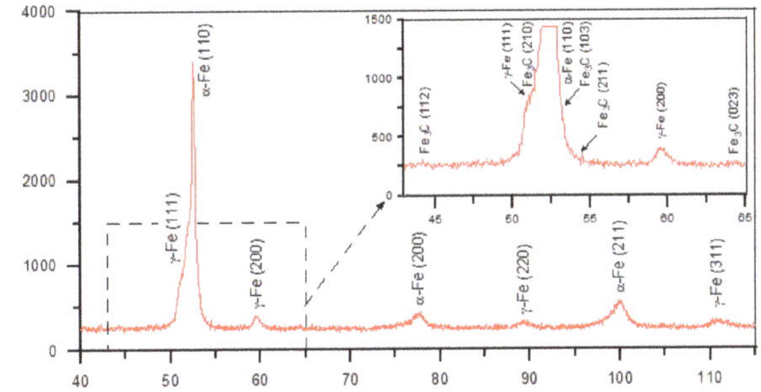

(c)

Figure 9. XRD patterns of the (**a**) as-received DCI and after the LSTH under conditions no. (**b**) H2 and (**c**) H8 (Table 2).

SEM investigations (Figures 10 and 11) indicated that the hardened layer can be divided into two sub-regions: the region of complete dissolution of cementite lamellae and the region of partial dissolution of cementite lamellae. Directly beneath the hardened surface, there was a region exhibiting a complete dissolution of cementite lamellae (Figure 11a). The thickness of this region was approx. 22 and 135 μm in the SHB no. H2 and H8, respectively. The microstructure of this region was composed of the graphite nodules and the martensitic matrix (the martensitic/austenitic matrix in the case of SHB no. H8). The region of the hardened layer adjacent to the core material had a matrix that was, in general, composed of partially dissolved cementite lamellae and fine martensite laths (Figure 11b,c). The thickness of the region of partial dissolution of cementite lamellae was approx. 39 and 61 μm in the SHB no. H2 and H8, respectively.

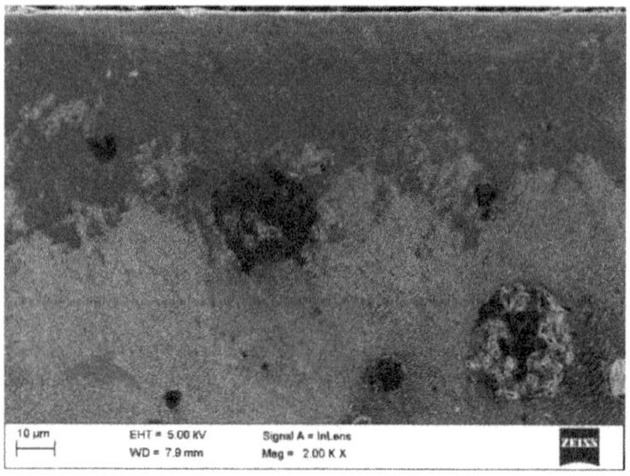

(a)

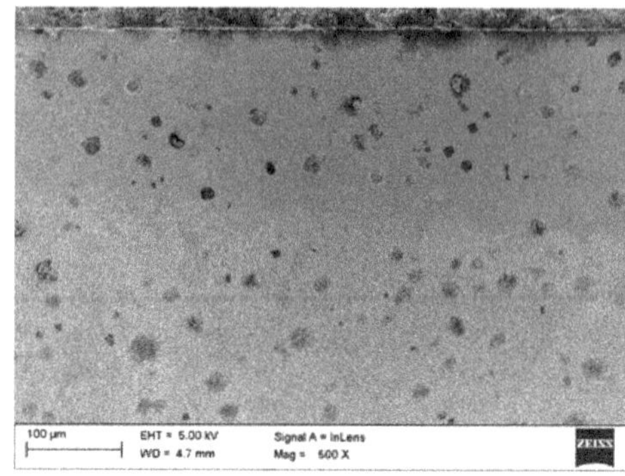

(b)

Figure 10. In-lens SEM images taken from the undersurface area of SHB no. (**a**) H2 and (**b**) H8 (Table 2).

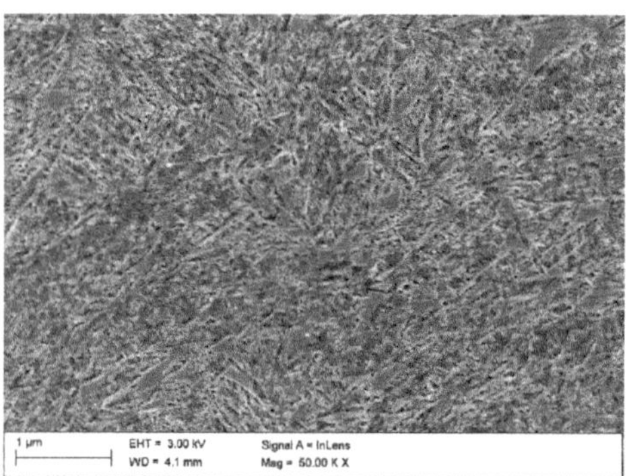

(a)

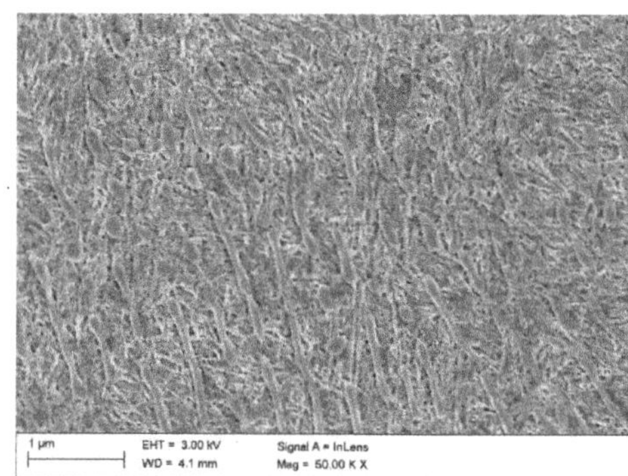

(b)

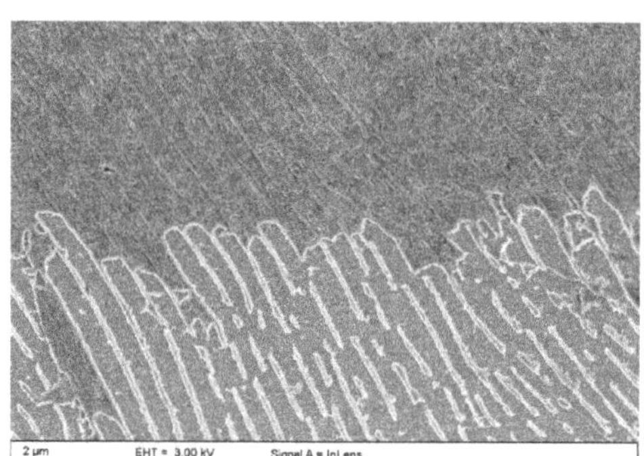

(c)

Figure 11. In-lens SEM images taken from the SHB no. H8 (Table 2) showing: (**a**) the region of complete dissolution of cementite lamellae; (**b**) the region of partial dissolution of the cementite lamellae; (**c**) a boundary between the region of partial dissolution of the cementite lamellae and the core material.

TEM investigations of the region of complete dissolution of the cementite lamellae in the hardened layer (Figures 12 and 13) revealed the presence of nanometric plate-like and spherical precipitates of cementite. The thickness of the cementite plates was close to

40 nm (Figure 12c). Both types of nanometric cementite precipitates were observed at the martensite lath boundaries. Further research will be required to clarify the mechanism of cementite precipitation during the investigated LSTH process.

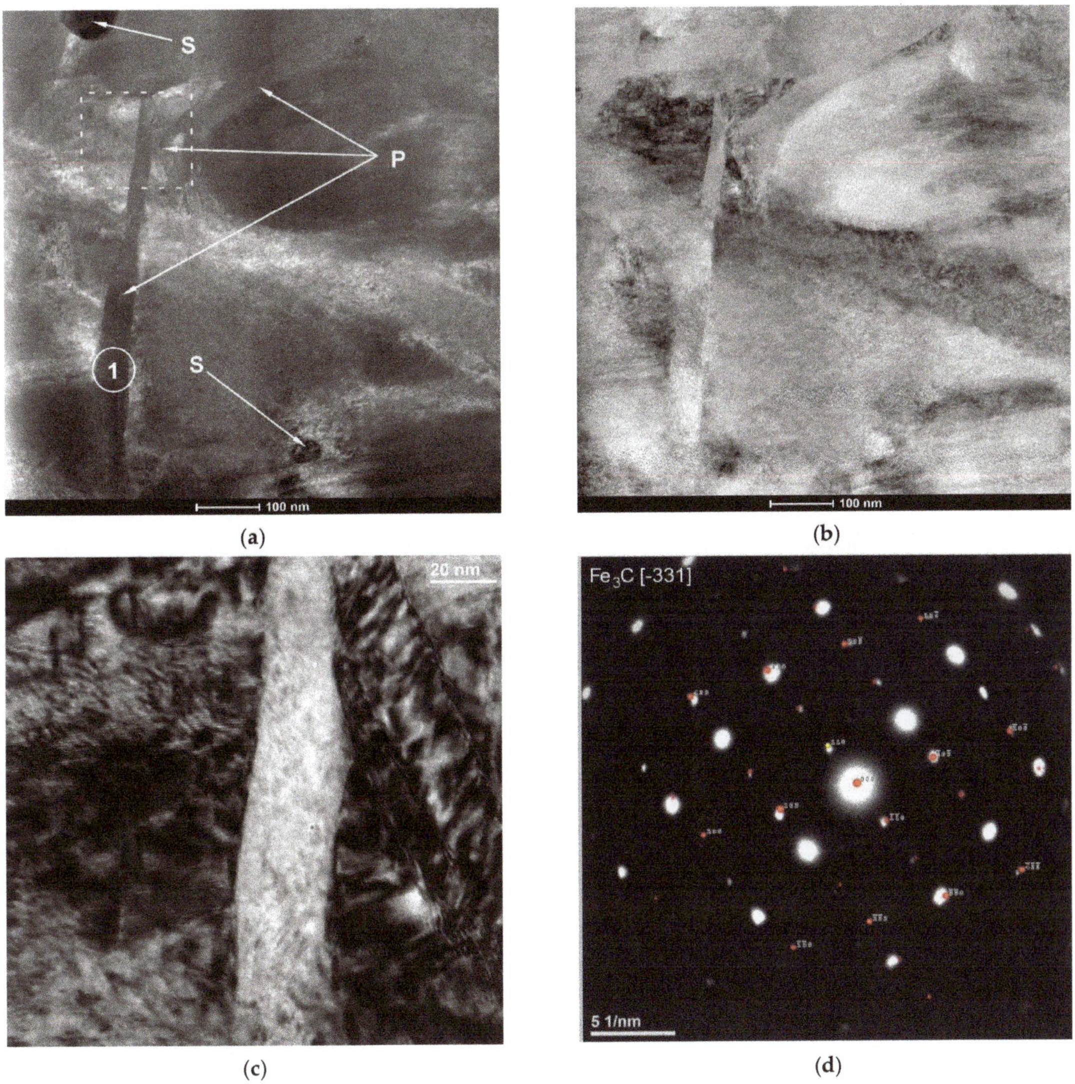

Figure 12. (**a**) Dark-field and (**b**) bright-field STEM HAADF images showing a distribution of plate-like (P) and spherical (S) cementite precipitates formed in the matrix of the SHB no. H8 (the region of complete dissolution of the cementite lamellae); (**c**) Bright-field TEM image taken from a marked area in (**a**); (**d**) SAED pattern obtained from region (1) in (**a**).

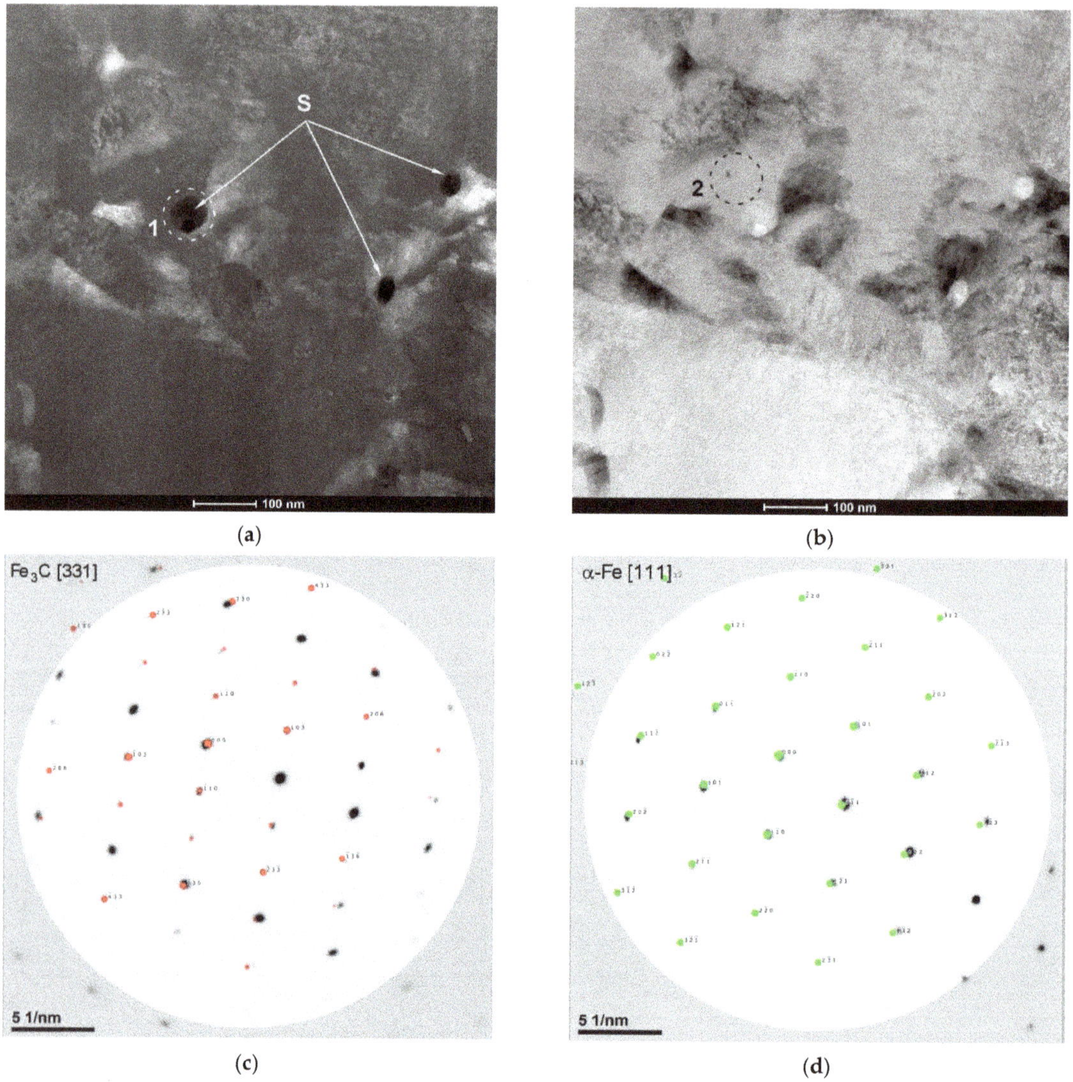

Figure 13. (**a**) Dark-field and (**b**) bright-field STEM HAADF images showing spherical (S) cementite precipitates formed in the matrix of the SHB no. H8 (the region of complete dissolution of the cementite lamellae); (**c**,**d**) SAED patterns obtained from regions (1) and (2), respectively, in (**a**).

3.3. Hardness Analysis

The hardness depth profiles of the SHBs no. H2 and H8 are reported in Figure 14. In general, the hardness profiles correspond well to the microstructural changes discussed in Section 3.2 and confirmed the previously mentioned thicknesses of the hardened layer. Moreover, the hardness test data confirmed the decrease in the surface layer hardness with increasing traverse speed mentioned in Section 3.1. Note that the subregions of the hardened layer, i.e., the regions of complete and partial dissolution of the cementite lamellae, have comparable hardness. This leads to a uniform hardness distribution throughout the hardened layer. The average hardness values for the SHBs no. H2 and H8 were 755 ± 5 and 651 ± 15 HV0.1, respectively.

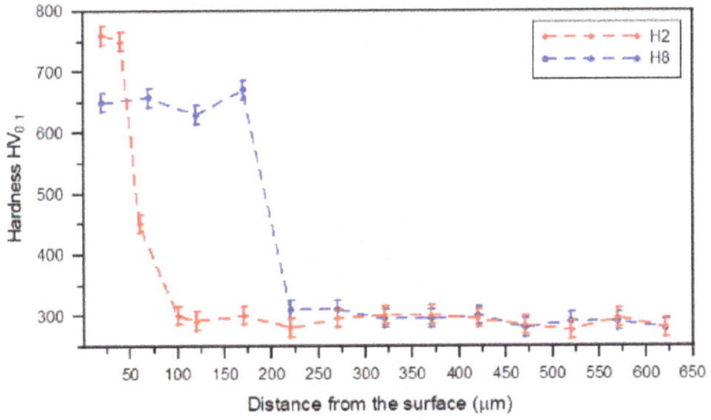

Figure 14. Hardness profiles of the SHBs no. H2 and H8 (Table 2).

3.4. Examination of Thermal Conditions in the Hardened Zone

The temperature distributions on the surface of the hardened specimens recorded by the IR thermography and calculated for processing conditions no. H2 and H8 (Table 2) are compared in Figure 15. The corresponding temperature profiles across the hardened zone are shown in Figures 16 and 17, respectively. Figures 16 and 17 show that the calculated temperature profiles match well with that recorded by the IR thermography, indicating the reliability of the numerical model used. Figure 18a,b present the calculated temperature fields in the cross-sections of the SHB no. H2 and H8, respectively. The calculated temperature distributions beneath the surface of the SHB under the above processing conditions are presented in Figure 19.

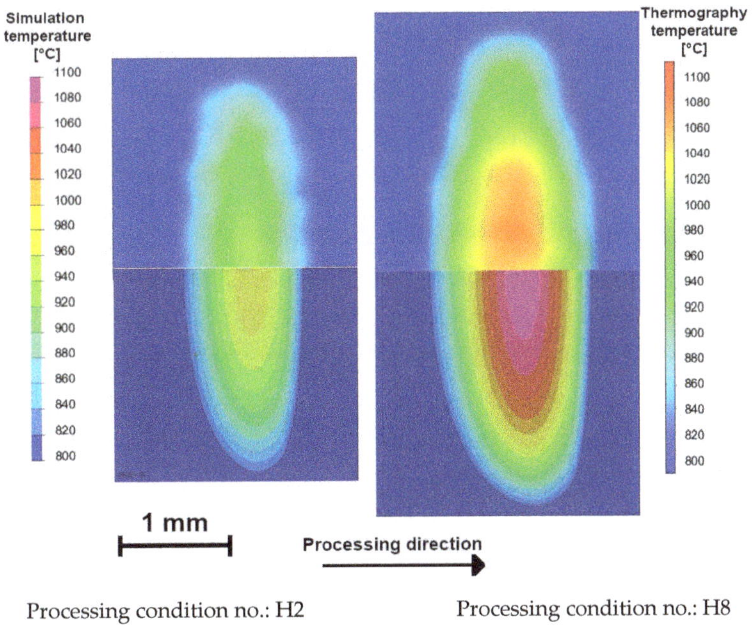

Figure 15. Temperature fields on the top surface of the laser surface-hardened specimens obtained via IR thermography (**upper part**) and thermal simulation (**lower part**). Processing conditions: H2 and H8 (Table 2). Selected emissivity value, $E = 0.17$.

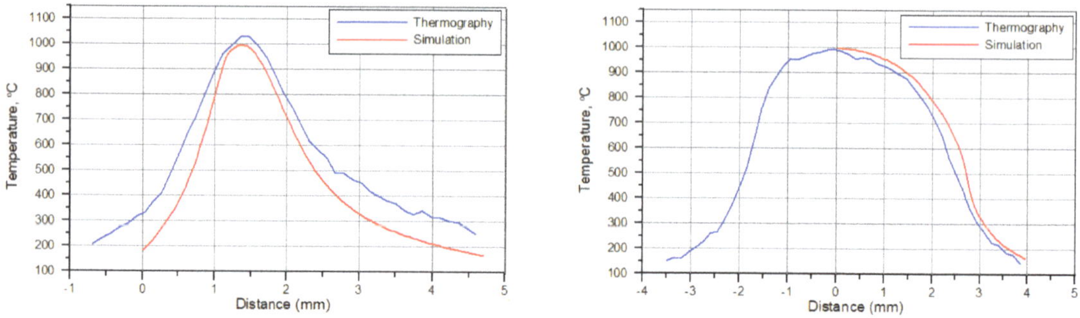

(**a**) (**b**)

Figure 16. Temperature profiles for the hardened surface in (**a**) the longitudinal (along the SHB centerline) and (**b**) traverse directions (along the slow-axis of the HPDDL beam). Processing conditions no. H2 (Table 2).

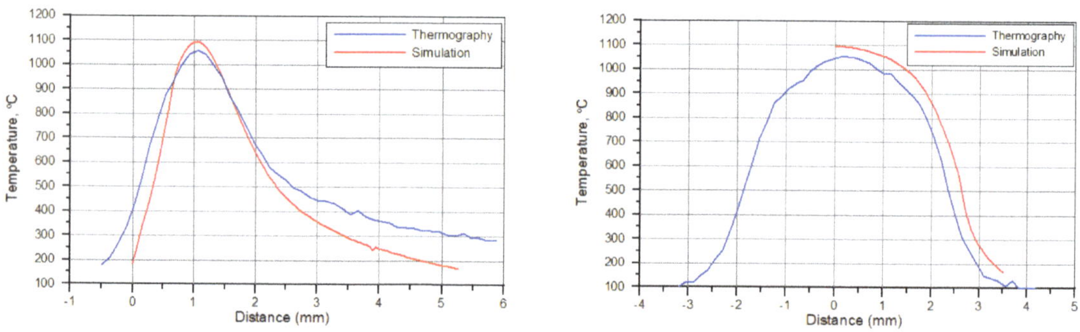

(**a**) (**b**)

Figure 17. Temperature profiles for the hardened surface in (**a**) the longitudinal (along the SHB centerline) and (**b**) traverse directions (along the slow-axis of the HPDDL beam). Processing conditions no. H8 (Table 2).

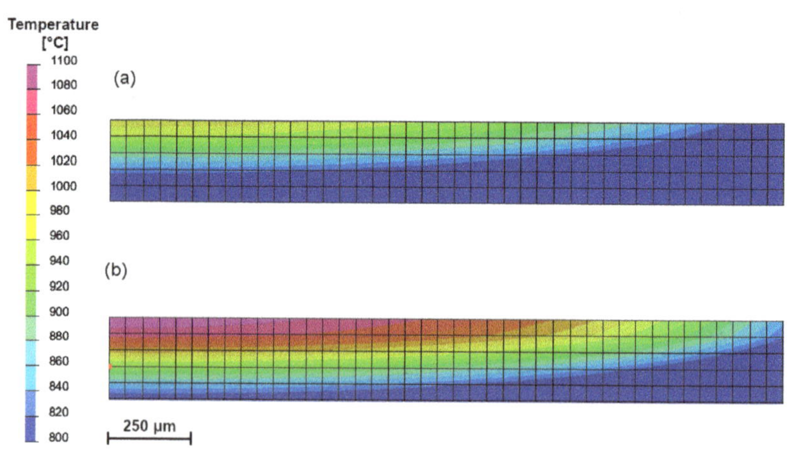

Figure 18. Calculated temperature fields in the cross-section of the SHB (a half of the bead) no.: (**a**) H2, (**b**) H8 (Table 2).

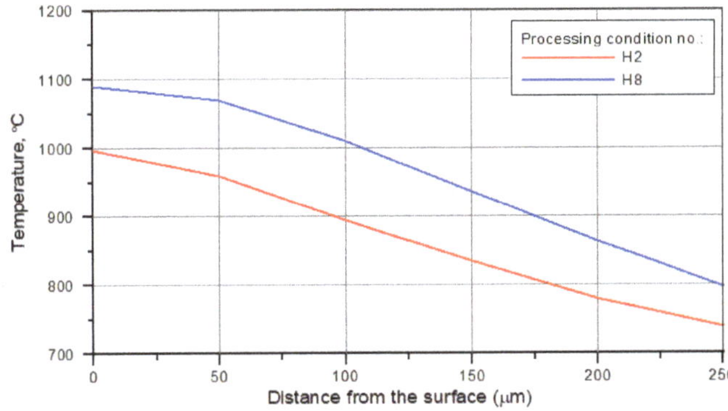

Figure 19. Calculated temperature distributions beneath the surface of the SHB (along the central vertical plane) under processing conditions no. H2 and H8 (Table 2).

The thermal data indicated that the maximum temperature on the hardened surface (on the SHB centerline) was about 1000 and 1100 °C for processing conditions no. H2 and H8, respectively. The average heating rate for the above two cases was 2300 and 8500 °C/s, respectively. The time spent above the critical temperature (i.e., the matrix structure becomes entirely austenitic; assuming approximately 800 °C) for processing conditions no. H2 and H8 was 0.33 and 0.14 s, respectively. The average cooling rate within the temperature range of 800–500 °C for processing conditions no. H2 and H8 was 1300 and 3400 °C/s, respectively. The thermal data errors due to simplification of the modelled process, limited mesh density, idealisation of the simulation and variety of emissivity of the sample can be present. Although small deviations between the simulation and thermographic experimental measurements are present, their convergence is high, which increases the confidence of investigated thermal parameters.

4. Discussion

The laser power density, determining the optimal range of the traverse speed (ensuring no surface melting), had a direct impact on the hardened layer thickness. The power density for processing conditions no. H2 and H8 was 40.4 and 80.8 W/mm^2, respectively. Despite a notably lower HI level (80 J/mm), the processing condition no. H8 provided an almost 3.5 times higher thickness of the hardened layers in comparison to that achieved under the processing condition no. H2 (HI = 120 J/mm). Similar relationships between the processing parameters and the hardening depth have been reported for LSTH of different grades of steel [23,24]. It should be noted that the maximum hardness produced in the hardened layer was affected by the traverse speed. The increase in the traverse speed led to a reduction in the maximum layer hardness. Comparing the surface hardness of the SHBs no. H2 and H8 shows that the lower traverse speed (0.2 and 0.6 m/min for H2 and H8, respectively) provided about 10% greater surface hardness (Table 2). This trend of decreasing hardness with increasing traverse speed is associated with an increase in the cooling rate and the resulting suppression of the martensitic transformation. The lower hardness of SHB no. H8 (651 HV0.1, Figure 14) in comparison to that of SHB no. H2 (755 HV0.1) is attributed to the presence of the retained austenite phase. The suppression of the martensitic transformation under non-equilibrium cooling conditions has been reported in several works on the casting and surface treatment of different cast iron grades [25,26]. It indicates that, from the point of view of the maximum surface hardness, the lower traverse speeds, ensuring lower values of the cooling rate, are beneficial in producing more wear-resistant layers [27,28]. On the other hand, to avoid surface melting with increasing power density, the elevation of traverse speed is required. The above results show that both the laser power density (determined by the laser power level and laser spot size used) and the traverse speed are equally important.

Hence, a proper combination of the above processing parameters is required to achieve optimization of the hardened layer hardness and depth.

Considering the notably shorter thermal cycle induced under processing condition no. H8 (the time spent above the critical temperature for processing condition no. H2 was almost 2.5 times longer than that for H8), it can be concluded the main factor controlling the hardening depth under the analysed two cases is the value of the temperature on the hardened surface. This relationship is consistent with that previously reported for LSTH of steel [23,29,30]. The explanation can be found in the fact that the carbon diffusion coefficient is a strong function of the temperature. The structural changes in the hardened layer are dependent on diffusion-controlled processes such as the transformation of perlite to austenite and the homogenization of the carbon content in austenite. Based on the data reported by Gegner [31], it can be assumed that there is a significant increase in the carbon diffusivity in the face-centered cubic Fe-C austenite lattice with increasing temperatures in the range of 1000–1100 °C.

Comparing the microstructure of the SHBs (Section 3.2) with the calculated temperature distributions beneath the surface of SHBs (Figures 18 and 19) suggests that the minimum temperature required for the complete dissolution of cementite lamellae in the hardened layer was about 970 °C. In turn, the formation of the hardened region that contained the partially dissolved cementite lamellae underwent treatment in the temperature range of 880–970 °C.

5. Conclusions

Surface-hardened layers were produced on ductile cast iron via laser surface transformation hardening. The laser source used was a high-power direct diode laser having a rectangular beam with a top-hat intensity distribution. The thermal conditions in the hardened zone were examined by using numerical simulations and infrared (IR) thermography. The results showed that the major factor controlling the thickness of the hardened layer is the laser power density, which determines the optimal range of the traverse speed and in consequence the maximum surface temperature. The hardening depth was found to increase with increasing power density. To avoid surface melting with increasing power density, the lower limit of traverse speed notably increases. The resulting increase in the cooling rate led to a suppression of the martensitic transformation and a decrease in the hardened layer hardness. In the investigated range of the laser power density (40.4–80.8 W/mm^2), the maximum hardening depth was approx. 200 μm. The hardness in the hardened layer was increased by almost three times in comparison with the as-received condition.

Author Contributions: Conceptualization, A.C. and D.J.; methodology, A.C., D.J., J.G., A.G., B.W. and K.M.; formal analysis, A.C., D.J., J.G., A.G. and B.W.; investigation, D.J., A.C., A.G., B.W., K.M. and W.K.; resources, A.C. and D.J.; writing—original draft preparation, A.C. and D.J.; writing—review and editing, A.C. and D.J.; visualization, A.C., D.J. and B.W.; supervision, A.C. and D.J.; project administration, A.C. and D.J.; funding acquisition, D.J. and A.C. All authors have read and agreed to the published version of the manuscript.

Funding: The research was funded by the Silesian University of Technology Rector's habilitation grant 10/050/RGH_20/1006 and pro-quality grant 10/050/RGJ_22/1029.

Institutional Review Board Statement: Not applicable.

Informed Consent Statement: Not applicable.

Data Availability Statement: Not applicable.

Conflicts of Interest: The authors declare no conflict of interest.

References

1. Grum, J.; Sturm, R. Microstructure analysis of nodular iron 400-12 after laser surface melt hardening. *Mater. Charact.* **1996**, *37*, 81–88. [CrossRef]
2. Janicki, D.; Górka, J.; Kwaśny, W.; Pakieła, W.; Matus, K. Influence of Solidification Conditions on the Microstructure of Laser-Surface-Melted Ductile Cast Iron. *Materials* **2020**, *13*, 1174. [CrossRef]
3. Hurtado-Delgado, E.; Huerta-Larumbe, L.; Miranda-Pérez, A.; Aguirre-Sánchez, A. Microcracks Reduction in Laser Hardened Layers of Ductile Iron. *Coatings* **2021**, *11*, 368. [CrossRef]
4. Jeyaprakash, N.; Yang, C.; Duraiselvam, M.; Prabu, G.; Tseng, S.; Kumar, D.R. Investigation of high temperature wear performance on laser processed nodular iron using optimization technique. *Results Phys.* **2019**, *15*, 102585. [CrossRef]
5. Janicki, D. Microstructure and sliding wear behaviour of in-situ TiC-reinforced composite surface layers fabricated on ductile cast iron by laser alloying. *Materials* **2018**, *11*, 75. [CrossRef] [PubMed]
6. Janicki, D. Effect of Chromium and Molybdenum Addition on the Microstructure of In Situ TiC-Reinforced Composite Surface Layers Fabricated on Ductile Cast Iron by Laser Alloying. *Materials* **2020**, *13*, 5750. [CrossRef]
7. Janicki, D. The friction and wear behaviour of *in-situ* titanium carbide reinforced composite layers manufactured on ductile cast iron by laser surface alloying. *Surf. Coat. Technol.* **2021**, *406*, 126634. [CrossRef]
8. Paczkowska, M.; Makuch, N.; Kulka, M. The influence of various cooling rates during laser alloying on nodular iron surface layer. *Opt. Laser. Technol.* **2018**, *102*, 60–67. [CrossRef]
9. Liu, Y.; Qu, W.; Su, Y. Reinforcement Composite Coating Produced Using Graphite of the Cast Iron by Laser Cladding. *Materials* **2016**, *9*, 815. [CrossRef]
10. Putatunda, S.K.; Bartosiewicz, L.; Hull, R.J.; Lander, M. Laser hardening of austempered ductile cast iron (ADI). *Mater. Manuf. Process.* **1997**, *12*, 137–151. [CrossRef]
11. Ocelik, V.; Tang, P.N.; De Boer, M.C.; De Oliveira, U.; De Hosson, J.T.M. Laser surface treatment of grey cast iron for automotive applications. In *Computer Methods and Experimental Measurements for Surface Effects and Contact Mechanics VII*; WIT Transactions on Engineering Sciences, 49; De Hosson, J.T.M., Brebbia, C.A., Nishida, S.I., Eds.; WIT Press: Ashurst, UK, 2005.
12. Liu, A.; Previtaili, B. Laser surface treatment of grey cast iron by high power diode laser. *Phys. Procedia* **2010**, *5*, 439–448. [CrossRef]
13. Janicki, D. Fabrication of high chromium white iron surface layers on ductile cast iron substrate by laser surface alloying. *Stroj. Vestn.* **2017**, *63*, 705–714. [CrossRef]
14. Xu, Y.; Liu, J. Laser surface hardening of ductile cast iron for vehicle die. *Proc. SPIE* **2010**, *7843*, 353–361. [CrossRef]
15. Soriano, C.; Leunda, J.; Lambarri, J.; García Navas, V.; Sanz, C. Effect of laser surface hardening on the microstructure, hardness and residual stresses of austempered ductile iron grades. *Appl. Surf. Sci.* **2011**, *257*, 7101–7106. [CrossRef]
16. Wanga, B.; Qiu, F.; Barber, G.C.; Pan, Y.; Cui, W.; Wang, R. Microstructure, wear behavior and surface hardening of austempered ductile iron. *J. Mater. Res. Technol.* **2020**, *9*, 9838–9855. [CrossRef]
17. Malek Ghaini, F.; Ameri, M.H.; Torkamany, M.J. Surface transformation hardening of ductile cast iron by a 600w fiber laser. *Optik* **2020**, *203*, 163758. [CrossRef]
18. Meijer, J.; van Sprang, I. Optimization of Laser Beam Transformation Hardening by One Single Parameter. *CIRP Ann.* **1991**, *40*, 183–186. [CrossRef]
19. Ion, J.C. *Laser Processing of Engineering Materials*; Elsevier Butterworth-Heinemann: London, UK, 2005.
20. Grum, J. Laser Surface Hardening. In *Encyclopedia of Tribology*; Wang, Q.J., Chung, Y.W., Eds.; Springer: Boston, MA, USA, 2013. [CrossRef]
21. Czupryński, A.; Wyględacz, B. Comparative Analysis of Laser and Plasma Surfacing by Nickel-Based Superalloy of Heat Resistant Steel. *Materials* **2020**, *13*, 2367. [CrossRef]
22. Kik, T. Heat Source Models in Numerical Simulations of Laser Welding. *Materials* **2020**, *13*, 2653. [CrossRef]
23. Hung, T.P.; Shi, H.E. Temperature Modeling of AISI 1045 Steel during Surface Hardening Processes. *Materials* **2018**, *11*, 1815. [CrossRef]
24. Yang, Y.S.; Na, J. A study on residual stresses in laser surface hardening of a medium carbon steel. *Surf. Coat. Technol.* **1989**, *38*, 311–324. [CrossRef]
25. Zarubova, N.; Kraus, V.; Cermak, J. Mechanisms of phase transformations during laser treatment of grey cast iron. *J. Mater. Sci.* **1992**, *27*, 3487–3496. [CrossRef]
26. Janus, A.; Stachowicz, M. Thermodynamic stability of austenitic Ni-Mn-Cu cast iron. *Metalurgija* **2014**, *3*, 353–356.
27. Czupryński, A.; Pawlyta, M. Influence of Preheating Temperature on Structural and Mechanical Properties of a Laser-Welded MMC Cobalt Based Coating Reinforced by TiC and PCD Particles. *Materials* **2022**, *15*, 1400. [CrossRef] [PubMed]
28. Sroka, M.; Jonda, E.; Pakieła, W. Laser Surface Modification of Aluminium Alloy AlMg9 with B4C Powder. *Materials* **2020**, *13*, 402. [CrossRef]
29. Tayal, M.; Mukherjee, K. Thermal and microstructural analysis for laser surface hardening of steel. *J. Appl. Phys.* **1994**, *75*, 3855. [CrossRef]

30. Tobar, M.J.; Álvarez, C.; Amado, J.M.; Ramil, A.; Saavedra, E.; Yáñez, A. Laser transformation hardening of a tool steel: Simulation-based parameter optimization and experimental results. *Surf. Coat. Technol.* **2006**, 362–6367. [CrossRef]
31. Gegner, J. Concentration-and temperature-dependent diffusion coefficient of carbon in FCC iron mathematically derived from literature data. In *Mathematical Modeling and Computer Simulation of Materials Technologies*; College of Judea and Samaria: Ariel, Israel, 2006; Volume 1, pp. 158–167.

materials

MDPI

Article

Numerical and Physical Simulation of MAG Welding of Large S235JRC+N Steel Industrial Furnace Wall Panel

Marek Mróz [1,*], Robert Czech [2], Bogdan Kupiec [1], Andrzej Dec [1], Marcin Spólnik [2] and Patryk Rąb [1]

[1] Department of Foundry and Welding, Faculty of Mechanical Engineering and Aeronautics, Rzeszow University of Technology, Al. Powstańców Warszawy 12, 35-959 Rzeszów, Poland
[2] Naftoremont-Naftobudowa Sp. z o.o., Branch Jedlicze, ul. Sikorskiego 17, 38-460 Jedlicze, Poland
* Correspondence: mfmroz@prz.edu.pl

Abstract: This paper presents the results of a study on the development of a Metal active gas (MAG) welding technology for an industrial furnace component made of steel S235JRC+N with respect to the minimizationof welding deformation. A numerical simulation of the welding process was performed in the first phase of the research. The numerical simulation was carried out with the SYSWELD software. For the numerical simulation of the welding process, the FEM method was used. In the simulation, four variants of restraint of the industrial furnace wall panel elements during the execution of the welding process were investigated. They differed in the number of restraints (model 1–4). It was found that the difference between the maximum mean strain in model 1 and the lowest mean strain in model 4 was only 11%. A physical simulation of the welding process was then performed with a restraint variant according to model 1. The displacement results obtained from the physical simulation of the welding process were compared with the displacement results from the numerical simulation. Discrepancies between numerical and physical simulation displacement values were found. The quality of selected welded joints was also evaluated. Visual testing (VT) and measurements of weld geometries were performed for this purpose. Metallographic tests and hardness measurements were performed to determine of influence of the welding process on the microstructure of the welded joint area, especially the heat affected zone (HAZ). The results obtained confirm the correctness of the assumptions made regarding the technology of manufacturing the furnace wall panels.

Keywords: MAG welding method; numerical and physical simulation; welding deformation; metallographic tests; hardness measurements

Citation: Mróz, M.; Czech, R.; Kupiec, B.; Dec, A.; Spólnik, M.; Rąb, P. Numerical and Physical Simulation of MAG Welding of Large S235JRC+N Steel Industrial Furnace Wall Panel. *Materials* **2023**, *16*, 2779. https://doi.org/10.3390/ma16072779

Academic Editors: Marcin Adamiak, Antonín Kříž, Tünde Anna Kovács and Artur Czupryński

Received: 1 March 2023
Revised: 22 March 2023
Accepted: 24 March 2023
Published: 30 March 2023

1. Introduction

Metal active gas (MAG) welding is a method of joining metal parts used in a variety of industries. Compared to the other methods [1], MAG produces highstrength welded joints at relatively low production costs.

In the MAG method, similar to other processes for manufacturing welded structures, the components to be joined are subjected to thermal cycles, i.e., heating and cooling, and phase transformations can occur in the structural material. Displacement of the heat source causes uneven heating and cooling of the component area around the weld, resulting in differential thermo-mechanical responses of the component. This results in stresses that, once exceeded, can cause deformation and strain in the joined components. This problem is particularly evident in complex, sophisticated welded structures [2,3].

Complete elimination of post-weld stress is difficult to achieve. However, efforts should be made to minimize their effect on the quality of the welded structure. Welding engineers have a number of solutions at their disposal. These include thermal and mechanical methods. Minimization of the influence of weld stress factors can also be achieved by proper selection of welding process parameters. These operations are often associated with increased financial outlay and extended production times [4,5].

With the development of computer technology, welding process simulation is increasingly used to optimize the production of welded structures. The use of the latest numerical simulation software for the welding process reduces the development time for the manufacturing technology of welded structures and the financial outlay [1,6].

Numerical simulations are used for the development of manufacturing processes for welded structures by means of various methods: MIG [2,4,7–11], MAG [7–9,12–14], TIG [1,7,15–18], NG-TIG [7], TIG-MIG [19], EBW [7], and WAAM [20,21].

Different types of software based on the Finite Element Method (FEM) are used for the numerical simulation of the welding process, in terms of welding stress and strain analysis. In works [1,5,13,14,18,22], the welding process is analyzed with the help of ABAQUS. In contrast, the authors of papers [3,8,10,12,15,16] used ANSYS for the numerical simulation of the welding process. Furthermore, in research on numerical simulations of the welding process, the SYSWELD program is used [4,7,17,23].

Numerical simulation of the welding process requires computers with high computing power. This is due to the large number of complex calculations involved. Therefore, various types of simplifications are used to reduce its implementation time. For example, to reduce the number of calculations in numerical simulations, axisymmetric elements are used, or 3D models are replaced by flat-transformed 2D elements [4,6,13,23].

The numerical simulation of a welding process should ensure that the results are as close as possible to the actual welding process (physical simulation). However, as the authors of papers [8,18] emphasize, this requires consideration of many variables describing the heat source, the welding method and its process parameters, the geometry of the welded joint, the welding technique, and the amount and type of restraint system (bracing).

The authors of works [5,13], analyzing the welding deformations and strains of large-scale welded structures, found a high correspondence between the results of the numerical simulation and those obtained under the conditions of the actual welding process (physical simulation).

The problem of welding stresses and deformations is particularly important in the fabrication of large welded structures consisting of a large number of welded joints made up of elements of varying thickness. In paper [12], numerical simulation in the ANSYS program was used to analyze the stress distribution arising in the welding process of a T-fillet joint of a Volvo truck chassis component. Some inconsistencies were found between the results of the stress distribution simulation and the results of residual stress measurements made in certain areas of the welded joint.

When welding large structures, the risk of weld distortion must be considered in the manufacturing technology. There are several methods used in the welding industry to minimize the distortion of welds. For example, the authors of paper [24] showed that the distribution and magnitude of weld distortion is strongly influenced by the method and sequence of weld layout. They analyzed six variants of weld placement in the welding process of a large plate with stiffening ribs made of the EH36 shipbuilding steel. They found that less weld distortion was achieved by placing the stiffener rib welds in the second variant, i.e., from the center to the edge of the plate.

The authors of paper [25] also found that the method and sequence of weld placement influences the amount of weld distortion produced during the welding process of large structures. Using numerical simulation, they analyzed the influence of the method and sequence of joint laying in the welding process of a vacuum vessel made of the 316LN austenitic steel, used in fusion processes.

High agreement between the results of numerical simulations of welding deformations and experimental results (physical simulation) in the case of welding of large constructions was demonstrated by the authors of paper [26]. They analyzed the welding process of a transport vehicle chassis beam made of the S235 steel using the MAG method. They also found a high correspondence between simulation results and experimental results in terms of temperature distribution in the welded joint area. They analyzed the influence of different methods for implementing the finite element method and found that the best results were obtained with the substructuring technique.

The aim of this work was to develop a MAG welding technology for a large construction furnace wall panel made of the S235JRC+N steel in terms of minimizing welding distortion, in accordance with the requirements of EN ISO 13920. The analysis of the welding technology was based on the results of numerical and physical simulation of the welding process, as well as visual inspections and measurements of the weld geometry. Metallographic tests and hardness measurements were performed to determine the effect of the welding process on the microstructure of the welded joint.

2. Materials and Methods

The subject of the study was an industrial furnace panel made of S235JRC+N steel with dimensions of 2000 mm × 4000 mm. The chemical composition of the S235JRC+N steel is shown in Table 1.

Table 1. Chemical composition of steel S235JRC+N (from metallurgical certificate) in wt.%.

Fe	C	Mn	P	S	Cu	N
balance	0.14	0.48	0.013	0.007	0.014	0.003

The furnace wall panel is a large welded structure made of sheet metal, flat bars, and a channel (Figure 1). The flat panel wall is made of 5 mm thick sheet metal, the outer frame of 200 mm × 80 mm channel sections with a wall thickness of 10 mm. The internal ribs are made from flat bars with a wall thickness of 6 mm and 80 mm × 80 mm angles with a wall thickness of 6 mm.

Figure 1. Furnace wall panel elements connected by spot welds.

For the numerical simulation of the stress and strain distribution, a CAD and FEM model of the industrial furnace wall panel was created (Figure 2).

The FEM numerical model utilized a mesh made of second-order (2D) planar elements with center nodes. Using the CAD geometry, the center surfaces of the plates were determined and surfaces representing the welds were added. In the FE mesh, triangular or quadrilateral finite elements were used. In the weld areas, the mesh was compacted.

The numerical modeling of the production of the welded joints was carried out using the shrinkage method implemented in the SYSWELD program (ESI Group, Paris, France).

This method consists of inducing shrinkage in the joint area with a value derived from the thermal expansion of the S235JRC+N steel. Sequential sections of the weld were defined by giving them a condition that induces shrinkage.

Figure 2. CAD and FEM model of the industrial furnace wall panel.

The welding sequence of the individual elements of the industrial furnace wall panel is shown in Figure 3. The welds were made in the back-step technique.

The MAG process is used in the production of furnace wall panels. The MAG welding process was performed using the technological parameters specified in the WPS (Welding Procedure Specification) developed in accordance with the WPQR document of the company manufacturing the industrial furnace wall panels.

The numerical simulation of the welding process of the industrial furnace wall panel was preformed assuming four models of restraint of the panel elements during the welding process (Figure 4).

All of the restraint system variants used the same values for the MAG welding process parameters and the same sequence of weld placement.

The boundary conditions of each restraint model took away degrees of freedom in selected directions in defined regions. These regions were consistent with the restraint model being analyzed.

Figure 5 shows model 1 of the distribution of restraint points.

A diagram of the distribution of restraints according to model 1 in the furnace wall panel construction during welding is shown in Figure 6.

For each restraint system model, the release of the panel from the fixture (restraint system) was realized by modification of the boundary conditions after completion of the welding process simulation. The post-weld restraint boundary conditions are shown in Figure 7.

The numerical simulation of the welding process of the furnace wall panel determined the distribution of the strain field and the deformation values at characteristic points on the furnace wall panel (Figure 8).

For the analysis of the numerical simulation results, deformations were specified in the direction of the X, Y, and Z axes. The values of these deformations were specified in order to be able to compare them with the results from the physical simulation.

The physical simulation consisted ofperforming the MAG welding process of the furnace wall panel. The welding process was performed using the 1st model of restraint. The view of the panel prepared for welding with the restraint system according to model 1 is shown in Figure 9.

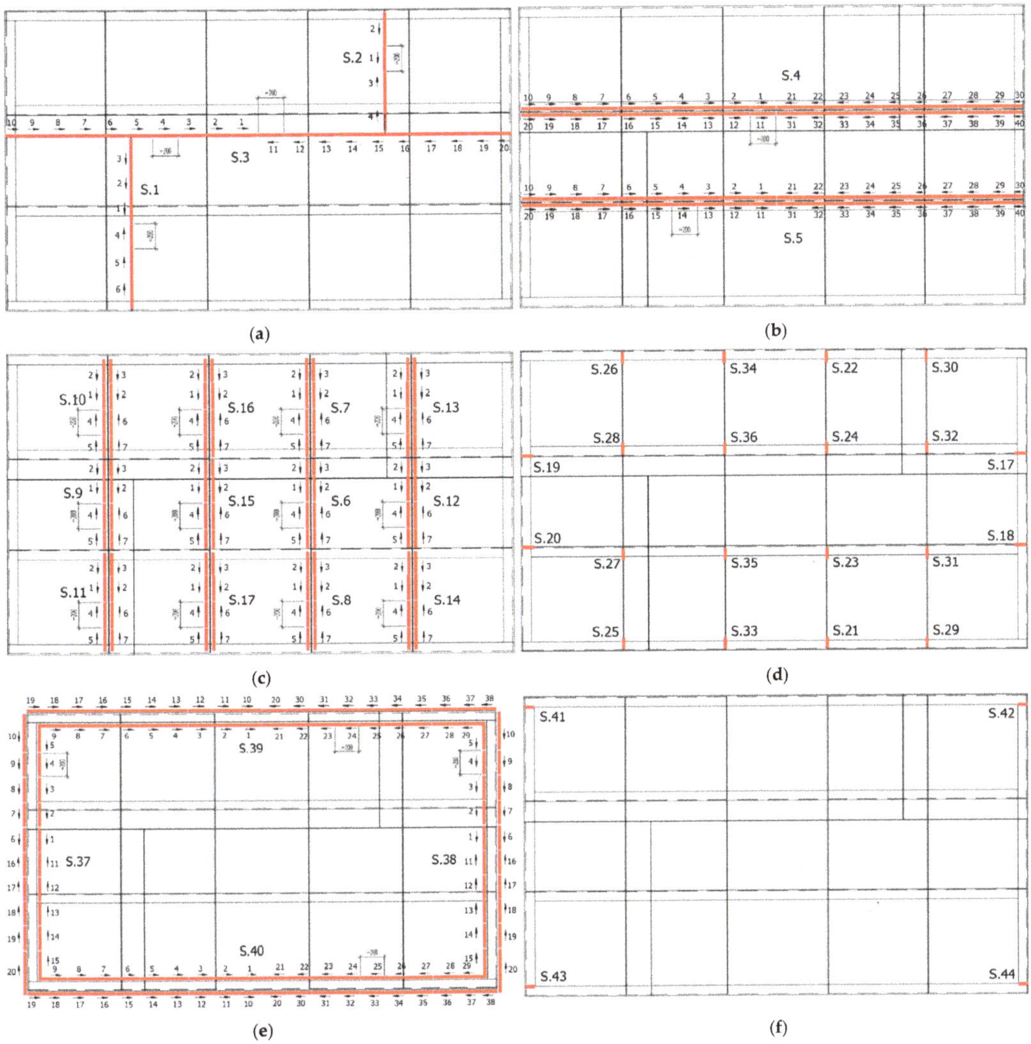

Figure 3. Sequence of welds in the numerical simulation of the welding process of the industrial furnace wall panel; (**a**) welds S.1–S.3 joining metal sheets (stage 1), (**b**) welds S.4 and S.5 joining ribs made of angles to the metal sheets (stage 2), (**c**) welds S.6–S.17 joining the ribs made of flat bars to the metal sheets (stage 3), (**d**) welds S.18–S.36 joining the ribs made of flat bars to the channel sections of the frame and ribs made of angles (stage 4), (**e**) welds S.37–S.40 joining the channel sections of the outer frame to the metal sheets (stage 5), and (**f**) welds S.41–S.44 joining the channel sections of the outer frame (stage 6).

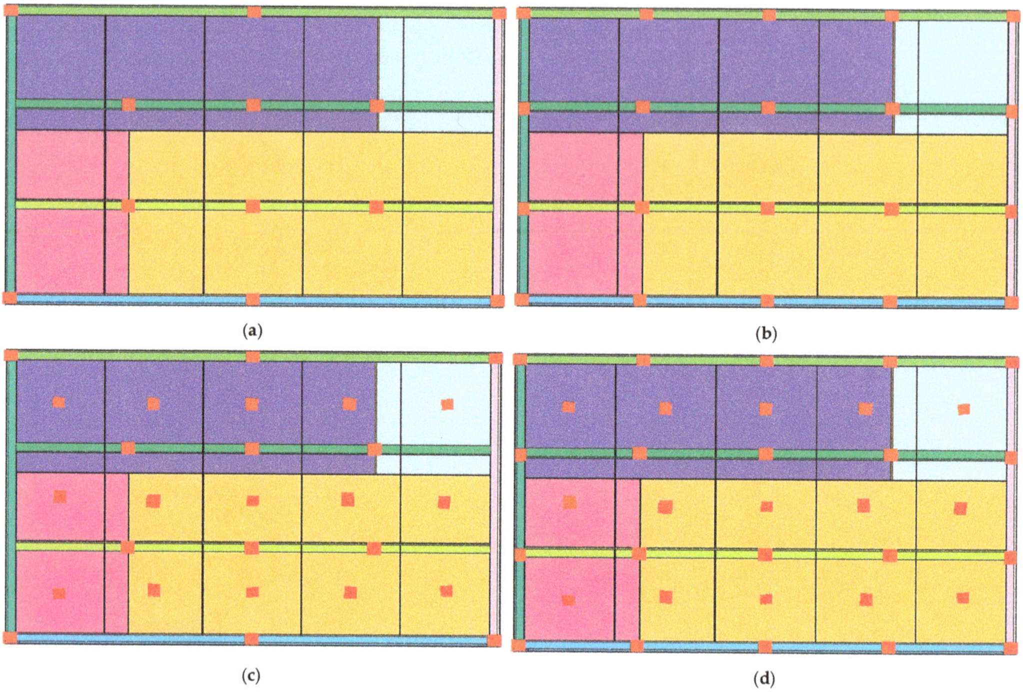

Figure 4. Diagram of the distribution of restraint points in four variants of the numerical simulation of the welding process of the industrial furnace wall panel; (a) model 1, (b) model 2, (c) model 3, and (d) model 4.

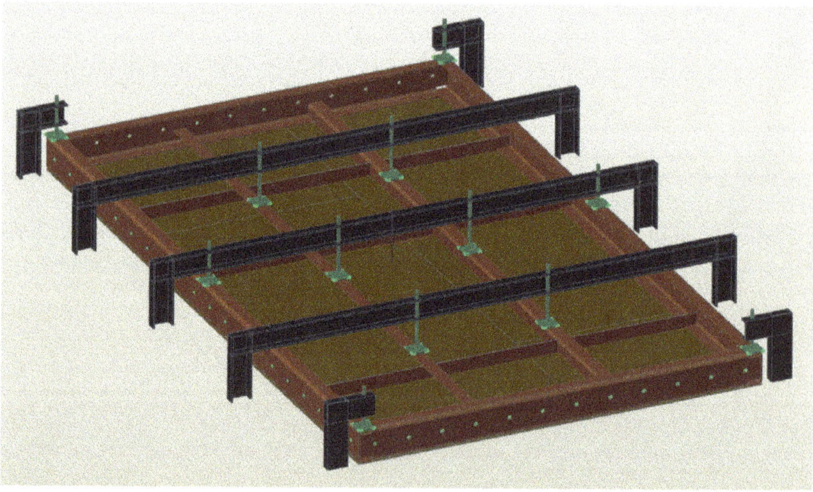

Figure 5. Diagram of the distribution of restraint points in model 1 (Figure 3a) of the welding process of the industrial furnace wall panel.

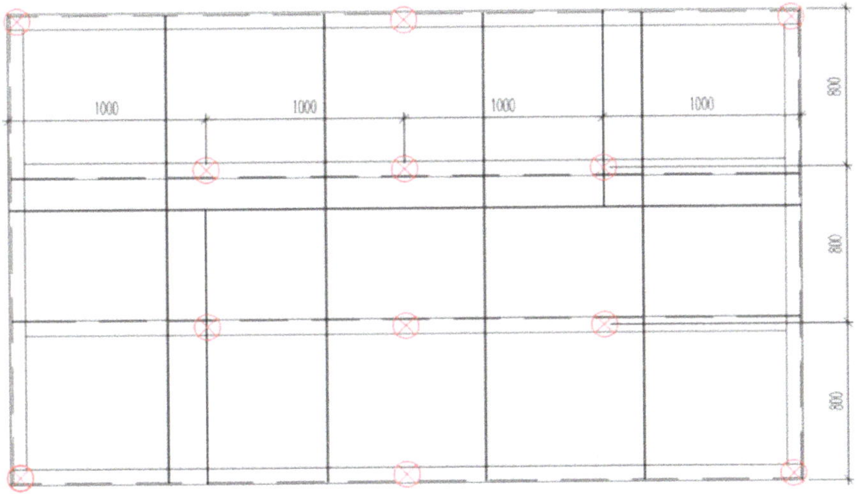

Figure 6. Diagram of the distribution of restraints in the furnace wall panel structure during welding.

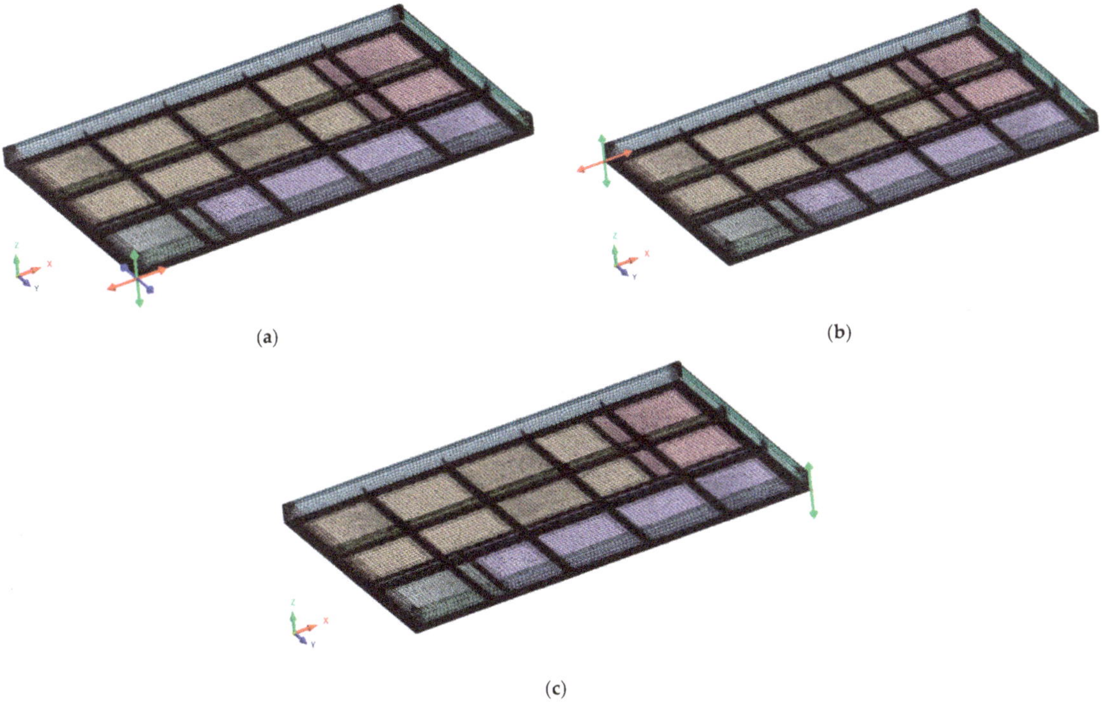

(a) (b) (c)

Figure 7. Diagrams of restraint boundary conditions after panel welding; (**a**) no node displacement in the X, Y, and Z directions, (**b**) no node displacement in the X and Y directions, and (**c**) no node displacement in the Y direction.

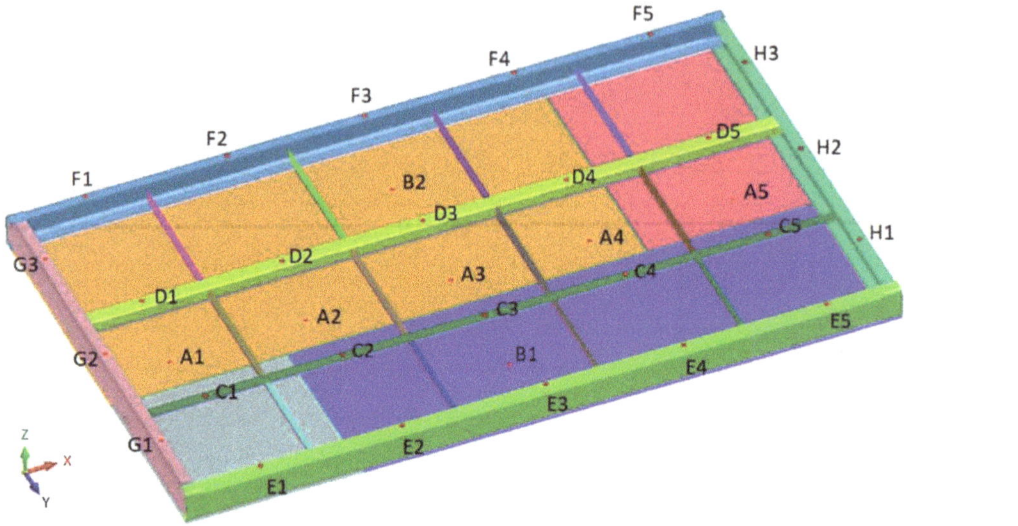

Figure 8. Distribution of selected measurement points where deformation values were determined in the numerical simulation of the furnace wall panel welding process.

Figure 9. View of the furnace wall panel elements connected by spot welds to the restraint system according to model 1.

Prior to the welding process, preliminary measurements were made. The measurements took into account the distances between characteristic points C1–C5, D1–D5, E1–E5, F1–F5 (direction of the X axis), and G1–G3 and H1–H3 (direction of the Y axis), as well as the values of relative height (in the direction of the Z axis) at characteristic measurement points. These measurements were carried out using a linear gauge (Stanley, Hartford, CT, USA) and a laser instrument (Bosch Power Tools, Warszawa, Poland). After the welding process, measurements were taken again.

Based on the deformation results in the numerical simulation and the results of measurements in the physical simulation of the welding process of the furnace wall panel, the tolerances for linear dimensions and tolerance for flatness were determined. These tolerances were calculated in accordance with EN ISO 13920.

To identify possible welding defects in the welded joints of the furnace wall panel, visual inspections were performed. These tests were carried out in accordance with the EN ISO 5817 standard. Measurements of the weld geometries (height and width of welds) were also taken.

Metallographic tests were carried out to reveal the microstructure of the welded joint area (weld, fusion zone, heat affected zone and parent material). Due to the large number of welds, one of the welds connecting the ribs made of flat bars to the plate (welds S.6–S.17 Figure 3c) was selected for metallographic examination.

In order to carry out metallographic examinations, it was necessary to prepare metallographic specimens (cutting, grinding, polishing) and to etch them properly, according to the commonly used methodology for steel specimens. Observations of the microstructure werecarried out using a NEOPHOT 2 optical microscope (Carl Zeiss, Jena, Germany).

Hardness measurements were carried out in accordance with the EN ISO 9015 standard, using the Vickers method at the HV5 load. On the basis of these measurements, the distribution of hardness in the weld, the heat affected zone, and the parent material weredetermined.

3. Results

3.1. Numerical and Physical Simulation

The distribution of the deformation fields of the industrial furnace wall panel in the direction of the X, Y, and Zaxes, resulting from the welding process with different restraint models, are shown in Figures 10–12.

The deformation values at the characteristic measurement points (according to Figure 8), determined in the numerical simulation, are presented in Tables 2–4.

The largest displacements in the analyzed structure occurred in the direction of the Z axis, i.e., the direction perpendicular to the main surface of the furnace wall (Figure 12). Displacements in the −Z and +Z directions were found in individual areas of the welded element. The furnace wall panel was characterized by localized metal bulges between the ribs. There was also a general bulge in the entire panel.

The displacements in the X and Y directions were about fourtimes lower than the displacements in the Z direction.

For all restraint systems of the panel structure (models 1–4), the differences in deformation values were small, despite the fact that the number of stiffening points in model 1 was more than twice as small as in model 4.

Table 5 shows the deformation of the linear dimensions specified in the physical and numerical simulation of the welding process of the furnace wall panel.

The results of measuring linear dimension deformation indicate that the tolerance for line dimensions in physical simulation is from −2.00 to +1.00 mm. In the case of numerical simulation, the tolerance for line dimensions is from −1.77 to +2.18 mm. Considering the specifics, assumptions, and conditions for both simulations, it can be said that these tolerances are similar. However, deformations specified in the physical and numerical simulations differ in the plus and minus sign. This shows a different character of deformation, because the plus sign means the tensile stress, while the minus sign means the occurrence of compressive stresses.

Table 2. Results, average value, and sum of displacements in the X direction at characteristic measurement points, obtained in the numerical simulation of the welding process of the furnace wall panel.

Model	A1	A2	A3	A4	A5	B1	B3	C1	C2	C3	C4	C5	D1	D2	D3	D4	D5	E1
1	−0.19	−0.49	−0.79	−1.02	−1.50	−0.89	−0.71	−0.41	−0.66	−0.85	−1.03	−1.24	−0.37	−0.59	−0.78	−1.00	−1.28	−0.88
2	−0.19	−0.48	−0.77	−1.00	−1.48	−0.87	−0.71	−0.41	−0.64	−0.83	−1.01	−1.20	−0.37	−0.58	−0.77	−0.98	−1.24	−0.85
3	−0.22	−0.50	−0.77	−1.00	−1.43	−0.86	−0.71	−0.42	−0.64	−0.83	−1.00	−1.18	−0.39	−0.58	−0.76	−0.97	−1.21	−0.85
4	−0.21	−0.49	−0.77	−0.99	−1.42	−0.85	−0.71	−0.42	−0.63	−0.82	−0.99	−1.16	−0.39	−0.58	−0.76	−0.97	−1.20	−0.84
Model	**E2**	**E3**	**E4**	**E5**	**F1**	**F2**	**F3**	**F4**	**F5**	**G1**	**G2**	**G3**	**H1**	**H2**	**H3**			
1	−0.85	−0.86	−0.88	−0.89	−0.75	−0.77	−0.79	−0.79	−0.77	−0.89	−1.15	−0.72	−0.81	−0.46	−0.80			
2	−0.83	−0.84	−0.86	−0.87	−0.72	−0.74	−0.76	−0.76	−0.75	−0.84	−1.15	−0.69	−0.81	−0.41	−0.77			
3	−0.83	−0.83	−0.85	−0.86	−0.73	−0.74	−0.76	−0.76	−0.75	−0.89	−1.19	−0.73	−0.76	−0.37	−0.74			
4	−0.82	−0.83	−0.85	−0.86	−0.71	−0.73	−0.75	−0.75	−0.74	−0.87	−1.20	−0.72	−0.76	−0.33	−0.72			

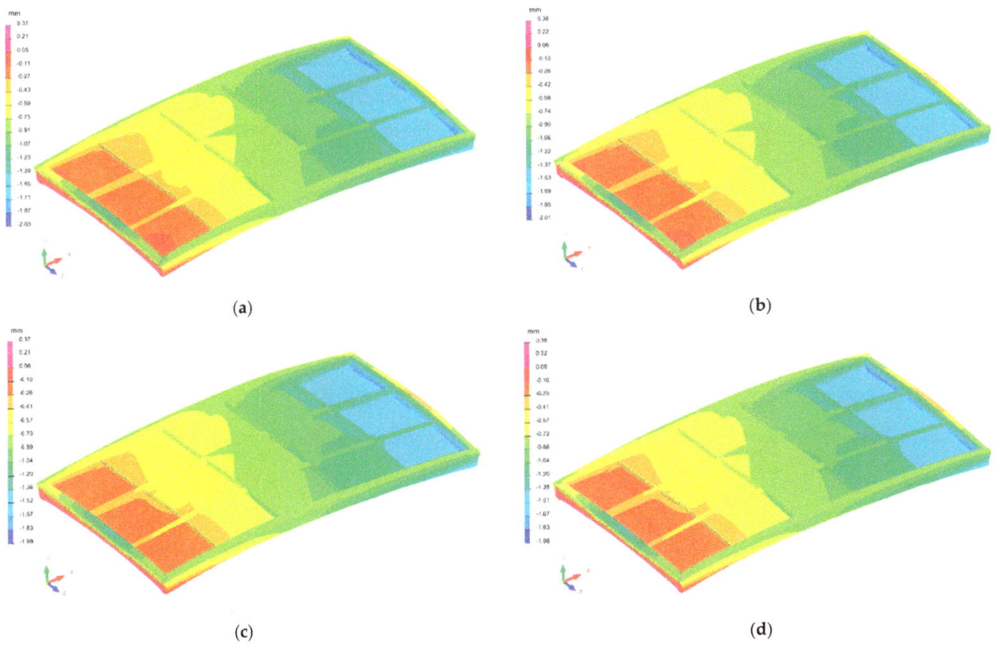

Figure 10. Distribution of the strain field of the industrial furnace wall panel in the X direction; (a) 1st restraint model, (b) 2nd restraint model, (c) 3rd restraint model, and (d) 4th restraint model.

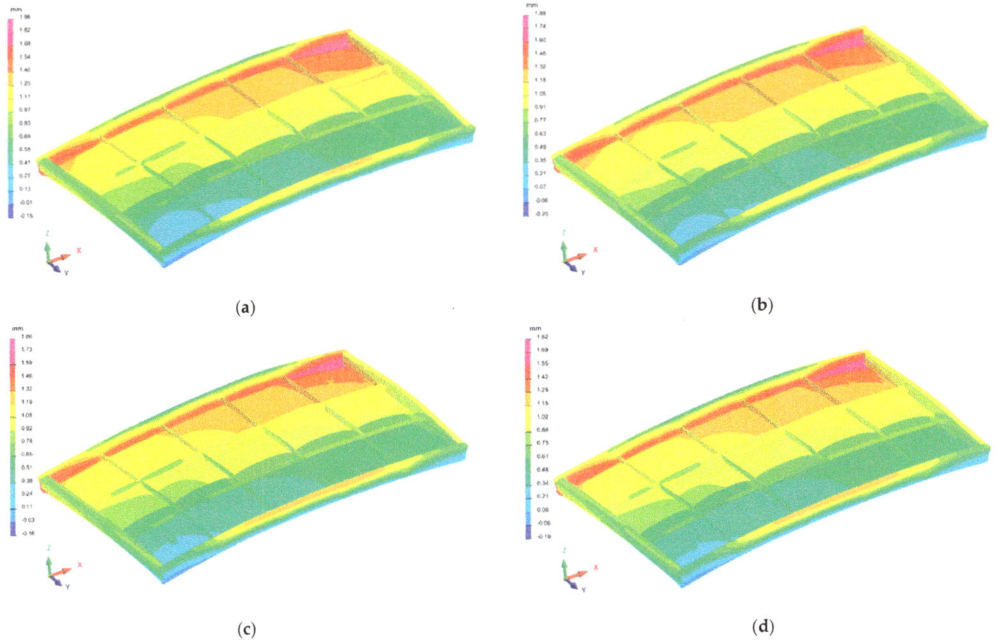

Figure 11. Distribution of the strain field of the industrial furnace wall panel in the Y direction; (a) 1st restraint model, (b) 2nd restraint model, (c) 3rd restraint model, and (d) 4th restraint model.

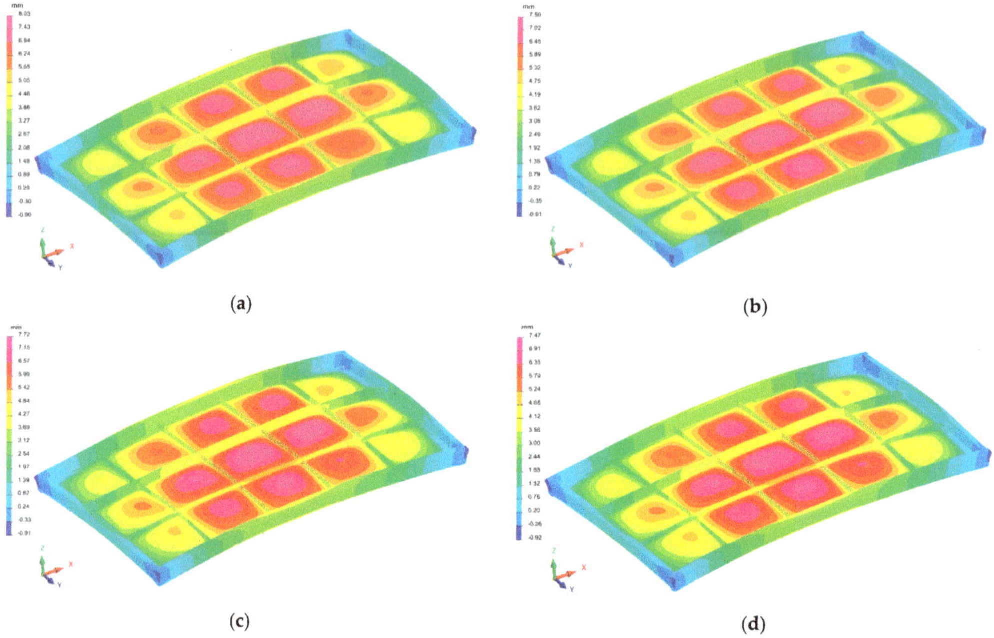

Figure 12. Distribution of the strain field of the industrial furnace wall panel in the Z direction; (**a**) 1st restraint model, (**b**) 2nd restraint model, (**c**) 3rd restraint model, and (**d**) 4th restraint model.

Table 3. Results, average value, and sum of displacements in the Y direction at characteristic measurement points, obtained in the numerical simulation of the welding process of the furnace wall panel.

Model	A1	A2	A3	A4	A5	B1	B3	C1	C2	C3	C4	C5	D1	D2	D3	D4	D5	E1
1	0.95	0.98	1.05	1.11	1.15	0.51	1.29	0.69	0.72	0.82	0.94	0.91	0.98	0.93	1.04	1.14	1.17	0.66
2	0.91	0.93	0.99	1.05	1.08	0.45	1.23	0.65	0.65	0.75	0.86	0.84	0.92	0.89	0.98	1.08	1.08	0.57
3	0.89	0.92	0.98	1.04	1.07	0.50	1.20	0.65	0.67	0.77	0.88	0.85	0.91	0.89	0.98	1.08	1.09	0.64
4	0.86	0.89	0.95	1.00	1.03	0.47	1.17	0.63	0.64	0.73	0.83	0.81	0.88	0.86	0.95	1.04	1.03	0.59
Model	**E2**	**E3**	**E4**	**E5**	**F1**	**F2**	**F3**	**F4**	**F5**	**G1**	**G2**	**G3**	**H1**	**H2**	**H3**			
1	1.21	1.25	1.24	0.88	1.08	0.77	0.74	0.68	1.23	0.83	0.82	0.82	1.09	1.09	1.09			
2	1.11	1.17	1.12	0.75	1.06	0.74	0.67	0.65	1.15	0.76	0.77	0.79	0.97	0.98	1.00			
3	1.22	1.26	1.22	0.81	1.01	0.66	0.62	0.56	1.12	0.78	0.78	0.78	1.00	1.00	1.00			
4	1.16	1.21	1.16	0.75	0.99	0.64	0.58	0.55	1.07	0.74	0.75	0.77	0.93	0.94	0.95			

Table 4. Results, average value, and sum of displacements in the Z direction at characteristic measurement points, obtained in the numerical simulation of the welding process of the furnace wall panel.

Model	A1	A2	A3	A4	A5	B1	B3	C1	C2	C3	C4	C5	D1	D2	D3	D4	D5	E1
1	5.59	7.43	8.01	7.94	6.14	7.58	7.36	2.82	4.68	4.98	4.56	2.94	2.40	3.83	4.55	4.67	2.86	1.56
2	5.23	7.05	7.57	7.48	5.64	7.27	6.94	2.44	4.32	4.58	4.15	2.46	2.03	3.44	4.09	4.18	2.28	1.42
3	5.25	7.13	7.70	7.56	5.68	7.28	6.93	2.81	4.72	5.00	4.57	2.87	2.38	3.85	4.53	4.60	2.68	1.50
4	5.05	6.90	7.45	7.29	5.41	7.10	6.69	2.57	4.51	4.78	4.33	2.57	2.15	3.63	4.27	4.32	2.34	1.43

Table 4. *Cont.*

Model	E2	E3	E4	E5	F1	F2	F3	F4	F5	G1	G2	G3	H1	H2	H3
1	3.59	3.80	3.02	1.14	1.21	3.08	3.91	3.78	1.85	0.95	1.78	0.70	1.01	2.11	1.28
2	3.32	3.58	2.82	1.05	1.09	2.78	3.51	3.23	1.39	0.65	1.34	0.46	0.65	1.47	0.72
3	3.54	3.80	3.01	1.10	1.18	3.02	3.81	3.57	1.58	0.87	1.67	0.64	0.90	1.87	1.02
4	3.38	3.67	2.89	1.06	1.11	2.85	3.58	3.26	1.33	0.68	1.39	0.50	0.67	1.49	0.69

Table 5. Deformation of the linear dimensions specified in the physical and numerical simulation of the welding process of the industrial furnace wall panel.

Distance	Deformation, mm				
	Physical Simulation	Numerical Simulation			
		Model 1	Model 2	Model 3	Model 4
C1–C5	−2.00	−1.65	−1.61	−1.60	−1.58
D1–D5	−2.00	−1.65	−1.61	−1.60	−1.59
E1–E5	0.00	−1.77	−1.72	−1.71	−1.70
F1–F5	+1.00	−1.52	−1.47	−1.48	−1.45
G1–G3	0.00	+1.65	+1.55	+1.56	+1.52
H1–H3	−2.00	+2.18	+1.97	+2.00	+1.88

According to EN ISO 13920, the tolerance for linear dimensions 2000–4000 mm in length (dimensions of the furnace wall panel) in class A should be ±4 mm. The values of tolerance for linear dimensions, specified in the physical and numerical simulations, are within this range.

The flatness tolerance was determined on the basis of the results of deformation measurements in the direction of the Z axis at characteristic points of the furnace wall panel. Table 6 presents the results of deformation measurements from the physical and numerical simulations of the welding process of the furnace wall panel. The mean absolute strain values obtained in the physical and numerical simulations are similar.

Table 6. Deformations in the Z direction at the measurement points of the furnace wall panel (physical and numerical simulation).

Point	Deformation in the Direction of the Z axis, mm				
	Physical Simulation	Numerical Simulation			
		Model 1	Model 2	Model 3	Model 4
A1	+4	+5.59	+5.23	+5.25	+5.05
A2	+6	+7.43	+7.05	+7.13	+6.90
A3	+7	+8.01	+7.57	+7.70	+7.45
A4	+5	+7.94	+7.48	+7.56	+7.29
A5	+13	+6.14	+5.64	+5.68	+5.41
B1	+6	+7.58	+7.27	+7.28	+7.10
B2	+6	+7.36	+6.94	+6.93	+6.69
C1	−2	+2.82	+2.44	+2.81	+2.57
C2	−2	+4.68	+4.32	+4.72	+4.51
C3	−1	+4.98	+4.58	+5.00	+4.78
C4	−8	+4.56	+4.15	+4.57	+4.33

Table 6. *Cont.*

Point	Deformation in the Direction of the Z axis, mm				
	Physical Simulation	Numerical Simulation			
		Model 1	Model 2	Model 3	Model 4
C5	−3	+2.94	+2.46	+2.87	+2.57
D1	−3	+2.40	+2.03	+2.38	+2.15
D2	−1	+3.83	+3.44	+3.85	+3.63
D3	+1	+4.55	+4.09	+4.53	+4.27
D4	+1	+4.67	+4.18	+4.60	+4.32
D5	−1	+2.86	+2.28	+2.68	+2.34
E1	+5	+1.56	+1.42	+1.50	+1.43
E2	+4	+3.59	+3.32	+3.54	+3.38
E3	+1	+3.80	+3.58	+3.80	+3.67
E4	0	+3.02	+2.82	+3.01	+2.89
E5	−2	+1.14	+1.05	+1.10	+1.06
F1	−4	+1.21	+1.09	+1.18	+1.11
F2	+1	+3.08	+2.78	+3.02	+2.85
F3	+1	+3.91	+3.51	+3.81	+3.58
F4	−1	+3.78	+3.23	+3.57	+3.26
F5	−4	+1.85	+1.39	+1.58	+1.33
G1	−1	+0.95	+0.65	+0.87	+0.68
G2	−3	+1.78	+1.34	+1.67	+1.39
G3	−3	+0.70	v0.46	+0.64	+0.50
H1	−3	+1.01	+0.65	+0.90	+0.67
H2	0	+2.11	+1.47	+1.87	+1.49
H3	5	+1.28	+0.72	+1.02	+0.69
Average of absolute values	**3.27**	**3.73**	**3.35**	**3.59**	**3.37**

According to EN ISO 13920, in class F for the length range from 1000 to 2000 mm, the flatness tolerance is t = 4.5 mm. The obtained average absolute values of deformations at the measurement points from both simulations are within this range.

However, it can be noticed that the deformations at the characteristic points of the panel determined in the physical and numerical simulations are different in terms of value and sign (plus or minus). For numerical simulation, strain values have a plus sign. This proves that the predicted deformations arising in the panel welding process are the result of tensile stresses. In the case of physical simulation, the deformations have a plus or minus sign, which indicates that the process of welding the furnace wall panel generates tensile or compressive stresses.

As welding practice shows, there will always be stress in awelded structure. Compressive stresses occur in the weld as a result of solidification of the molten pool. On the other hand, tensile stresses occur in the heat-affected zone and the area of the welded material in this zone. The distribution of these stresses depends on many factors, including the welding method (heat source), the complexity of the welded structure, the thickness of the joined elements, and the welding process.

The methods used in welding processes differ in the source of heat. In the case of arc welding, the size, shape, and concentration of the heat source are determined by the values of the technological parameters of the welding method. This affects the course of welding thermal cycle, which results in the creation of a weld. This cycle includes heating and cooling of the welded joint area. The use of a wide electric arc with high power is associated with the risk of creating a wide heat-affected zone (HAZ) and a wide thermal impact zone on the parent material. In the case of the HAZ, a microstructure with low fracture toughness is formed in Fe-C alloys. In the areas further away from the weld, the thermal influence of the heat source may result in welding stresses due to cyclical heating and cooling. If the value of these stresses exceeds the value of the elastic limit, then plastic deformations, i.e., welding deformations, will occur. These deformations may also occur as a result of stresses arising in the adjacent welded joints (inhibition of weld shrinkage in the structures composed of a large number of welded joints). This is the case in the analyzed furnace wall panel; there are 38 welds.

Reducing the energy value of the arc and its width (high concentration) will result in a lower thermal impact on the material of the joined elements. This solution is used in the case of making welds of thick elements, which are joined with a multi-pass weld—the weld is created as a result of several passes of the electric arc. In addition, maintaining the appropriate temperature between the stitches reduces the adverse effect of the thermal cycle on the material of the joined elements. This reduces the risk of welding deformations.

The selection of technological parameters of the welding method, on the one hand, determines the energy value of the heat source, and on the other hand, the value of these parameters depends on the type and thickness of the elements. In the case of welded structures containing a large number of welds, these parameters also depend on the complexity of the structure and the method and sequence of arranging the welds.

The method and sequence of making the welds proposed in this paper are the result of research conducted as part of the optimization of the welding process of the furnace wall panel. Figure 3 shows the order in which the welds should be made and how they should be made with a step-back stitch. The choice of such a welding concept was the result of numerical simulations of deformations of many variants of the order and method of arranging the welds, which were carried out in the initial period of the optimization process.

The welding sequence is very important in the welding process. Welding in the correct order may result in partial relaxation of welding stresses. In the publication, the idea of the proposed welding sequences and methods of making welds of elements of the furnace wall panel (back-step technique—Figure 3) was to limit the possibility of generation of tensile and compressive stresses that could cause plastic deformations. Further minimization of welding deformations was obtained as a result of the use of a restraint system.

3.2. Visual Inspection and Measurement of Weld Geometry

The objective of visual testing (VT) was to assess the welded joints for the presence of welding defects. The results of the observations of the welded joints indicate the absence of welding defects. No excessive face overflow wasdetected. Analysis of the cross-section of the joint indicates the presence of full fusion. Small amounts of welding spatter were found on the joint surface.

Based on the measurement of the weld geometry, the average value of the weld face width over a length of 1 m was found to be 9.0 mm, while the average value of the weld face height over the analyzed length was 4.0 mm. These values areconsistent with the weld face width and height variation ranges adopted in the WPS manual It should be noted that the WPS technological instructions were developed in accordance with the existing WPQR procedures, because one of the criteria in the WPQR qualification process is the thickness of the joined elements. The thickness of these elements determines the thickness of the weld. For example, for butt welds made with full melting, the thickness of the weld is equal to the thickness of the joined elements. Depending on the thickness of the weld, the range of values of the other geometric parameters of the weld, i.e., the width and height of the weld face, is determined. The measured average values of the width and height of the weld face are within the ranges specified in the WPS and WPQR.

The microstructure of the welded joint was analyzed in individual areas of the joint (A–E), according to the diagram shown in Figure 13.

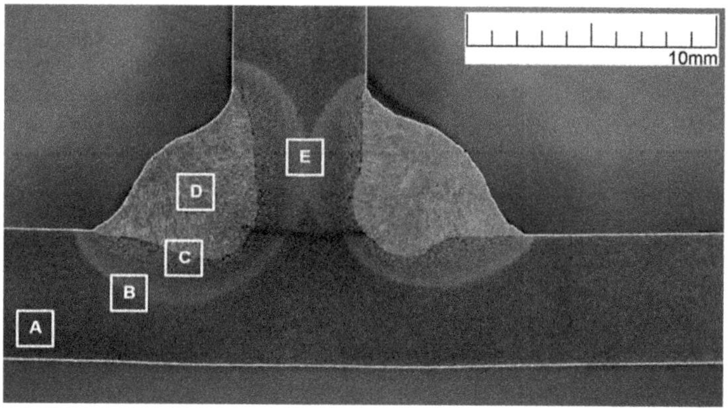

Figure 13. Macrostructure of the test joint with marked areas of the microstructure test: A—parent material, B—heat-affected zone (HAZ), C—weld and heat-affected zone boundary, D—weld, and E—heat-affected zone (HAZ).

Area A (Figure 14) shows the microstructure of the parent material, i.e., S235JRC+N steel. This steel is structural steel in which the microstructure is obtained by hot rolling. The microstructure of S235JRC+N steel consists of ferrite and perlite (cementite and ferrite) distributed along the grain boundaries of the ferrite. The cementite in the perlite can take a lamellar or fibrous form.

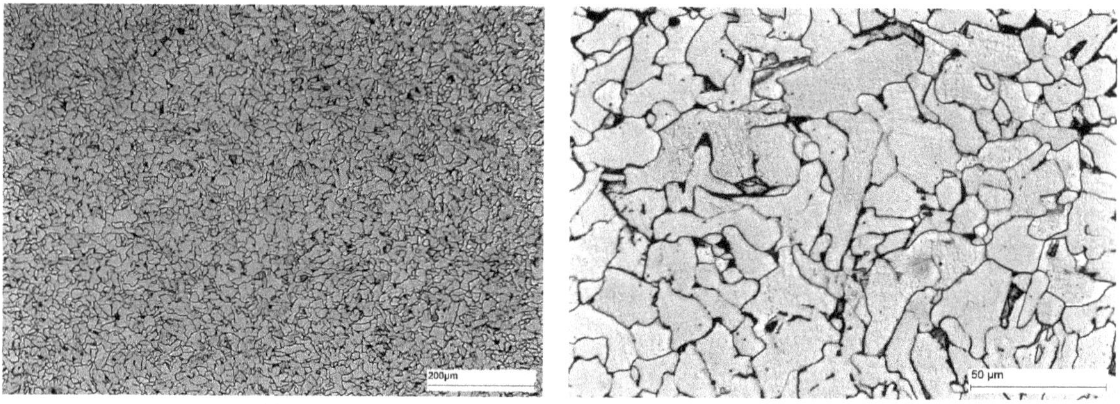

(**a**) (**b**)

Figure 14. Microstructure of area A of the welded joint—parent material: (**a**) 100× magnification, and (**b**) 500× magnification. Ferrite—white precipitates, and perlite—dark precipitates at the ferrite grain boundaries.

Area B (Figure 15) shows the microstructure of the part of the HAZ furthest from the fusion boundary. In this part of the heat-affected zone, the effect of the heat source is finer gradation of the ferrite grains compared to the parent material.

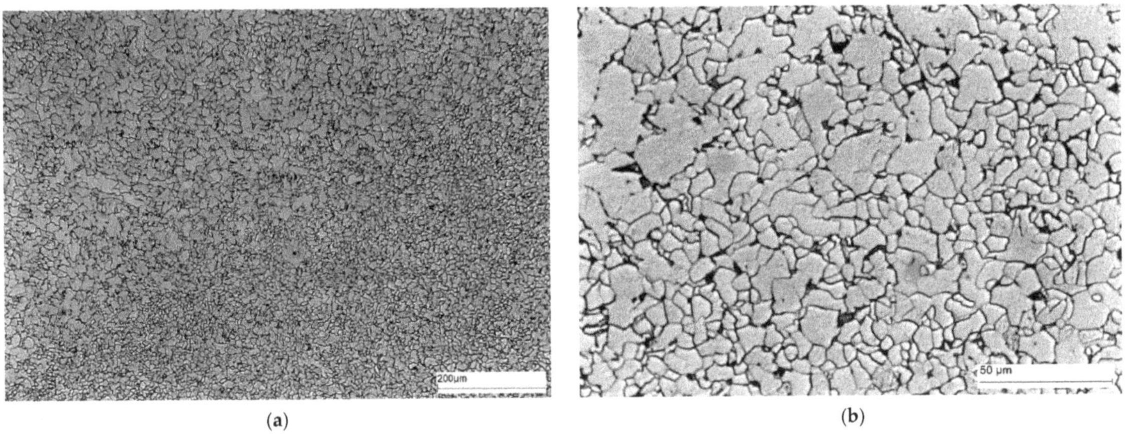

Figure 15. Microstructure of area B of welded joint—HAZ: (**a**) magnification 100×, and (**b**) magnification 500×. Ferrite—white precipitates, and perlite—dark precipitates at the ferrite grain boundary.

The microstructure of area C, i.e., the fusion boundary, is shown in Figure 16. In this area, Widmanstätten ferrite (light) and bainite (dark) precipitates occur on the weld side. Ferrite (light) and bainite (dark) precipitates with different morphology are also present on the HAZ side.

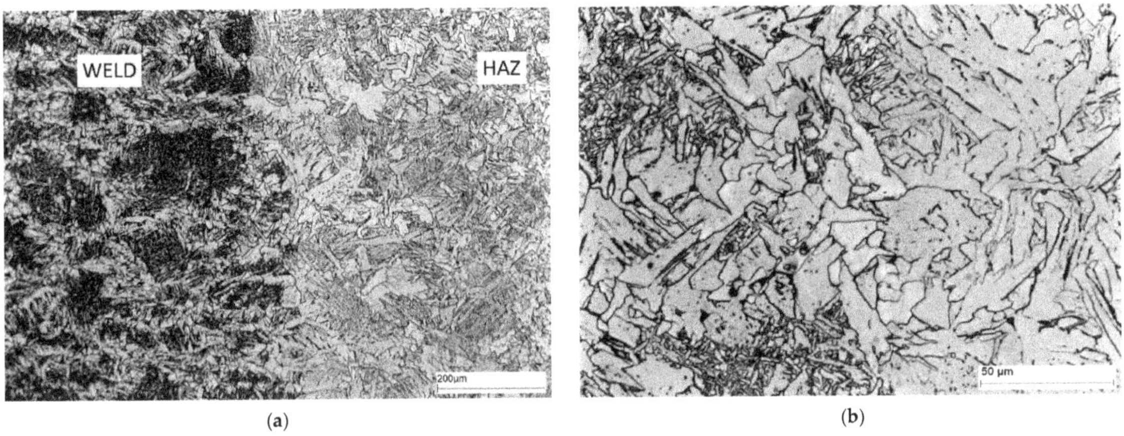

Figure 16. Microstructure of area C—fusion boundary: (**a**) 100× magnification, and (**b**) 500× magnification. Widmanstätten ferrite—white separations, hardening products, mainly bainite—dark separations.

The microstructure of the weld (area D) is shown in Figure 17. The weld contains Widmanstätten ferrite (light) and bainite (dark) precipitates.

Figure 18 shows the microstructure of area E, i.e., the location of the dual influence of the heat source. The microstructure of this area, as in Figure 15, is characterized by a greater degree of fragmentation of the ferrite grains compared to the parent material.

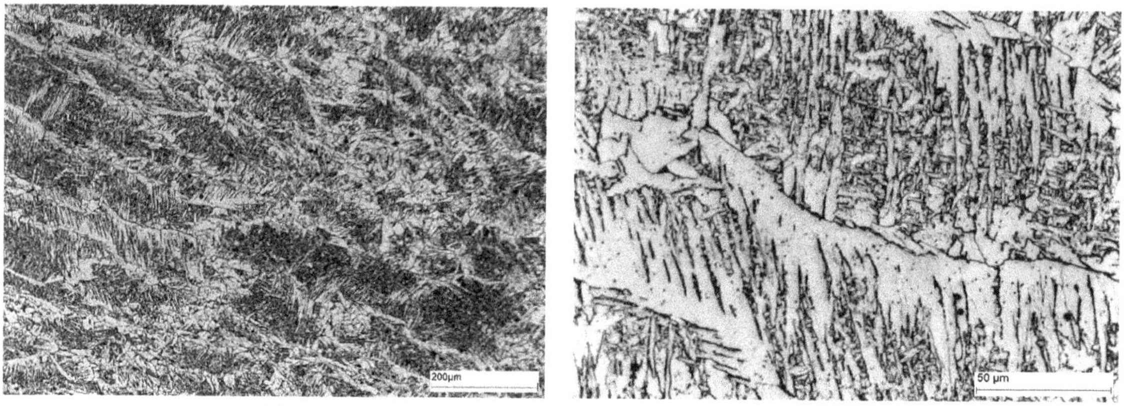

(**a**) (**b**)

Figure 17. Microstructure of area D—weld: (**a**) 100× magnification, and (**b**) 500× magnification. Widmanstätten ferrite—white precipitates with different morphology, quenching products, mainly bainite—dark precipitates.

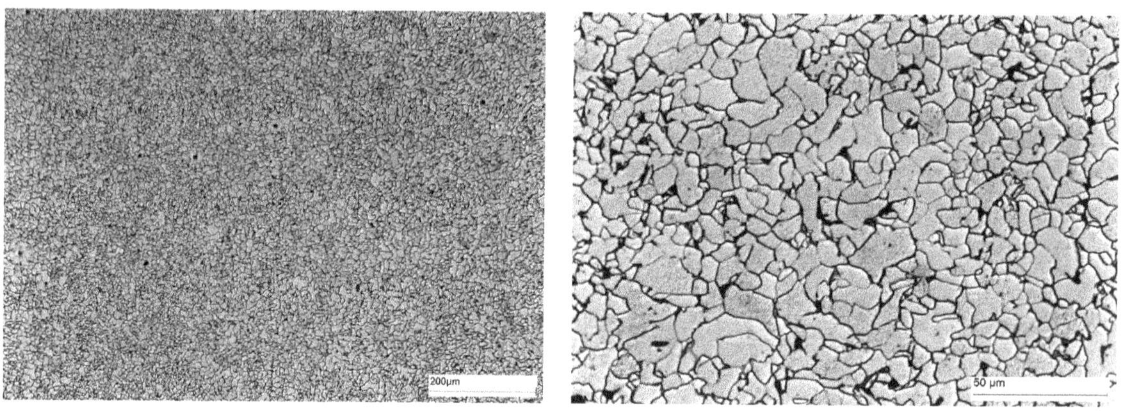

(**a**) (**b**)

Figure 18. Microstructure of area E of welded joint—SWC: (**a**) magnification 100×, and (**b**) magnification 500. Ferrite—white precipitates, and perlite—dark precipitates at grain boundaries.

3.3. Hardness Measurements

The macrostructure of the welded joint area with the HV5 hardness measurement lines marked is shown in Figure 19.

The results of HV5 hardness measurements along measurement lines 1 and 2 are shown in Figure 20.

The results obtained show a similar distribution of the HV5 hardness on lines 1 and 2. The highest value of the HV5 hardness was found in the weld area, slightly lower in the HAZ and the lowest in the parent material.

The hardness value does not exceed 350HV5, which proves the lack of tendency of the welded joint material to form hard and brittle hardening products such as martensite.

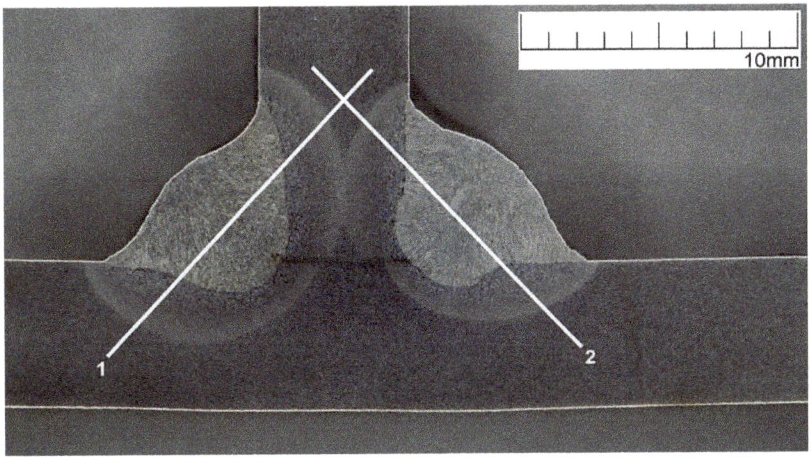

Figure 19. Macrostructure of welded joint with HV5 hardness measurement lines marked.

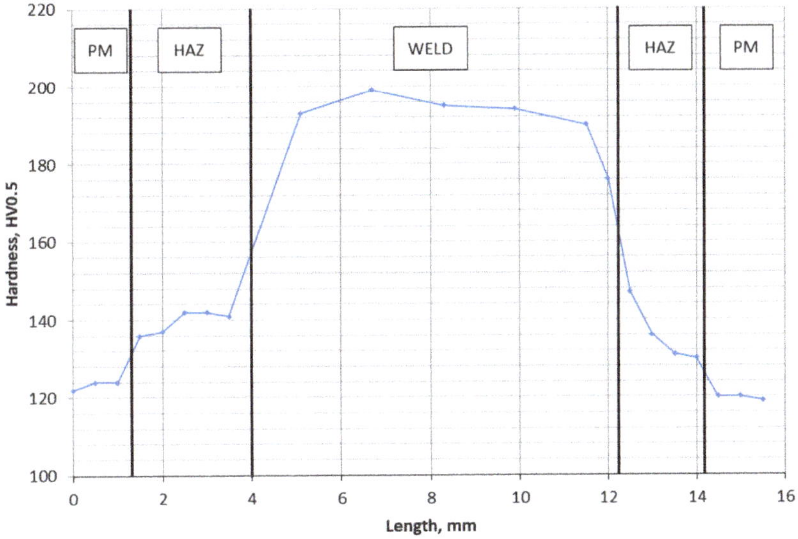

(a)

Figure 20. *Cont.*

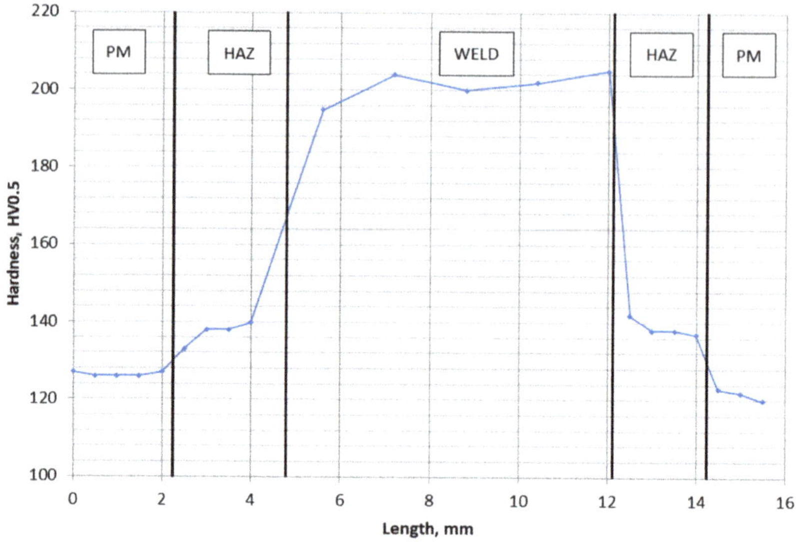

(**b**)

Figure 20. HV5 hardness distribution in the welded joint area; (**a**) measurement line 1, and (**b**) measurement line 2. PM—parent material, HAZ—heat-affected zone.

4. Conclusions

As a conclusion of the research on the development of welding technology for a large industrial furnace wall panel, using numerical and physical simulation of the MAG welding process, it can be stated that:

- The use of different variants of the restraint system in the numerical simulation of the welding process of the industrial furnace wall panel gives a similar character of the deformation distribution, but with different values. The difference between the maximum mean strain in model 1 and the lowest mean strain in model 4 was found to be only 11%.
- The results of deformation measurements in the physical simulation of the furnace wall panel welding process differ from the results of the numerical simulation. These differences are visible in the values and the signs of the strains. This confirms the observations of other authors that numerical simulation of the welding process of alarge structures is difficult.
- Due to the requirements of EN ISO 13920 standard, all restraint models provide acceptable linear dimensional and flatness tolerances. Therefore, for the technology of welding a large wall panel of an industrial furnace, a variant of the constraint system according to model 1 is recommended.
- The results of microstructural tests and hardness measurements confirm the correctness of the developed technology, taking into account the recommended system of restraining the elements of the furnace wall panel during the welding process and the proposed sequence and method of welding.

Author Contributions: Conceptualization, M.M. and R.C.; Data curation, P.R. and B.K.; Formal analysis, M.M., R.C., B.K. and A.D.; Investigation, M.M., A.D., B.K., R.C. and M.S.; Methodology, M.M., R.C. and M.S.; Project administration, M.M. and P.R.; Resources, M.M.; Supervision, M.M.; Validation, M.M. and R.C.; Writing—original draft, M.M. and P.R.; Writing—review & editing, M.M. All authors have read and agreed to the published version of the manuscript.

Funding: Project co-financed by the European Union from the European Regional Development Fund under the RPOWP for the years 2014–2020, under Measure 1.2 Industrial research, development and their implementation, Project type: R&D works. Project title "Development of an innovative prefabrication technology for steel structures, pipelines and pressure equipment components based on the results of numerical analysis".

Institutional Review Board Statement: Not applicable.

Data Availability Statement: Data is contained within the article.

Conflicts of Interest: The authors declare no conflict of interest.

References

1. Domański, T.; Piekarska, W.; Saternus, Z.; Kubiak, M.; Stano, S. Numerical Prediction of Strength of Socket Welded Pipes Taking into Account Computer Simulated Welding Stresses and Deformations. *Materials* **2022**, *15*, 3243. [CrossRef]
2. Potukutchi, R.; Agrawal, H.; Perumalswami, P. *Fatigue Analysis of Steel MIG Welds in Automotive Structures*; SAE Technical Paper Series; SAE World Congress: Detroit, MI, USA, 2004.
3. Barsoum, Z.; Barsoum, I. Residual stress effects on fatigue life of welded structures using LEFM. *Eng. Fail. Anal.* **2009**, *16*, 449–467. [CrossRef]
4. Lee, S.H.; Kim, E.S.; Park, J.Y.; Choi, J. Numerical analysis of thermal deformation and residual stress in automotive muffler byMIG welding. *J. Comput. Des. Eng.* **2018**, *5*, 382–390.
5. Perić, M.; Seleš, K.; Tonković, Z.; Lovrenić-Jugović, M. Numerical simulation of welding distortions in large structures with a simpli fied engineering approach. *Open Phys.* **2019**, *17*, 719–730. [CrossRef]
6. Ilam, M.; Buijk, A.; Rais-Rohani, M.; Motoyama, K. Simulation based numerical optimization of arc welding process for reduced distortion in welded structures. *Finite Elem. Anal. Des.* **2014**, *84*, 54–64. [CrossRef]
7. Prakash, R.; Gangradey, R. Review of high thickness welding analysis using SYSWELD for a fusion grade reactor. *Fusion Eng. Des.* **2013**, *88*, 2581–2584. [CrossRef]
8. Farias, R.M.; Teixeira, P.R.F.; Araújo, D.B. Thermo-mechanical analysis of the MIG/MAG multi-pass welding process on AISI304L stainless steel plates. *J. Braz. Soc. Mech. Sci. Eng.* **2017**, *39*, 1245–1258. [CrossRef]
9. Ogino, Y.; Hirata, Y.; Murphy, A.B. Numerical simulation of GMAW process using Ar and an Ar–CO_2 gas mixture. *Weld World* **2016**, *60*, 345–353. [CrossRef]
10. Zhu, W.; Xu, C.; Zeng, L. Coupled finite element analysis of MIG welding assembly on auto-body high-strength steel panel and door hinge. *Int. J. Adv. Manuf. Technol.* **2010**, *51*, 551–559. [CrossRef]
11. Skri, A.; Guidara, M.; Elhalouani, F. Numerical simulation of MIG type arc welding induced residual stresses and distortions in the sheets of S235 steel. *Mater. Sci. Eng.* **2010**, *13*, 012020. [CrossRef]
12. Barsoum, Z.; Lundbäck, A. Simplified FE welding simulation of fillet welds–3Deffects on the formation residual stresses. *Eng. Fail. Anal.* **2009**, *16*, 2281–2289. [CrossRef]
13. Li, W.; Yu, R.; Huang, D.; Wu, J.; Wang, Y.; Hu, T.; Wang, J. Numerical simulation of multi-layer rotating arc narrow gap MAG welding formed in steel plate. *J. Manuf. Process.* **2019**, *45*, 460–471. [CrossRef]
14. Perić, M.; Tonković, Z.; Rodić, A.; Surjak, M.; Garašić, I.; Boras, I.; Švaić, S. Numerical analysis and experimental investigation of welding residual stresses and distortions in a T-joint fillet weld. *Mater. Des.* **2014**, *53*, 1052–1063. [CrossRef]
15. Vemanaboina, H.; Akella, S.; Buddu, R.K. Welding process simulation model for temperature and residual stress analysis. *Procedia Mater. Sci.* **2014**, *6*, 1539–1546. [CrossRef]
16. Varma Prasad, V.M.; Joy Varghese, V.M.; Suresh, M.R.; Siva Kumar, D. 3D simulation of residual stress developed during TIG welding of stainless-steel pipes. *Procedia Technol.* **2016**, *24*, 364–371. [CrossRef]
17. Koňár, R.; Patek, M. Numerical simulation of dissimilar weld joint in Sysweld simulation software. *Tehnički Vjesnik* **2017**, *24*, 137–142.
18. Deng, D. FEM prediction of welding residual stress and distortion in carbon steel considering phase transformation effects. *Mater. Des.* **2009**, *30*, 359–366. [CrossRef]
19. Han, Y.; Chen, J.; Ma, H.; Zhao, X.; Wu, C.; Gao, J. Numerical simulation of arc and droplet behaviors in TIG-MIG hybrid weling. *Materials* **2020**, *13*, 4520. [CrossRef]
20. Casuso, M.; Veiga, F.; Suárez, A.; Bhujangrao, T.; Aldalr, E.; Artaza, T.; Amondarain, J.; Lamikiz, A. Model for the prediction of deformations in the manufacture of thin-walled parts by Wire Arc Additive Manufacturing technology. *Metals* **2021**, *11*, 678. [CrossRef]
21. Ding, J.; Colegrove, P.; Mehnen, J.; Ganguly, S.; Sequeira Almeida, P.M.; Wang, F.; Williams, S. Thermo-mechanical analysis of Wire and Arc Additive Layer Manufacturing process on large multi-layer parts. *Comput. Mater. Sci.* **2011**, *50*, 3315–3322. [CrossRef]
22. Finch, D.M.; Burdekin, F.M. Effects of welding residual stresses on significance of defects in various types of welded joint. *Eng. Fract. Mech.* **1992**, *41*, 721–735. [CrossRef]

23. Giętka, T.; Ciechacki, K.; Kik, T. Numerical simulation of duplex steel multi pass welding. *Arch. Metall. Mater.* **2016**, *61*, 1975–1984. [CrossRef]
24. Chen, Z.; Chen, Z.; Ajit Shenoi, R. Influence of welding sequence on welding deformation and residual stress of a stiffened plate structure. *Ocean Eng.* **2015**, *106*, 271–280. [CrossRef]
25. Zhang, J.; Yu, L.; Liu, Y.; Li, H.; Liu, C.H.; Wu, J.; Ma, J.; Li, Z. Effect of welding sequences on the welding stress and distortion in the CFETR vacuum vessel assembly using finite element simulation. *Int. J. Press. Vessel. Pip.* **2019**, *175*, 103930. [CrossRef]
26. Bhatti Ayjwat, A.; Bersoum, Z.; Khurshid, M. Development of finite element simulation frame work for the prediction of residual stresses in large welded structures. *Comput. Struct.* **2014**, *133*, 1–11. [CrossRef]

materials

MDPI

Article

Calorimetric Method for the Testing of Thermal Coefficients of the TIG Process

Marek Mróz [1,*], Antoni Władysław Orłowicz [1], Magdalena Lenik [1], Andrzej Trytek [2] and Mirosław Tupaj [2]

[1] Department of Foundry and Welding, Rzeszow University of Technology, Powstancow Warszawy 12, 35-959 Rzeszow, Poland

[2] Faculty of Mechanics and Technology, Rzeszow University of Technology, Kwiatkowskiego 4, 37-450 Stalowa Wola, Poland

* Correspondence: mfmroz@prz.edu.pl; Tel.: +48-17-865-16-83

Abstract: This paper presents an original design of a test apparatus for calorimetric measurements of arc efficiency η and melting efficiency η_m in welding processes. The construction and principle of operation of a new flow calorimeter are described, as well as the method for determining the η and η_m values in the process of the surface melting of aluminium–silicon alloy casting surfaces with a concentrated heat flux generated by the TIG (Tungsten Inert Gas) method. The results obtained indicate the advisability of using calorimetric testing to assess the arc efficiency of welding processes. It was demonstrated that changing the welding current and arc scanning speed, as well as changing the chemical composition of the silumin, has an effect on the arc efficiency value η. This has the effect of introducing a different amount of heat into the area of the heated material. The consequence of this is a change in the value of the melting efficiency η_m, which results in a change in the width and depth of the surface melting areas, through this, the cooling conditions of the material. As is well known, this will affect the microstructure of the welds and the width and microstructure of the heat-affected zone, and thus the performance of the welded joints.

Keywords: TIG method; aluminium–silicon alloys; calorimeter; arc efficiency η; melting efficiency η_m

Citation: Mróz, M.; Orłowicz, A.W.; Lenik, M.; Trytek, A.; Tupaj, M. Calorimetric Method for the Testing of Thermal Coefficients of the TIG Process. *Materials* **2022**, *15*, 7389. https://doi.org/10.3390/ma15207389

Academic Editors: Marcin Adamiak, Artur Czupryński, Antonín Kříž and Tünde Anna Kovács

Received: 13 October 2022
Accepted: 19 October 2022
Published: 21 October 2022

1. Introduction

The performance of welded joints depends on the chemical composition and microstructure of the base material, the geometry and microstructure of the welds and the width and microstructure of the heat-affected zone (HAZ). The weld geometry and the width of the heat-affected zone are determined by the amount of heat introduced into the joint during the welding process. The amount of heat introduced has an effect on the primary crystallisation of the welds and the course of phase transitions in the heat-affected zone during the cooling of the welded joints [1–3].

As the microstructure of the alloys used in the different types of structures determines their mechanical properties and fatigue strength, it is important to know how it can be modelled in the welded joints in the weld and heat-affected zone areas. The welding process parameters used in the manufacturing of structures should be such that, with the set cooling conditions, enough heat can be introduced into the welded joint to ensure that, in the weld and the HAZ, a microstructure is formed, which provides levels of mechanical properties, and a fatigue strength not inferior to those of the base material [3–5].

Consequently, the development of the welding procedure for the respective structures includes the study of the influence of the welding process parameter values on the amount of heat introduced into the welded joint and its effect on the weld geometry and the width of the HAZ, the determination of the type of microstructure components formed, their volume proportions and material property values, followed by the study of the mechanical properties and fatigue strength. Their results are the basis for the development of empirical relationships linking the amount of heat introduced to the weld geometry and its microstructure, the width of the HAZ and its microstructure, and the mechanical properties and fatigue strength. It is sometimes necessary to carry out such tests for cooling conditions with isothermal endurance, for example, when there is a need to block the martensitic transformation in the weld and HAZ [6,7].

Such studies constitute the know-how of leading responsible construction manufacturers and are not usually published. Sometimes these companies agree to publish partial research results or to patent them.

Since the amount of heat introduced into a welded joint depends on many factors—for example, the chemical composition of the parts to be joined and their microstructure, the current waveforms obtained from the current sources, the set values of the current parameters, the length of the arc, the speed of the arc scanning, and the protective atmosphere used—the precise determination of the amount of heat introduced into the heated material requires the use of experimental methods. The results of such tests allow the value of an important parameter of the electric arc welding process to be calculated, namely the arc efficiency η. The arc efficiency of the welding process indicates how much of the energy supplied to the heated material was used to form a surface melting area or weld [8].

To clarify the issue of heat generation and heat flows in TIG (Tungsten Inert Gas) welding, Du Pont and Marder [9] proposed the heat balance diagram shown in Figure 1.

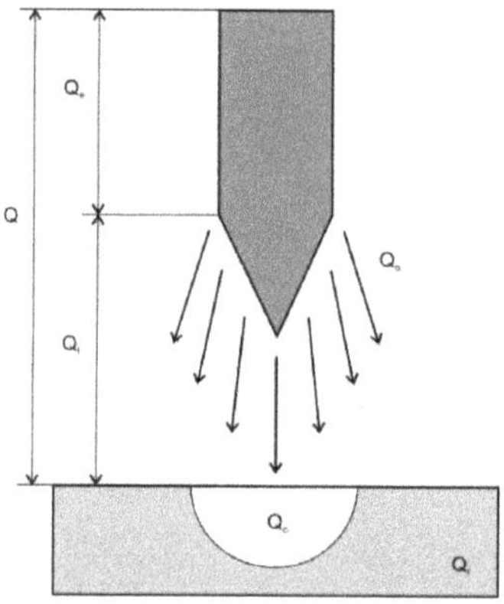

Figure 1. Energy distribution in the TIG process: Q, total heat generated in the electric arc Q_{arc} and on the electrode Q_e; Q_o, amount of heat transferred to the surrounding area; Q_c, amount of heat required to heat and partially melt the material and then overheat the liquid metal; $Q_r + Q_c$ is that part of the heat absorbed by the heated element; Q_r, amount of heat transferred by conduction to the parent material.

The heat balance of the process shown in Figure 1 can be expressed by the equation:

$$Q = Q_{arc} + Q_e + Q_o + Q_c + Q_{r,} \tag{1}$$

where Q is the total heat generated in the welding process (J), Q_{arc} is the amount of heat generated in the electric arc (J), Q_e is the amount of heat generated in the electrode (J), Q_o is the amount of heat transferred to the surrounding area (J), Q_c is the amount of heat needed to heat and melt the metal and further overheat the liquid metal (J), and Q_r is the amount of heat transferred by conduction to the parent material (J).

The main amount of heat is generated in the arc (Q_{arc}), and only a small amount (Q_e) is generated in the electrode. Part of the heat (Q_o) goes directly into the surrounding space, and the remainder ($Q_r + Q_c$) penetrates the heated material. Heat (Q_c) is used to form a pool of liquid metal and overheat the liquid within it. Heat (Q_r) diffuses in a direction from the liquid metal pool into the parent material, heating it up. In the area heated to temperatures where phase transformations take place, microstructural changes occur. This area is called the heat-affected zone (HAZ). The width of this zone increases with an increase in the time of the spot heat effect of the electric arc.

For the calculation of the theoretical amount of heat Q_{teor} generated in an electric arc, the following expression is used:

$$Q_{teor} = Q_{arc} = U \cdot I \cdot t \tag{2}$$

where U is the arc voltage (V), I is the welding amperage (A), and t is the scanning time (t).

The degree to which the heat released in the electric arc is used to heat the material is characterised by the arc efficiency of the process η. The arc efficiency of the process is described by an expression involving the quotient of the measured amount of heat taken up by the heated sample $Q_{cal} = Q_r + Q_c$ and the theoretical amount of heat generated in the electric arc Q_{teor}:

$$\eta = \frac{Q_{cal}}{Q_{teor}} = \frac{Q_c + Q_r}{Q_{teor}} \tag{3}$$

The degree of utilisation of the heat absorbed by the material to produce surface melting is characterised by the melting efficiency η_m. The melting efficiency is described by the following expression:

$$\eta = \frac{Q_c}{Q_c + Q_r} \tag{4}$$

From the expressions presented, it can be seen that the selection of parameters for the surface melting process should be carried out in such a way as to maximise the value of the arc efficiency of the process and the melting efficiency. It should be carried out in such a way as to limit energy losses and minimise the width of the heat-affected zone.

The arc efficiency η of the TIG process is calculated from the following expression:

$$\eta = \frac{Q_{cal}}{U \cdot I \cdot t} \tag{5}$$

where Q_{cal} is the amount of heat absorbed by the heated material, determined experimentally using a calorimeter (J), U is the arc voltage (V), I is the amperage (A), and t is the arc scanning time (t).

The melting efficiency is calculated from the following expression:

$$\eta_m = \frac{V_n \cdot \rho \cdot Q_H}{Q_{cal}} \tag{6}$$

$$Q_H = \Delta Q_f + \int_{T_0}^{T_1} c_p dT \tag{7}$$

where V_n is the surface melting area volume (mm^3), ρ is the alloy density (g/mm^3), Q_H is the heat required to heat a unit volume of the alloy from the temperature T_0 to the temperature T_1 and melt it (J/g), Q_f is the melting heat (J/g), and c_p is the specific heat (J/gK).

The main problem in calculating arc efficiency is determining the amount of heat taken up by the heated element.

Two lines of research on this issue can be distinguished in the literature. The first, using methods of mathematical analysis of the heat transfer process, involves searching for the temperature distribution in the material obtained as a result of the action of concentrated heat flux. The temperature field equations presented in the literature, based on a heat source model and a heat flux model founded on a Gaussian distribution function, were obtained by assuming that the values of the material constants of the heated workpieces are invariant with temperature. This approximation distorts the true picture of the phenomena occurring between the arc and the scanned surface. Taking into account the variation of these properties as a function of temperature complicates the calculation flow considerably. Progress on this issue has been made through the use of numerical methods. The application of these methods to describe the heat conduction process during the arc plasma scanning of the workpiece surface has been addressed by a number of researchers [10–14]. Numerical methods allow complex and lengthy calculations, which in turn make it possible to take into account the effect of temperature on the change in the physical properties of the heated material and to analyse the phenomenon as a sum of instantaneous states. Thermal arc efficiency values, calculated in this way, are in the range of 75–80%. They raise doubts primarily concerning two issues. Firstly, there is insufficient experimental verification of them. Secondly, the adopted values of the conductivity coefficient of the liquid metal (pool) are up to four times higher than the value of this coefficient at the melting temperature. It would, therefore, appear that the arc efficiency values quoted are an overestimate for results close to the experimental results. In summary, it can be said that in studies on the amount of heat applied to the superheated surface, the results obtained, depending on the assumptions made, are more or less close to reality and often differ [15].

Therefore, the final results of this modelling should be treated with due caution. For this reason, the results of calculating the heat introduced into the heated material must be verified by experimental methods [16].

This second line of research is based on calorimetric measurements of the amount of heat taken up by the superficially melted workpiece [10,15–23]. There are few studies in the technical literature on calorimetric measurements of welding processes. Various measurement techniques are used for this purpose. For example, Havalda [18] placed the sample to be hardfaced above the surface of the water filling the calorimetric vessel (Figure 2).

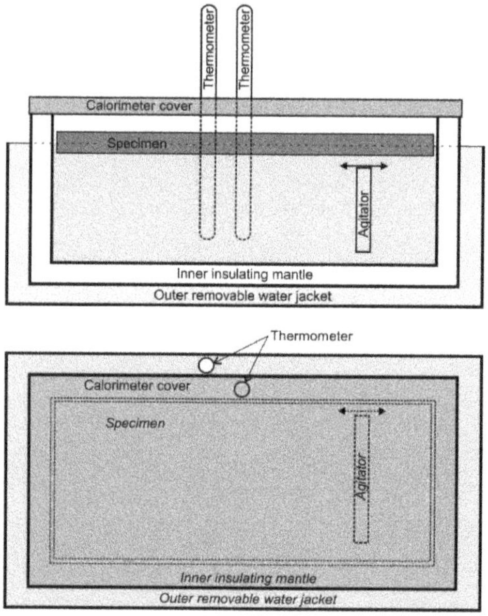

Figure 2. Havaldas's calorimeter [18].

The appropriate design of the holder allowed it to be quickly submerged in water after the bead was made. Once the sample was immersed, the vessel was closed with a lid. Based on the predetermined water constant of the calorimeter (the amount of heat that corresponds to a 1 °C increase in the temperature of the water in the calorimetric vessel) and on measurements of the increase in water temperature, the amount of heat introduced into the calorimeter was determined. The measurement technique used required taking into account the amount of heat used to evaporate the water as a result of the electric arc and during the immersion of the sample.

Giedt, Knorowski and other authors [8,22] determined the arc efficiency of the TIG process using a Seebeck calorimeter (Figure 3).

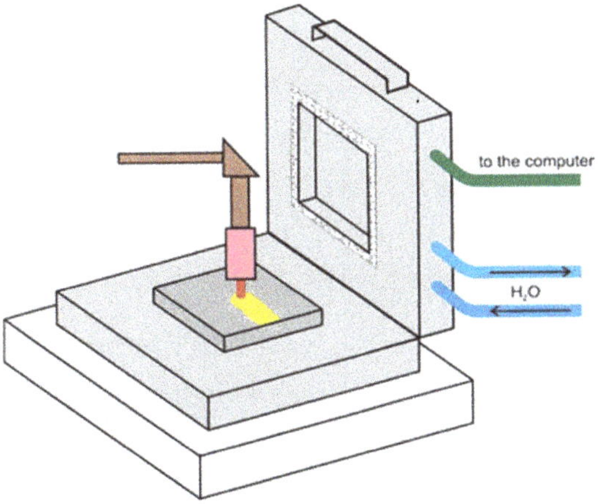

Figure 3. The Seebeck calorimeter [8].

With the calorimeter lid open, a melting section was performed on the sample placed on the base of the calorimeter. The lid was then closed. The correct orientation of the heat flow in the lid was enforced by the flow of water in the cooling pipes. The temperature drop across the walls of the calorimeter was also assessed and measured using a set of thermocouples.

Heat loss due to radiation and convection was estimated to be about 1% of the energy supplied by the arc. The inconvenience of the calorimetric measurements with a Seebeck calorimeter is the long measurement cycle time, reaching six hours in some cases.

Another design of apparatus for calorimetric measurements in welding processes is the Krzyzanowski calorimeter [19] in the form of a rotating cylinder made of brass sheet (Figure 4). During measurements, the outer surface of the cylinder is scanned with concentrated heat flux. The amount of heat input is assessed by measuring the flow rate of water through the cylinder and the difference in temperature of the water at the calorimeter's outlet and inlet lines.

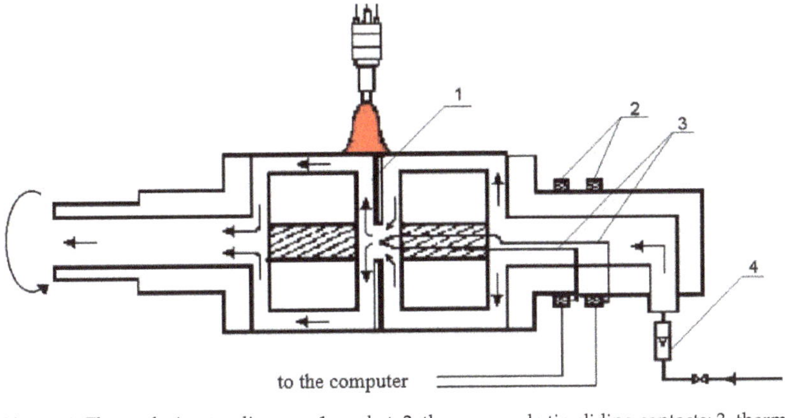

Figure 4. Flow calorimeter diagram; 1, gasket; 2, thermocouple tip sliding contacts; 3, thermocouple; 4, flow meter [19].

The Havalda calorimeter and the Seebeck calorimeter present some disadvantages that affect the measurement results. First, the sample is superficially melted and then immersed in water. This involves heat loss before the actual measurement. The Seebeck calorimeter has the problem of unrecorded heat flow into the body on which the superficially melted sample rests. In the Krzyzanowski calorimeter, the samples would have to be tube-shaped. The calorimeter presented in this study does not suffer from such disadvantages.

A new design of flow calorimeter for measuring thermal quantities characterising welding processes is the subject of Polish patent no. 211283 [23]. The flow calorimeter takes the form of a rectangular vessel. The design of the flow calorimeter is shown in Figure 5.

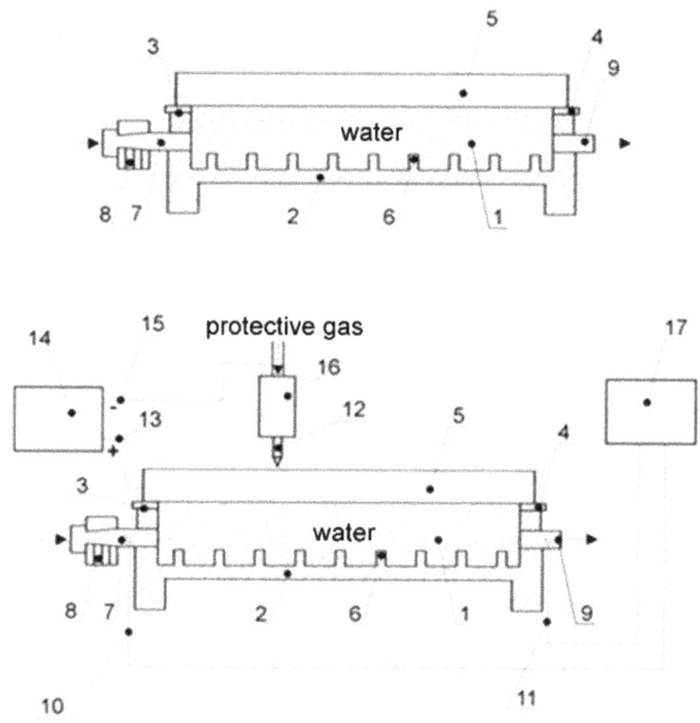

Figure 5. Structural design of the flow calorimeter; 1, water; 2, flow chamber; 3, top edge of the chamber; 4, gasket; 5, superficially melted sample; 6, partitions; 7, water inlet pipe; 8, flow rotameter; 9, water drainage pipe; 10, 11, thermocouples; 12, tungsten electrode; 13, positive pole; 14, power source; 15, negative pole; 16, holder with protective gas; 17, temperature recording [23].

The sample is placed at the top. Water flowing through the calorimeter rinses the lower surface of the sample. The water flow rate on the flow rotameter is chosen so that neither turbulence nor gas bubbles appear at the sample–water interface. Water is supplied via a slotted inlet, and the water temperature is measured in both the inlet and outlet pipes. The actual measurements start once the temperature of the outflowing water has stabilised, which usually occurs a few seconds following the start of the melting of the sample surface.

A diagram of the test system using the flow calorimeter is shown in Figure 6.

The aim of this study was to present the measurement capabilities and test methodology using the new calorimeter and to present example results of tests of the effects of TIG welding current and arc scanning speed on the arc efficiency of the process, the melting efficiency and the geometry of the surface melting areas of AK 51 and AK 20 alloy samples.

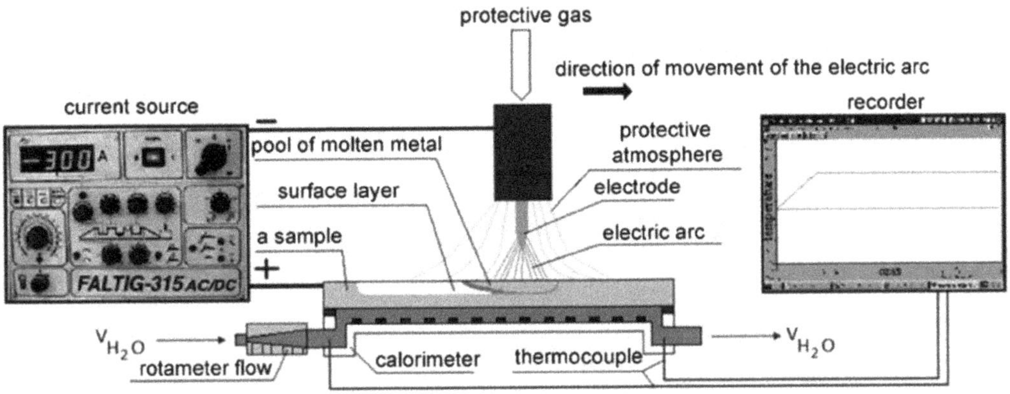

Figure 6. Diagram of the flow calorimeter test system.

2. Materials and Methods

2.1. Material and Casting Procedure

The test material consisted of die-cast plates measuring 260 × 55 × 15 mm made from an aluminium–silicon alloy. AK 51 and AK 20 alloys were used in the study. The chemical composition of the alloys is given in Table 1.

Table 1. Chemical composition of the alloy.

Alloy	Elemental Content, % by Weight							
	Si	Mg	Mn	Cu	Fe	Ti	Cr	Al
AK 51	5.56	0.38	0.34	0.24	0.26	0.110	0.009	remainder
AK 20	21.65	0.04	0.02	0.10	0.28	0.020	0.010	remainder

Smelting was carried out in a 300 kg electric resistance furnace. In the case of the AK 51 alloy, the metal feedstock was scrap metal from the AK 51 alloy castings. A refining treatment was carried out at 710 °C with Dursalit EG 281 at 500 g per 100 kg of metal. Five minutes after the refining procedure and the removal of oxides from the liquid metal surface, the metal moulds of the test plates were flooded with the alloy at 710 °C. The AK 20 alloy was prepared on the basis of the AK 20 pre-alloy. Refining was carried out at 740 °C with Dursalit EG 281 at 500 g per 100 kg of metal. Five minutes into the treatment, the metal moulds of the test plates were flooded with the alloy at 750 °C.

The chemical composition of the calorimetric samples was analysed using a Q4 Tasman Bruker spectrometer, Bruker Poland, Poznań.

2.2. Arc and Melting Efficiency

After mounting in the flow calorimeter, the plate casts were superficially melted with electric arc plasma using a Faltig 315 AC/DC device, Ozas company, Opole, Poland. The test bench is shown in Figure 7. A 4 mm diameter tungsten electrode was used. The surface melting was performed in an argon atmosphere. The shielding gas output was 20 l/min. Amperage levels of I = 100, 150, 200, 250 and 300 A and arc scanning speeds of vs. = 200 and 800 mm/min were used.

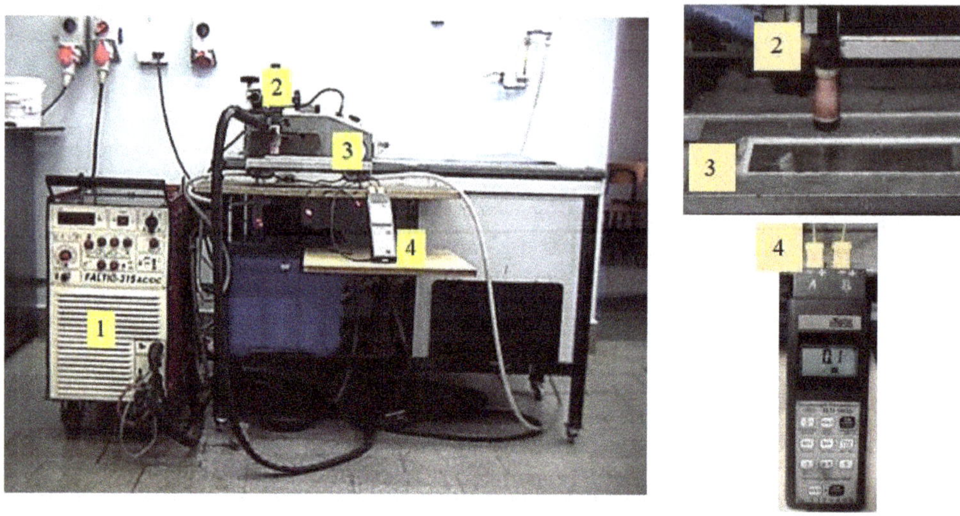

Figure 7. Calorimetric test bench; 1-TIG welding machine, 2—welding torch with tungsten electrode, 3—flow calorimeter with sample, 4—temperature recorder.

The calorimeter was in the shape of a cuboidal closed tub (Figure 5) through which water flowed. The water was supplied via a slotted inlet. The water temperature was measured at the inlet and outlet pipes. From the top, the calorimeter was covered with a superficially melted sample. The water flowing through the calorimeter rinsed its lower surface. With a water outflow of 7.5 L/min, no turbulence or vapour bubbles were observed during the surface melting cycle of the samples. The water temperature stabilised after 10 s. Calorimetric measurements were made for 20 cm long surface melting areas.

Each measurement point was developed for three non-overlapping beads made on the same sample plate. The automated travel of the welding torch ensured a repeatable scanning speed. Precise welding current settings were used. Precision control of the water flow in the calorimeter and computerised recording of water temperature values were used. This preparation of the test material and the conduct of the experiment ensured almost 100% reproducibility of the test results.

The amount of heat released in the arc Q_{teor} was determined from expression (2).

The amount of heat absorbed by the heated sample Q_{cal} was determined from the following relationship:

$$Q_{cal} = m_W \cdot c_{pW} \cdot \Delta T_W \quad (8)$$

where Q_{cal} is the amount of heat absorbed by the calorimeter, m_W is the water mass, c_{pW} is the specific heat of water in the temperature range under investigation, and ΔT_W is the water temperature difference: $\Delta T_W = T_{W_k} - T_{W_p}$, T_{W_p} is the initial water temperature, T_{W_k} is the final water temperature.

The melting efficiency was calculated from relation (6).

2.3. *Fusion Geometry Examination*

Measurements of the width and depth of the surface melting areas were taken using a Nephot 2 optical microscope, Carl Zeiss, Jena, Germany. The cross-sectional area of the surface melting areas in a plane perpendicular to their longitudinal axis was assessed using a planimetric method based on their photographs. The results of these measurements were the arithmetic average of the ten analysed welding cross-sections perpendicular to their longitudinal axis.

2.4. Microstructure Parameters

Due to the strong fragmentation of the microstructure in the surface melting areas, the value of the eutectic structural parameter λ_E was assessed on images of the microstructures taken with a scanning microscope at sufficiently high magnifications.

3. Results and discussion

3.1. Arc efficiency and melting efficiency

The results of calorimetric tests of the surface melting process of AK 51 and AK 20 alloy castings are shown in Tables 2 and 3.

Table 2. Results of calorimetric tests of the surface melting process of AK 51 alloy plate castings.

No.	GTAW Process Parameters		Efficiency	
	I, A	v_s, mm/min	arc, η	Melting, η_m
1	100	200	0.74	0.017
2	150	200	0.77	0.018
3	200	200	0.80	0.020
4	250	200	0.82	0.022
5	300	200	0.84	0.024
6	100	800	0.56	0.059
7	150	800	0.58	0.087
8	200	800	0.60	0.111
9	250	800	0.61	0.138
10	300	800	0.62	0.169

Table 3. Results of calorimetric tests of the surface melting process of AK 20 alloy plate castings.

No.	GTAW Process Parameters		Efficiency	
	I, A	v_s, mm/min	arc, η	Melting, η_m
1	100	200	0.81	0.046
2	150	200	0.83	0.048
3	200	200	0.84	0.051
4	250	200	0.86	0.054
5	300	200	0.88	0.059
6	100	800	0.57	0.207
7	150	800	0.61	0.234
8	200	800	0.62	0.258
9	250	800	0.64	0.286
10	300	800	0.65	0.328

The results of the arc efficiency tests of the process of TIG melting indicate that, for both materials tested, the values of this coefficient increase with increasing welding current and decreasing arc scanning speed. This results in a similar nature of the effect of welding parameter values on the amount of heat introduced into both test materials.

It turned out that alloy AK 20 had slightly higher arc efficiency values and significantly higher melting efficiency values. This suggests its greater susceptibility to forming surface melting areas.

Figure 8 illustrates graphically the test results (nature) of the effect of welding current and arc scanning speed on the TIG melting efficiency of the AK 51 alloy and AK 20 alloy samples.

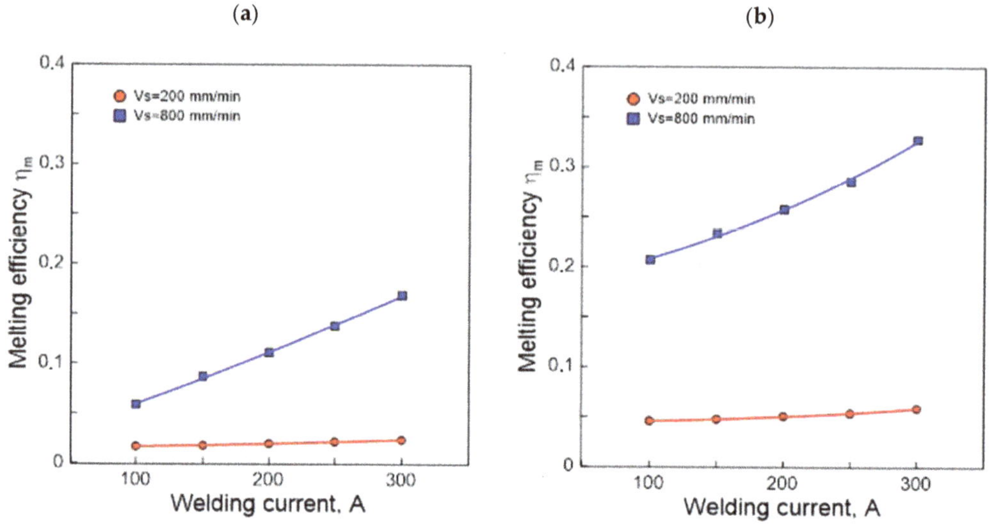

Figure 8. Effect of welding current and scanning speed of TIG arc welding in an argon atmosphere on melting efficiency in calorimetric testing of AK 51 (**a**) and AK 20 (**b**) alloys samples.

The results show that higher values of arc efficiency and melting efficiency were obtained for higher welding current and lower arc scanning speeds. Alloy AK 20 was found to have a higher susceptibility to the formation of surface melting areas.

The results of the tests indicate that, due to the achievement of surface melting areas with a large depth and width, the selection of the values of the TIG process parameters should be carried out in such a way as to maximise the value of the arc efficiency of the welding process.

On the other hand, it is known that low-depth low-width surface melting areas have a lower heat capacity compared to high-depth, high-width surface melting areas. Consequently, their cooling rate is higher, resulting in a more strongly fragmented microstructure. It is known that in the case of silumins, the fragmentation of the microstructure provides higher levels of mechanical properties and fatigue strength.

The importance of the arc efficiency value of the welding process in scientific work on developing relationships between the amount of heat input and the performance of welded joints is not always appreciated, as shown in study [24]. In calculating the amount of heat introduced into the welded joints of aluminium alloy sheets, it adopted one and the same efficiency value from the literature, $\eta = 0.8$, although the welding current values varied from 135 to 180 A and the linear welding speed from 5.9 to 7.32 mm/s.

Du Pont and Marder [9], Smartt et al. [23] and Giedt et al. [8] determined the thermal efficiency of TIG welding steel. For the range of the welding current values used, similar to those presented in this study, the thermal efficiencies were 0.67, 0.7 and 0.8.

Fuerschbach and Knorowsky [22] determined a thermal efficiency of 0.3.

The flow calorimeter used in the present paper was used in study [25], in which surface melting was performed on the surface of AZ 91 alloy castings using the TIG method. The amperage varied from 100 to 400 A, and the welding speed from 200 to 400 mm/min. The authors found that an increase in arc scanning speed is accompanied by a decrease in the arc efficiency of the process, while an increase in welding current increases the arc efficiency value of the process. The range of variation in thermal efficiency values was 0.63–0.88. The authors of paper [23] state that the effect of an increase in the arc efficiency of the process is an increase in the amount of heat introduced into the surface melting areas, thus increasing its width and depth. The amount of heat introduced into the surface melting areas influences the conditions for their crystallisation through which their microstructure is determined.

It has also been used to evaluate the geometry of the surface melting areas and the arc efficiency of the surface melting process of aluminium–silicon alloy [26–28], cast iron [29–31], and cobalt alloy castings [32,33].

3.2. Fusion Geometry

The results of measuring the width and depth of the surface melting areas and the results of calculating the cross-sectional area and volume of the surface melting areas on the AK 51 and AK 20 alloy samples are shown in Tables 4 and 5.

Table 4. Width a, depth h, cross-sectional area F and volume V of surface melting areas on AK 51 alloy samples.

No.	GTAW Process Parameters		Surface Melting Area Indicators			
	I, A	v_s, mm/min	a, mm	h, mm	F, mm^2	V_n, cm^2
1	100	200	3.54	0.43	1.37	0.27
2	150	200	5.05	0.69	2.62	0.48
3	200	200	6.57	0.96	4.31	0.73
4	250	200	8.09	1.21	6.34	1.10
5	300	200	9.57	1.45	8.62	1.57
6	100	800	1.04	0.19	0.26	0.06
7	150	800	2.48	0.45	0.98	0.19
8	200	800	3.94	0.69	2.06	0.35
9	250	800	5.24	0.92	3.35	0.57
10	300	800	6.45	1.11	4.73	0.85

Table 5. Width a, depth h, cross-sectional area F and volume V of surface melting areas on AK 20 alloy samples.

No.	GTAW Process Parameters		Surface Melting Area Indicators			
	I, A	vs, mm/min	a, mm	h, mm	F, mm^2	Vn, cm^2
1	100	200	5.63	2.10	7.26	0.84
2	150	200	6.47	2.65	10.24	1.41
3	200	200	7.28	3.25	13.86	1.98
4	250	200	8.17	3.82	17.99	2.81
5	300	200	8.97	4.38	22.38	4.11
6	100	800	3.17	1.37	2.97	0.32
7	150	800	3.88	1.90	4.79	0.49
8	200	800	4.55	2.49	7.13	0.72
9	250	800	5.36	2.99	9.80	1.07
10	300	800	6.25	3.57	13.32	1.55

The results of the melt geometry tests indicate that for both materials tested, the depth and width of the surface melting areas increase with increasing welding current and decreasing arc scanning speed, with the width of the surface melting areas being greater than their depth. This results in a similar nature to the effect of welding parameter values on the cross-sectional area of the surface melting areas and their volume assessed over the same length of 20 cm.

The AK 20 alloy was found to be more prone to surface melting. That is, at the same welding current and arc scanning rate, the width and depth of surface melting on the AK 20 alloy samples were greater than on the AK 51 alloy samples. This results in higher values for the volume of surface melting obtained in the case of alloy AK 20. Figure 9 shows the results (nature) of the effect of welding current and arc scanning speed on the volume of 200 mm of surface melting for the AK 51 and AK 20 alloy samples.

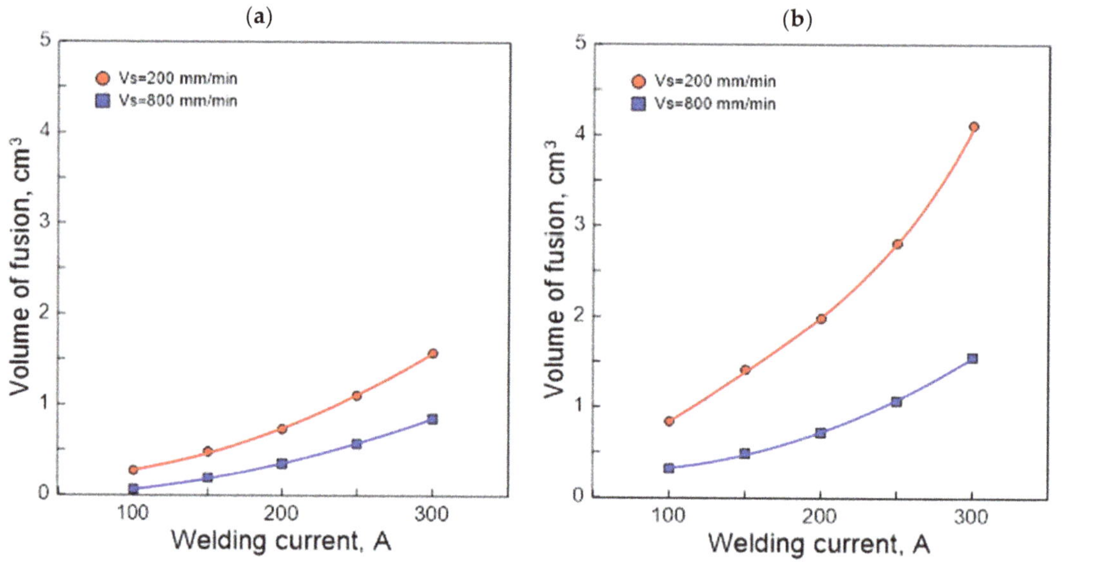

Figure 9. Effect of welding current and arc scanning speed values using the TIG method in an argon atmosphere on the volume of 200 mm long surface melting obtained from calorimetric measurements of AK 51 (**a**) and AK 20 (**b**) alloys samples.

3.3. Microstructure

The microstructure of the AK 51 and AK20 alloys in the areas adjacent to the fusion line on the surface melting side and on the native material side are shown in Figures 10–15.

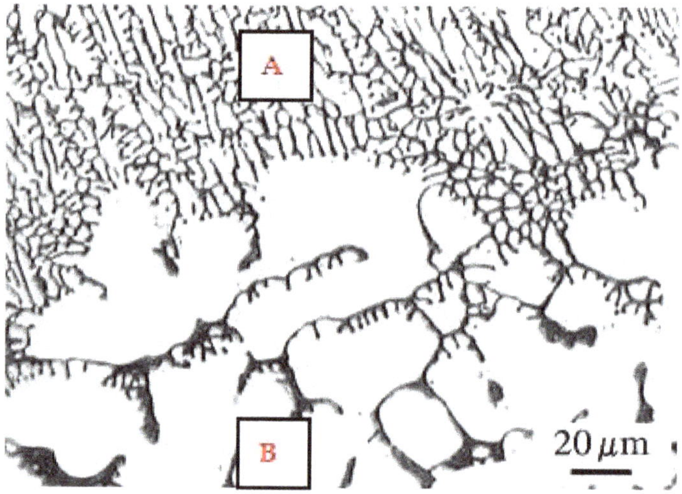

Figure 10. Microstructure of the AK 51 alloy in the surface melting area (top—area A) and in the native material area (bottom—area B). In the surface melting area, there is eutectic α(Al) + β(Si) and very fine dendrites of the α(Al) phase. Etch with Dix and Keith reagent.

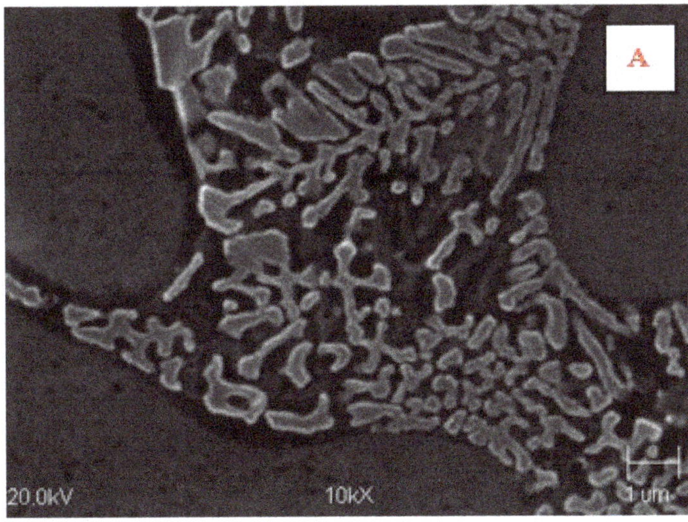

Figure 11. Microstructure of AK 51 alloy surface melting using the TIG method in an argon atmosphere at a welding current of I = 150 A and a scanning speed of vs. = 800 mm/min. A is the area in Figure 10. An eutectic characterised by a distance between lamellae and silica fibres λ_E of less than 1 μm. Etch with Dix and Keith reagent.

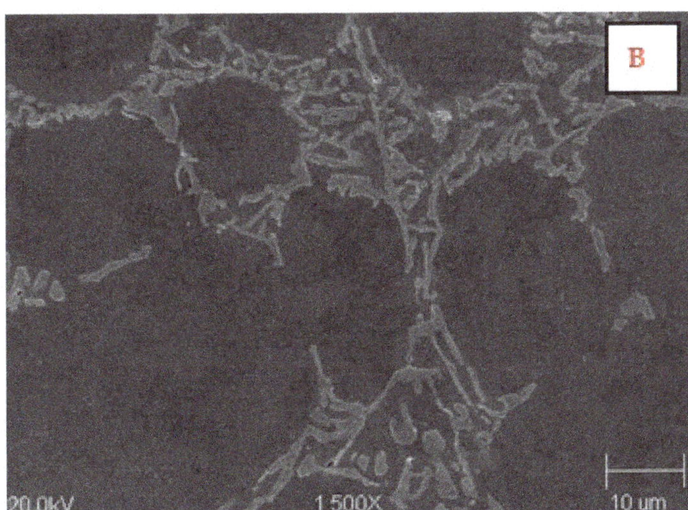

Figure 12. Microstructure of the AK 51 alloy in the native material area. B is the area in Figure 10. Visible lamellar and fibrous eutectics with the value of the parameter λ_E on the order of a few micrometers and the dendrites of α(Al) phase. Etch with Dix and Keith reagent.

In the microstructure of the surface melting area, made on castings of AK 51 alloy, there is eutectic α(Al) + β(Si) and very fine dendrites of the α(Al) phase, compared to the native material.

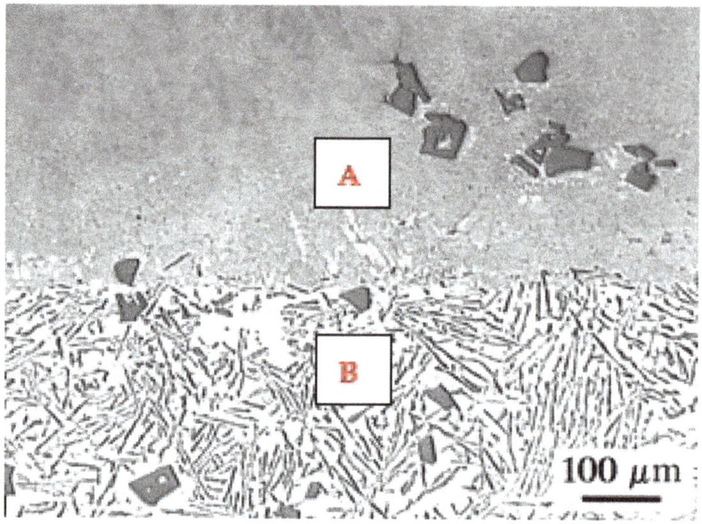

Figure 13. Microstructure of the AK 20 alloy in the surface melting area (top—area A) and in the native material area (bottom—area B). In the surface melting area, there is a eutectic, undissolved primary silicon crystals and very fine dendrites of the α(Al) phase at the super-melting boundary and around the primary silicon crystals. Etch with Dix and Keith reagent.

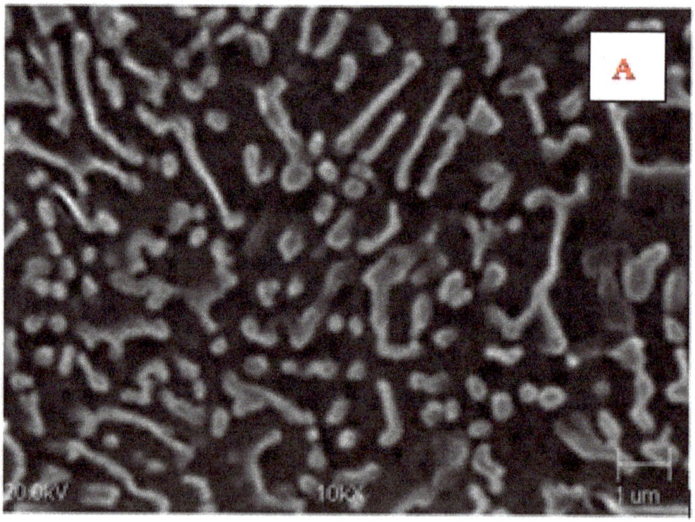

Figure 14. Microstructure of AK 20 alloy surface melting using the TIG method in an argon atmosphere at a welding current of I = 150 A and a scanning speed of vs. = 800 mm/min. A is the area in Figure 10. An eutectic characterised by a distance between lamellae and silica fibres λ_E of less than 1 μm. Etch with Dix and Keith reagent.

Compared to the substrate material, the value of the structural parameter λ_E of the α + β eutectic is several times smaller. The λ_E parameter is very sensitive to the temperature gradient in the liquid metal pool [34,35]. The different values of this parameter in the case of the base material, which was a die casting, and in the case of surface melting are indicative of the different values of the temperature gradients in the liquid melt. Therefore, for surface melting differing in geometry, different values of temperature gradients are expected to be present, resulting in an eutectic with different values of the structural parameter λ_E during the crystallisation process. On the other hand, it is known that the geometry of surface melting areas depends on the flows of liquid metal in the welding pool as a result of electromagnetic interactions (Lorentz forces) and as a result of Marangoni convection [36].

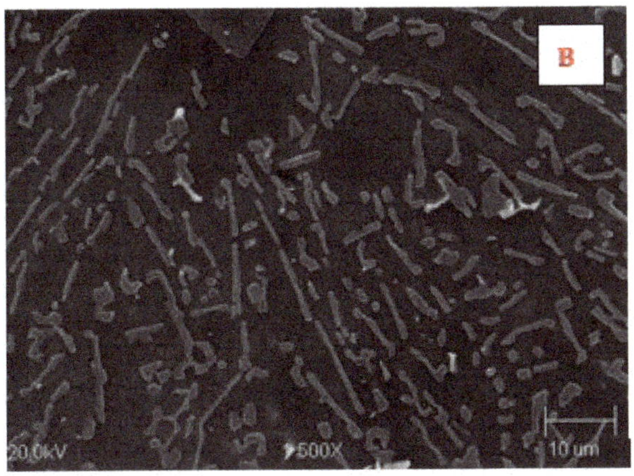

Figure 15. Microstructure of the AK 20 alloy in the native material area. B is the area in Figure 10. A visible lamellar and fibrous eutectic with an λ_E-value of a few micrometres and a primary silicon crystal (top). Etch with Dix and Keith reagent.

Aluminium–silicon alloys with a structure containing thick silicon lamellae have a low level of mechanical properties. These properties improve as the shape of the Si precipitates changes from lamellar to fibrous. This can be achieved as a result of rapid crystallisation.

4. Conclusions

The results obtained indicate the suitability of the newly developed calorimeter for testing the arc efficiency of the welding process. Correct data on the arc and melting efficiency values allow the necessary amount of heat to be designed that needs to be introduced into the joint in order to achieve the assumed weld geometry and its microstructure, as well as the width of the HAZ and its microstructure.

The effect of the surface melting of Al-Si alloy castings using concentrated heat flux is the rapid crystallisation of the superficially melted area. On the other hand, the effect of rapid crystallisation is a highly fragmented microstructure (low value of the λ_E parameter). The consequence will be an increase in performance.

Variations in amperage and arc scanning speed affect arc efficiency, melting efficiency and the amount of heat introduced into the heated material. This provides the opportunity to control the geometry of surface melting areas. In turn, changes in temperature gradients and the cooling conditions of the liquid metal pool affect the structure of the surface melting area.

The thermal efficiency of the surface melting process decreases as the arc scanning speed increases and the welding current decreases. The highest arc efficiency values are obtained when low arc scanning speeds and high welding current values are used.

The effect of the microstructure of the base material on the value of arc efficiency and melting efficiency was found. This is related to the amount of heat introduced into the heated material, which affects the geometry of the surface melting areas. In turn, the amount of heat introduced into the heated material depends on the presence in the microstructure of varying volume proportions of its individual components and their morphology, which differ in the value of the thermal conductivity coefficient and the value of the temperature and heat of fusion.

Author Contributions: Conceptualization, M.M. and A.W.O.; Data curation, M.T.; Formal analysis, M.M. and A.W.O.; Investigation, M.M., A.T., M.L. and M.T.; Methodology, M.M., A.T. and M.T.; Project administration, A.W.O. and M.L.; Resources, M.L.; Supervision, A.W.O. and A.T.; Validation, A.W.O. and A.T.; Writing—original draft, M.L. and M.T.; Writing—review & editing, M.M. and A.W.O. All authors have read and agreed to the published version of the manuscript.

Funding: This research received no external funding.

Institutional Review Board Statement: Not applicable.

Informed Consent Statement: Not applicable.

Data Availability Statement: Not applicable.

Conflicts of Interest: The authors declare no conflict of interest.

References

1. Ferec, K.; Ferenc, J. *Konstrukcje spawane. Połączenia*; Wydanie WNT: Warsaw, Poland, 2003.
2. Klimpel, A. *Technologia spawania i cięcia metali*; Wydawnictwo Politechniki Śląskiej: Gliwice, Poland, 1998.
3. Pilarczyk, J.; Adamiec, P.; Klimpel, A.; Włosiński, W.; Banasik, M.; Brózda, J.; Dębski, E.; Dobaj, E.; Dziuba, S.; Ferenc, K.; et al. *Poradnik inżyniera*; Wydawnictwo WNT: Warszawa, Poland, 2021.
4. Tasak, E. *Metalurgia spawania*; Wydawnictwo JAK: Kraków, Poland, 2008.
5. Ferenc, K. *Spawalnictwo*; Wydawnictwo WNT: Warszawa, Poland, 2007.
6. Orłowicz, A.W.; Mróz, M. Sposób zmniejszania udziału ferrytu Widmanstättena w złączu spawanym elementów konstrukcyjnych wykonanych ze stali niskowęglowej. International Patent PL 237300, 11 September 2019.
7. Orłowicz, A.W.; Mróz, M. Sposób zmniejszania udziału martenzytu oraz ferrytu Widmanstättena w dwuimiennym złączu spawanych elementów konstrukcyjnych wykonanych ze stali niskowęglowej. International Patent PL 236933, 11 September 2019.
8. Giedt, W.H.; Tallerico, L.N.; Fuerschbach, P.W. GTA Welding efficiency: Calorimetric and temperature field measurements. *Weld. J. Res. Suppl.* **1989**, *68*, 28–32.
9. Du Pont, J.N.; Marder, A.R. Thermal efficiency of arc welding processes. *Weld. Res. Suppl.* **1995**, *74*, 406–416.
10. Glikstein, S.S.; Friedman, E. Temperature transients in gas tungsten arc weldments. *Weld. Rev.* **1983**, *62*, 72–75.
11. Kou, S.; Sun, D.K. Three-dimensional heat flow and solidification during the autogenous GTA welding of aluminium plates. *Metall. Trans. A* **1983**, *14*, 2245–2253. [CrossRef]
12. Kou, S.; Sun, D.K. Fluid flow and weld penetration in stationary arc welds. *Metall. Trans. A* **1985**, *16*, 203–213. [CrossRef]
13. Han, Y.; Chen, J.; Ma, H.; Zhao, X.; Wu, C.; Gao, J. Numerical simulation of arc and droplet behaviors in TIG-MIG hybrid welding. *Materials* **2020**, *13*, 4520. [CrossRef]
14. Karganroudi, S.S.; Maradi, M.; Attar, M.A.; Rasouli, S.A.; Ghoreishi, M.; Lawrence, J.; Ibrahim, H. Experimental and numerical analysis on TIG arc welding of stainless steel using RSM Approch. *Materials* **2021**, *11*, 1659. [CrossRef]
15. Okada, A. Applications of melting efficiency and its problem. *J. Jpn. Weld. Soc.* **1977**, *46*, 53–61. [CrossRef]
16. Glikstein, S.S.; Friedman, E. Weld modeling applications. *Weld. J.* **1984**, *63*, 38–42.
17. Glikstein, S.S.; Friedman, E. An investigation of thermal response of stacionary gas tungsten arc welds. *Weld. J.* **1976**, *55*, 408–420.
18. Havalda, A. *Thermal Processes in Electric Welding*; Wydawnictwo WNT: Warszawa, Poland, 1963.
19. Krzyżanowski, M. *Iron Alloy Surface Overwork with the Use of Plasma Heating*; Wydawnictwo AGH: Cracow, Poland, 1995; pp. 287–297.
20. Smartt, H.B.; Stewart, J.A.; Einerson, C.J. Heat transfer in gas tungsten arc welding. In Proceedings of the ASM metals congress/TMS-AIME fall meeting, Toronto, ON, Canada, 13 October 1985; Series nr 8511–011. pp. 1–14.
21. Knorowski, G.A.; Fuerschbach, P.W. Calorimetry of Pulsed Versus Continous Gas Tungsten Arc Welding. In Proceedings of the ASM International Trends in Welding Research Meeting, Gatlinburg, TN, USA, 18 May 1986; pp. 393–400.
22. Fuerschbach, P.W.; Knorovsky, G.A. A study of melting efficiency in plasma arc and gas tungsten arc welding. *Weld. J. Res. Suppl.* **1991**, *70*, 287–297.
23. Orłowicz, A.W.; Mróz, M.F.; Trytek, A.; Tupaj, M.; Betlej, J. A Flow Calorimeter for Thermal Measurements on Welding Processes. Polish Patent No. 211283, 11 May 2012.
24. Grączewski, A. Wytrzymałość i Trwałość Zmęczeniowa Złączy Spawanych Wykonanych Ze Stopu EN AW-5754. Ph.D Thesis, Bydgoszcz University of Technology, Bydgoszcz, Poland, 2022.
25. Braszczyńska-Malik, K.; Mróz, M. Gas-tungsten arc welding of AZ 91 magnesium alloy. *J. Alloy. Compd.* **2011**, *509*, 9951–9958. [CrossRef]
26. Orłowicz, A.W.; Mróz, M. Structure and operational characteristics of casting of LM25 alloy with surface layer refined by rapid solidification. *Int. J. Cast Met.* **2002**, *15*, 85–93. [CrossRef]
27. Orłowicz, A.W.; Mróz, M. Study on susceptibility of AL-Si alloy castings to surface refinement with TIG arc. *Z. Mtallkunde* **2004**, *12*, 1391–1397. [CrossRef]
28. Mróz, M. Orłowicz, A.W.; Tupaj, M. Współczynniki termiczne procesu GTAW. *Arch. Foundry Eng.* **2005**, *5*, 217–224.

29. Orłowicz, A.W.; Trytek, A.; Malik, J. *Formation of Microstructure and Service Properties of Cast Iron Castings by Surface Improvement with Electric arc Plasme*; Rzeszow University of Technology: Rzeszów, Poland, 2013.
30. Orłowicz, A.W.; Trytek, A. Susceptibility of iron castings to heat absorption from an electric arc and to hardened-layer shaping. *Metall. Trans. A* **2003**, *34A*, 2973–2984. [CrossRef]
31. Orłowicz, A.W.; Trytek, A. Structural and operating aspects of rapid solidification of surface layer of spheroidal graphite cast iron. *Int. J. Cast Met.* **2003**, *16*, 496–505. [CrossRef]
32. Opiekun, Z.; Orłowicz, A.W. Kształtowanie struktury warstwy wierzchniej odlewów ze stopów kobaltu plazmą łuku elektrycznego. *Bull. Weld. Inst.* **2004**, *5*, 43–49.
33. Orłowicz, A.W.; Trytek, A. Nadtapianie powierzchniowe odlewów z superstopu na bazie kobaltu MAR-M509. *Arch. Foundry Eng.* **2004**, *4*, 348–355.
34. Hogan, L.M.; Song, H. Interparticle spacings and undecooling in Al-Si eutectic microstructure. *Metall. Trans. A* **1987**, *18*, 707–713. [CrossRef]
35. Cáceres, C.H.; Wang, Q.G. Solidifications conditions, heat treatment and tensile ductility of Al-Si-0.4Mg casting alloys. *AFS Trans.* **1996**, *104*, 1039–1043.
36. Oroper, G.M.; Eagar, T.W.; Szekely, J. Convection in arc weld pools. *Weld. J.* **1983**, *11*, 307–312.

materials

MDPI

Article

Problems of HLAW Hybrid Welding of S1300QL Steel

Bogdan Kupiec [1], Michał Urbańczyk [2], Magdalena Radoń [1] and Marek Mróz [1,*]

1 The Department of Casting and Welding, Faculty of Mechanical Engineering and Aeronautics, Rzeszow University of Technology, Al. Powstancow Warszawy 12, 35-959 Rzeszow, Poland
2 Lukasiewicz Resarch Network—Institute of Welding, Błogoslawionego Czeslawa 16-18, 44-100 Gliwice, Poland
* Correspondence: mfmroz@prz.edu.pl

Abstract: This paper presents the results of tests on the fabrication of welded joints in S1300QL steel according to the requirements of ISO 15614-14 and ISO 12932. The butt-welded joint without bevel was made from 350 × 150 × 8 mm sheets. The welding process was carried out at the hybrid welding (laser–MAG) station. MAG means metal active gas. The test welded joints were subjected to non-destructive and destructive testing. Visual and radiographic examinations were carried out. The distribution of HV10 hardness was determined in the weld, the heat-affected zone, and the base material. The microstructure of these areas was also analysed for the presence of hard and brittle hardening products and non-metallic inclusions. Tensile strength and yield strength, as well as bending strength, were assessed in the mechanical property tests. The impact test was performed in accordance with ISO 9016.

Keywords: hybrid welding; S1300QL steel; microstructure; mechanical properties

Citation: Kupiec, B.; Urbańczyk, M.; Radoń, M.; Mróz, M. Problems of HLAW Hybrid Welding of S1300QL Steel. *Materials* **2022**, *15*, 5756. https://doi.org/10.3390/ma15165756

Academic Editor: Daolun Chen

Received: 25 July 2022
Accepted: 18 August 2022
Published: 20 August 2022

1. Introduction

Increasing demands on welded structures for high durability and reliability, as well as operational safety, are forcing the use of modern construction materials with improved strength parameters.

Modern structural materials should be characterised by high yield strength, good ductility and low brittleness, good weldability, and good impact strength, while maintaining a relatively low price. These materials can be used for heavily loaded machine parts and welded structures, such as cranes, gantries, bridges, tanks, mining machinery, and equipment [1–3].

Modern structural materials include structural steel with improved yield strength in the tempered state, classified in EN 10025-6 [4]. In the cases of S960QL and S1100QL steel, which haves minimum yield strengths of 960 and 1100 MPa, respectively, the fine-grained microstructures consist of martensite or martensite and bainite, i.e., the microstructural components obtained as a result of the tempering treatment.

Further development of the material group of higher-yield-strength steel has led to the development of ultra-high-strength materials with yield strengths of up to 1300 MPa, for example steel S1300QL [5], and even up to 1700 MPa [6]. These steels belong to the group of UHSS steels (ultra-high-strength steels) [7]. The Swedish company SSAB, which produces HARDOX, WELDOX, ARMOX, or TOOLOX steel, as well as TyssenKrupp [8–11], have become pioneers among the world's producers of this type of material.

The occurrence of hard and brittle structural components (hardening products) in high-yield-strength steel is the reason for the limited weldability of these materials, especially in terms of microstructural changes occurring in the heat-affected zone (HAZ) [12,13]. During the welding process, the action of the heat source results not only in the formation of a pool of liquid metal with subsequent solidification of the weld area, but also in changes to the microstructure of the HAZ as a result of phase transformations in the solid state. The austenitisation of the base material, which takes place during the welding process, results in the growth of austenite grains and the occurrence of under-hardened areas, which can lead to the nucleation and propagation of cracks. In the dimension of mechanical properties, this results in a reduction in ductility and impact strength, as shown in the work [14,15].

Research into the development of welding technology for higher-yield-strength steel seeks to minimise the HAZ by modelling the welding process through the selection of appropriate process parameters. The tangible result of these measures should be a welded joint with high plastic properties. For example, the authors of [16] show that the use of a gas mixture of 98% Ar + O_2 results in a welded joint of S690QL and S960MC steel sheets with a yield strength of 435 MPa (R_e) and good strength properties (R_m) of 717 MPa.

Recently, hybrid laser-arc welding (HLAW) has attracted considerable interest [17–19]. A schematic diagram of the HLAW method is shown in Figure 1 [20].

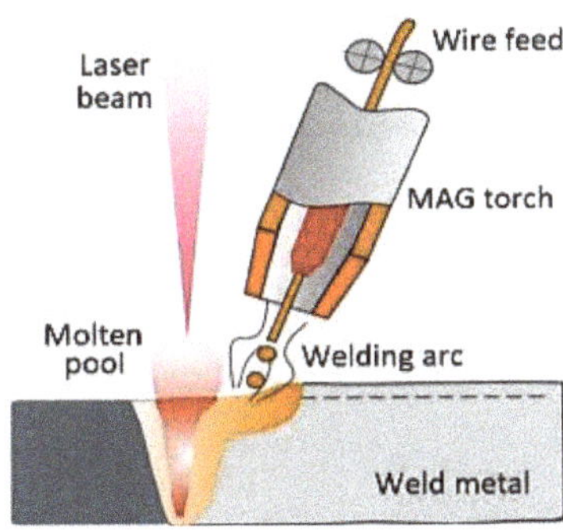

Figure 1. Schematic diagram of HLAW [19].

Urbańczyk M. et al. [20] analysed the mechanical properties and microstructure of a welded joint made of S960QL steel using the HLAW method. Turichin G. et al. [21] analysed the effect of the chemical composition of the binder on the microstructure and mechanical properties of welded joints of ARMOX, WELDOX, and HARDOX steel fabricated using the HLAW method. Mazar Atabaki M. et al. [22] investigated the feasibility of using HLAW to join advanced high-strength steel (AHSS) components with yield strength Re = 1034 MPa. The authors of the paper [23] indicate that the HLAW method can be successfully used to produce welded joints in S960QL steel in terms of meeting the requirements of EN ISO 15614-14. The author of the paper [24] presents the possibility of fabricating welded joints from steel grades of different thickness values by hybrid welding using a 32 kW laser.

It is now becoming a major challenge to develop correct welding technology for steel with a yield strength exceeding 1300 MPa (steel grade S1300QL). Many authors have tried to apply methods such as HPAW (hybrid plasma-arc welding) [25], TIG, A-TiG and MAG [26], or Laser [10,11] to weld parts made of this steel.

The aim of the study was to assess the feasibility of welded joints in S1300QL steel using the hybrid welding method (HLAW) in terms of meeting the requirements of ISO 15614-14 and ISO 12932.

2. Materials and Methods

In this study, 8 mm-thick sheets of high-strength tempered S1300QL steel were selected for testing. The chemical composition and mechanical properties according to EN 10025-6, the manufacturer's certificate, and that determined by control chemical analysis are given in Table 1. The control analysis of the chemical composition of the steel was carried out using Bruker Q4 TASMAN spark spectrometer (AXS, Kalkar, Germany).

Table 1. Chemical composition and mechanical properties of steel grade S1300QL.

Chemical Composition, (%)										
	C	Si	Mn	Cr	Ti	Ni	Mo	Nb	V	Cu
Standard	0.25	0.50	1.40	0.80	0.02	2.0	0.70	0.04	0.08	0.10
Certificate	0.24	0.22	0.89	0.49	0.004	1.3	0.392	0.021	0.02	0.01
Control test	0.20	0.212	0.87	0.50	0.002	1.23	0.391	0.021	0.019	0.01
Mechanical properties										
R_m [MPa]				R_e [MPa]					A_5 [%]	
1400–1700				min. 1300					min. 8	

The supplementary material used during welding was a wire with the trade name BÖHLER Union X96 with a diameter of 1.2 mm (Vienna, Austria). This welding wire made a low-alloyed solid wire of high-strength, fine-grained steel. The chemical composition of the wire according to the certificate and control analysis is shown in Table 2.

Table 2. Chemical composition of the Union X96 welding wire.

	C	Si	Mn	Cr	Mo	Ni
Certificate	0.10	0.80	1.80	0.35	0.60	2.30
Control test	0.11	0.78	1.93	0.43	0.55	2.31

Based on the chemical inventory of the EN test, it is equivalent to CEV ISO 1011-2 Formula (1). This pattern is also provided by the MIS (International Welding Institute, Paris, France).

$$CEV = C + \frac{Mn}{6} + \frac{Cr + Mo + V}{5} + \frac{Ni + Cu}{15}\% \quad (1)$$

From the point of view of carbon content, carbon steels can be divided into: well weldable (maximum 0.25% C), sufficiently weldable (0.25–0.35% C), with limited weldability (0.35–0.45% C), and badly weldable (>0.45% C). The calculated equivalent was ≈ 0.44%, which means that this steel is characterised by limited weldability, which was also demonstrated by the authors of research [27–30].

The test joints were made from sheets of S1300QL steel measuring 350 × 150 × 8 mm. The sheets were not bevelled ("I" butt joint, no clearance/gap). Once the sheets were cleaned, tack welds were made, and the welding process began. The welding process was carried out after preheating to 200 °C. According to [24], this temperature is defined as the maximum interpass temperature when welding S1300QL steel.

The welding process was carried out using a robotised Laser–MAG hybrid welding station, which is owned by the Łukasiewicz Research Network—Welding Institute (Figure 2). The station consists of a 12 kW TruDisk 12,002 disk laser (TRUMPF, Stuttgart, Germany), a KRC30HA welding robot (KUKA, Augsburg, Germany) equipped with a hybrid welding head and a PHOENIX 452 RC PULS MIG/MAG welding machine (Shanghai, China), which allows welding with a maximum current of 450A (EWM Hightec Welding GmbH, Mündersbach, Germany).

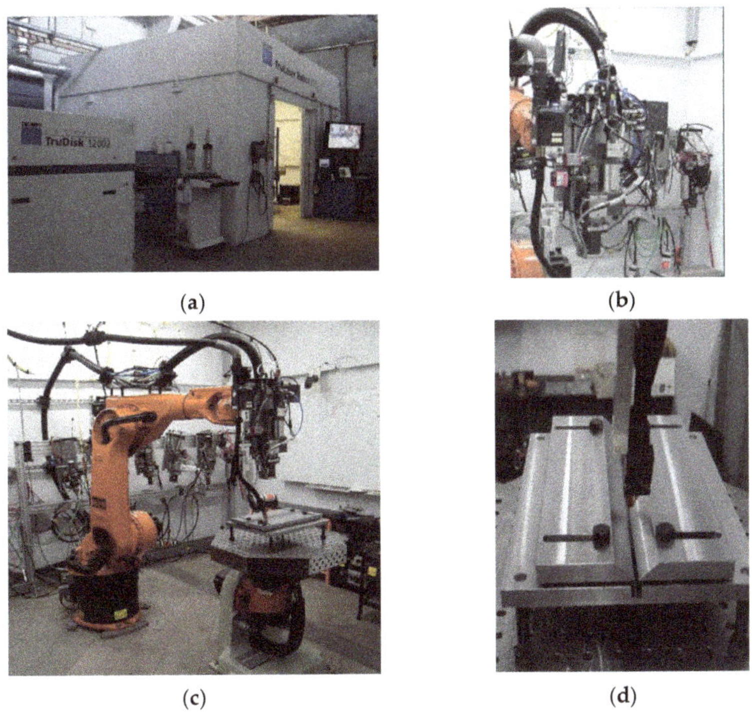

Figure 2. Robotic laser welding station (TruLaser Robot 5120) with a TruDisk 12,002 disk laser and a D70 hybrid welding head (Trumpf): (**a**) view of the booth and laser source, (**b**) hybrid head, (**c**) view of the robot with welding table, (**d**) how the welding sample is fixed.

The welding process was carried out in the flat position (PA) (Figure 3). The welding process parameters are shown in Table 3.

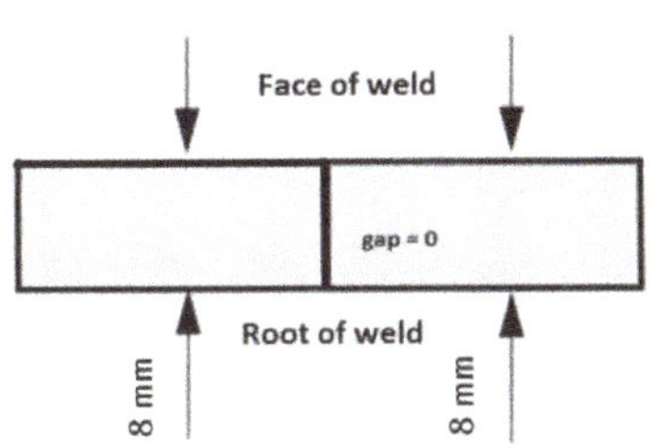

Figure 3. Position of the plates in the flat position (PA) according to the positions of the ISO standard.

Table 3. Parameters of hybrid welding of S1300QL steel.

P, kW	V_s, m/min	V_d, m/min	I, A	U, V	kor. U, V	Q, kJ/mm	T_p, °C
4.5	1.2	8.5	275	27	3	0.59	200

P—laser power, V_s—welding speed, V_d—wire feed speed, I—current, U—voltage, cor. U, V—correction, Q—heat input, T_p—sheet pre-heating temperature.

The S1300QL steel welded joints were subjected to non-destructive and destructive testing. Non-destructive testing is visual and radiographic. Visual testing was carried out in accordance with the requirements of ISO 17637. The macroscopic metallographic examination was performed using a Zeiss Neophot 2 microscope (Zeiss, Jena, Germany). To determine the macrostructure, the specimens were etched with Adler's reagent (Chmes, Poznań, Poland). Radiographic testing of the welded joints was carried out in accordance with the requirements of ISO 17636-1. An Eresco 65 MF3 X-ray machine (GE Sensing&Inspection Technologies; Ahrensburg, Germany) was used.

Metallographic tests to analyse the microstructure were carried out in accordance with the requirements of ISO 17639 on metallographic specimens taken from sample welded joints. In the preparation of the metallographic specimens, the samples were ground with 800- and 1000-grit sandpaper, polished using an ATM Saphir 320E polisher with an ATM Rubin 500 head (Mammelzen, Germany) and Struers' M-type monocrystalline diamond suspension (3 μm), and etched in 5% Nital (5% HNO3 in ethanol). The metallographic specimens were observed using a Zeiss Neophot 2 light microscope (Zeiss Jena, Germany) and a scanning electron microscope (SEM- VEGA 3 TESCAN, Brno, Czech Republic).

Hardness distribution tests were carried out in accordance with the requirements of ISO 9015-1 using a GNEHM DIGITAL BRICKERS 220 hardness tester. Vickers hardness (HV10) measurements were carried out along two measuring lines 2 mm from the top and bottom edges of the specimen. Impressions were made in the base material, the heat-affected zone (HAZ), and the weld.

A static tensile test was carried out in accordance with the requirements of ISO 6892-1 using 300 mm × 25 mm × 5.0 mm specimens. Excessive root and face height was rectified during the sample preparation process. The tests were carried out on an MTS 810 TEST SYSTEMS testing machine (Eden Prairie, MN, USA).

The weld-face tensile bend test (FBB) and weld-root tensile test (RBB) were performed according to the requirements of ISO 5173. The tests included four specimens—two specimens on each side. The tests were conducted on specimens measuring 300 mm × 20 mm × 5.0 mm. The tests were carried out on an LOS12126 testing machine (Losenhausenwerk AG; Düsseldorf, Germany).

Impact tests were performed in accordance with ISO 9016. Two sets of specimens were used in the tests. The first set was taken from the weld area and the second set from the heat-affected zone. There were three specimens in each set. The thickness of the test joint sheets made it impossible to take impact test specimens of standard dimensions (10 × 10 × 55mm), so specimens with reduced cross sections (2.5 × 8.0 × 55mm) were taken. V-notched specimens were used with a notch depth of 2 mm. The specimens were cooled to −40 °C before testing. The cooling process was carried out using an FP89 cooling circulator (Julabo, Stamford, UK). Impact tests were carried out using an RKP 300 impact machine (Amsler, Colombo, Sri Lanka).

3. Results and Discussion

3.1. Visual and Macroscopic Examination

Visual examination showed that the welded joint was characterised by a smooth face without splatter and a correctly executed root. Non-conformity in No. 502 was observed, i.e., there was excess weld metal according to EN ISO 6520-1. The height of the weld face was 2.3 mm. The completed welded joint based on ISO 12932 met the relevant criteria and represented quality level B, while the excess weld metal met the requirements of quality level D.

The macrostructure and dimensions of the welded joint in S1300QL steel are shown in Figure 4.

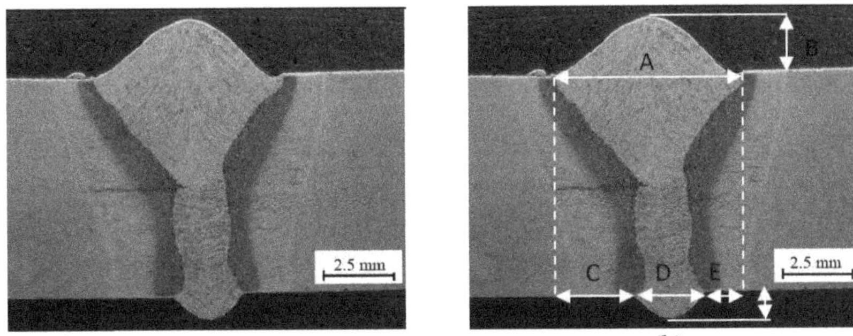

(**a**) (**b**)

Figure 4. Joint in S1300QL steel: (**a**) macrostructure, (**b**) weld dimensions: A—width of the weld face, D—width of the weld root, C, E—shift of the weld root relative to the weld face, F—height of the weld root.

The metallographic macroscopic examination revealed misalignment of the weld. A slight offset was found at the top of the weld (weld face), visible in Figure 4b, which may be due to non-axial feeding of the electrode wire during welding. The width of the weld face (A) of the joint made in the PA position was 7 mm, while its face height (B) was 2.3 mm. The width of the weld root (D) is 2.5 mm, and its height (F) is 0.8 mm. A photo of the macrostructure revealed a dendritic weld and a clear heat-affected zone. No cracks were observed in this joint.

After visual and macroscopic examinations, the analysed welded joint was found to comply with quality level B, while the excess weld metal met the requirements of quality level D according to the requirements of ISO 12932 for the qualification of the hybrid welding procedure.

3.2. *Radiographic Examinations*

Examples of radiographic results are shown in Figure 5. Analysis of the radiographs, taken along the entire length of the welded joint, makes it possible to conclude that there are no internal defects in the area.

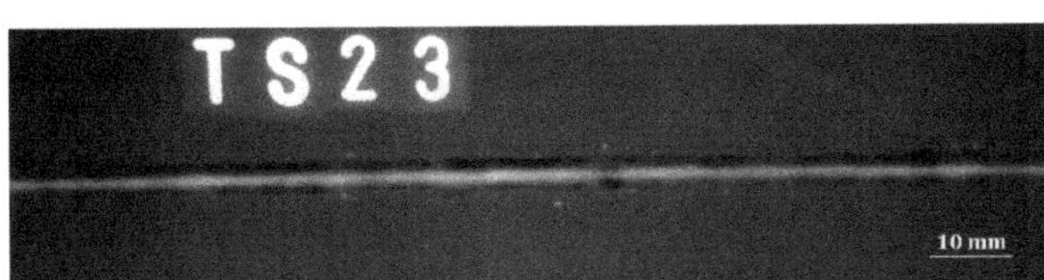

Figure 5. Photo of a test of a hybrid welded joint in S1300QL steel.

3.3. *Hardness Distribution*

The results of the hardness distribution in the different zones of the welded joint are shown in Figure 6. A maximum hardness of 458 HV10 was observed in the HAZ (point 11 of measuring line A). The hardness in the weld axis was 344 HV10. The hardness at the base material–HAZ transition line (point 12 of measuring line A) was 373 HV10. The hardness of the base material ranged from 436 ÷ 455 HV10. The average results of the hardness test are presented in Table 4.

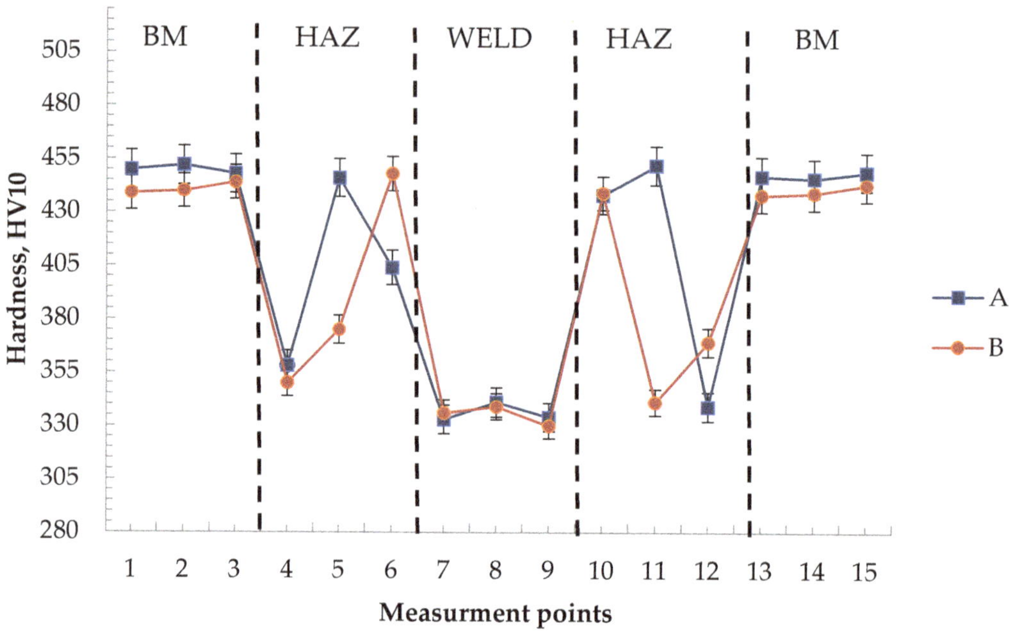

Figure 6. Hardness distribution in the tested welded joint—average result.

Table 4. Average hardness test results.

	BM			HAZ			WELD			HAZ			BM		
	1	2	3	4	5	6	7	8	9	10	11	12	13	14	15
A	450	452	448	358	446	404	333	341	334	438	452	339	447	446	449
B	439	440	444	350	375	448	336	339	330	439	341	369	438	439	443

According to the requirements of PN-EN ISO 15614-14 [31], the maximum permissible joint hardness for the material group according to ISO/TR 15608 [32] is 450 HV10, with special values to be agreed with the customer for Re > 890 MPa steel.

When qualifying the welding technology of S1300Ql steel, it is proposed to take the maximum hardness of the base material as a criterion for the allowable hardness value of this joint. The maximum HV10 hardness of the base material is 452 HV10. By analysing the hardness distributions in the weld and in the heat-affected zone, it was found that this criterion is fulfilled.

3.4. Welded Joint Microstructure

Figure 7 shows the microstructure of the base material (BM), the HAZ, and the weld joint of S1300QL steel.

The base material is characterised by a fine-grained martensitic and martensitic–bainitic structure, with dispersive carbide and carbide–nitride precipitates, as shown in Figure 7c, occurring at the borders of the martensite strips and in their interior. Bainite mostly occurs within grain boundaries. Similar conclusions were reached by the authors [2].

The HAZ area is characterised by a microstructure with fine martensite precipitates and an area heated in the Ac1 ÷ Ac3 range (partial recrystallization area), in which bright martensite fields are observed against the dark etching structure of highly moderated martensite.

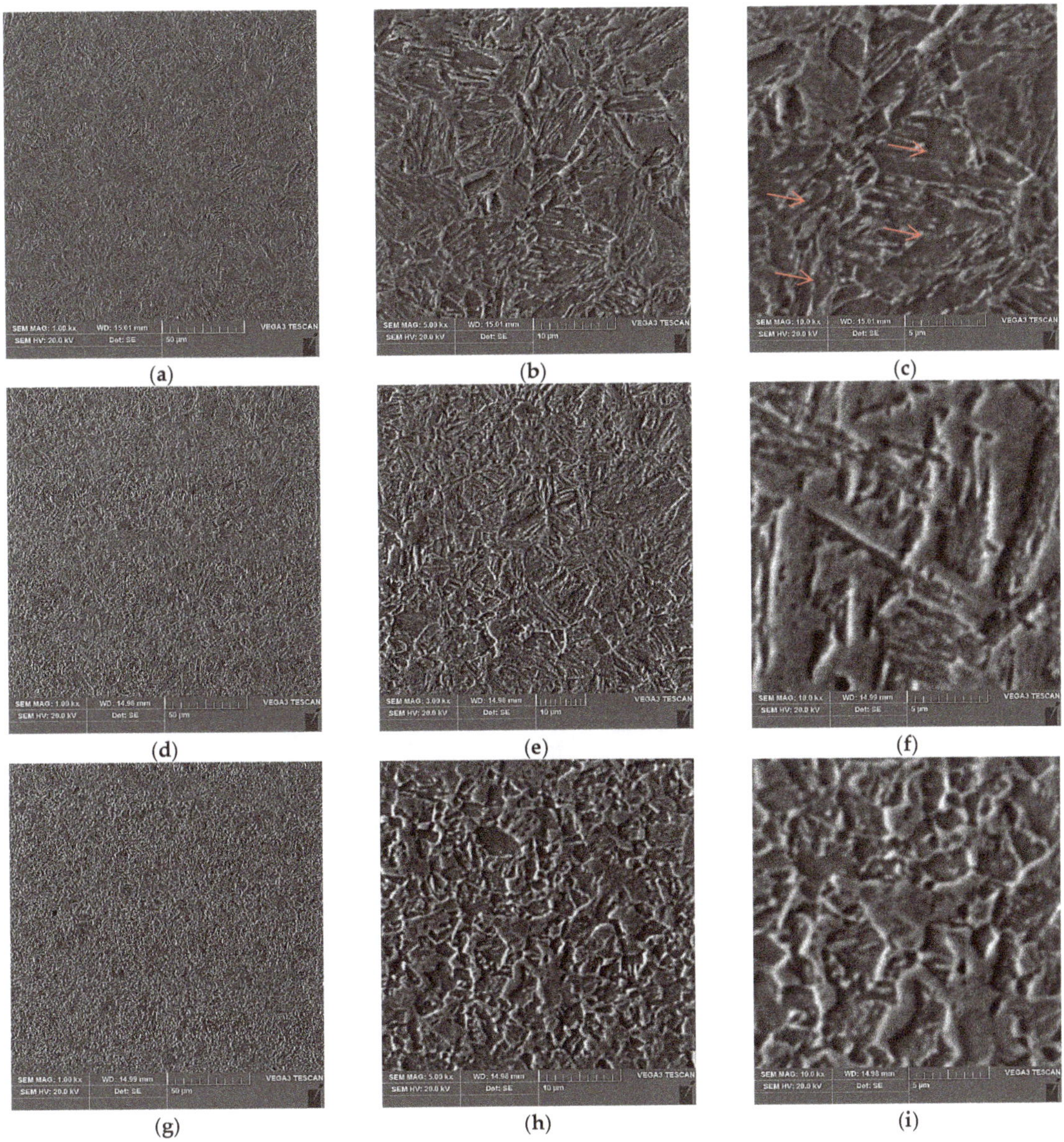

Figure 7. Microstructure of the welded joint in the following characteristic areas: (**a**–**c**) base material; (**d**–**f**) weld; (**g**–**i**) HAZ. The red arrows indicate an example dispersive carbide and carbide–nitride precipitates.

A martensitic structure is observed throughout the weld. From the middle of the weld to the face, the columnar crystals are arranged from the fusion line to the face, as shown in Figure 4a. This is due to the large width of the joint and slower cooling compared to the other half of the joint. On the other hand, the lower half of the weld (to the root), in Figure 4a, is characterised by columnar crystals arranged from the fusion line to the weld axis. This is due to the small width of the joint and the rapid dissipation of heat to the base material. Due to the high purity of the steel (low S and P content), no impurity band is observed in the area where the crystals meet.

3.5. Static Tensile Test

The results of the static tensile test of the samples taken from the S1300QL steel joint are shown in Table 5. The basis for accepting the conformity of a welded joint is the achievement of the required minimum tensile strength of the base material, Rm ≥ 1400 MPa. In the case of test joints R1 and R2 analysed, tensile strength values of 171 and 114 MPa less were obtained, respectively, which is the basis for concluding that the analysed joint does not comply with the tensile strength requirements.

Table 5. Static tensile test results.

No.	Determination of Sample	Tensile Strength R_m, MPa	Comments
1	R1	1229	rupture in HAZ
2	R2	1286	

A view of the specimens after tensile testing is shown in Figure 8. Rupture of the specimens occurred in the heat-affected zone.

Figure 8. View of the specimens after tensile testing.

The observational results of the fracture surfaces of the specimens after the tensile tests are shown in Figure 9.

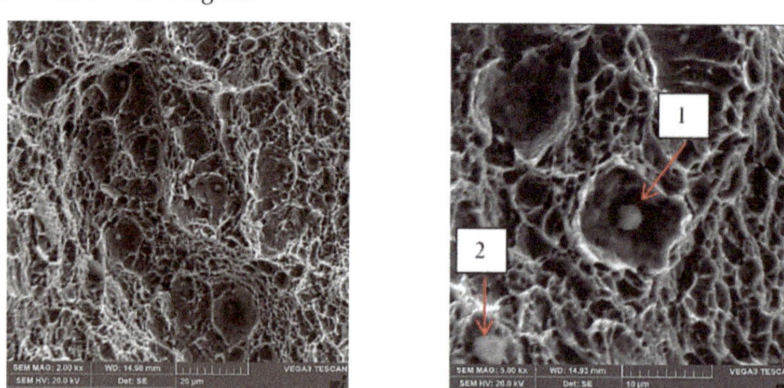

(**a**) (**b**)

Figure 9. SEM image of the fracture surface of specimen R1 after tensile testing: (**a**) magnification 2.00 kx; (**b**) points 1 and 2 mark the places where a quantitative point X-ray microanalysis was performed, magnification 5.00 kx.

The fracture surface was found to have the characteristic features found in ductile fracture. On the surface of the fracture, there are areas with spherical particles of intermetallic phases Figure 9a,b. Similar precipitates in the fracture were also found by the authors of the paper [1], who analysed the fractures of welded joint specimens in Weldox 1100 steel. For phase identification, spot quantitative X-ray microanalysis was performed according to the designation given for Figure 9b. The results of this analysis are shown in Table 6.

Table 6. Chemical composition of particles.

Spot	C	O	Al	Ca	Fe	Si	Balance
1	11.37	38.07	18.38	15.15	13.70	2.63	0.70
2	11.81	38.41	17.57	15.63	12.76	2.99	0.83

The presence of calcium and aluminium in the spherical particles of the intermetallic phase, according to the results of [1], suggests that the spherical particles are calcium aluminate (CaO-Al2O3). These particles have a detrimental effect on the fatigue strength of the material.

3.6. Bending Test

The results of the bending test on the S1300QL steel joints are shown in Table 7.

Table 7. The results of the bending test on the S1300QL steel joints.

No.	Designation of Sample	Angle of Bend, °	Comments
1	FBB/I	35	Brittle cracking in HAZ
2	FBB/II	70	
3	RBB/I	65	
4	RBB/II	70	

FBB—face bending test, RBB—root bending test.

According to the criterion adopted, the value of the maximum bending angle should be 90° in the absence of visible signs of specimen fracture. In the case of three specimens, the angle value exceeded 60°. The bending angle for the FBB/II and RBB/II samples was 70°, while for the RBB/I sample it was 65°. Observation of the specimen surface after bending revealed the presence of brittle cracks in the heat-affected zone. If an angle value of less than 90° is achieved and brittle cracks are present, the basic criterion for the bending test is not met. Figure 10 shows all 4 samples after the bending test.

Figure 10. View of specimens after bending.

3.7. Impact Testing

The results of the impact tests at −40 °C are shown in Table 8.

Table 8. Results of impact testing at −40 °C.

No.	Designation	Notch Type	Breaking Work, J	Impact Strength J/cm^2	Fracture Characteristics
1	VWT/I	Charpy V	17	42.5	Mixed fracture
2	VWT/II		18	45.0	
3	VWT/III		15	37.5	
4	VHT/I		18	45.0	
5	VHT/II		24	60.0	
6	VHT/III		20	50.0	

VWT—impact strength in the weld, VHT—impact strength in the HAZ.

The results show that the criterion that the breaking work should be ≥18 J was only met for specimens 4, 5, and 6 where the impact strength of the heat-affected zone was assessed. For the specimens where the notch was in the weld area material, the breaking work only for specimen 2 was within the acceptable range. It must therefore be concluded that the impact test results are not satisfactory and do not meet the requirements of ISO 15614-14. This is also confirmed by observations of the fractures of the specimens, the results of which are shown in Figure 11.

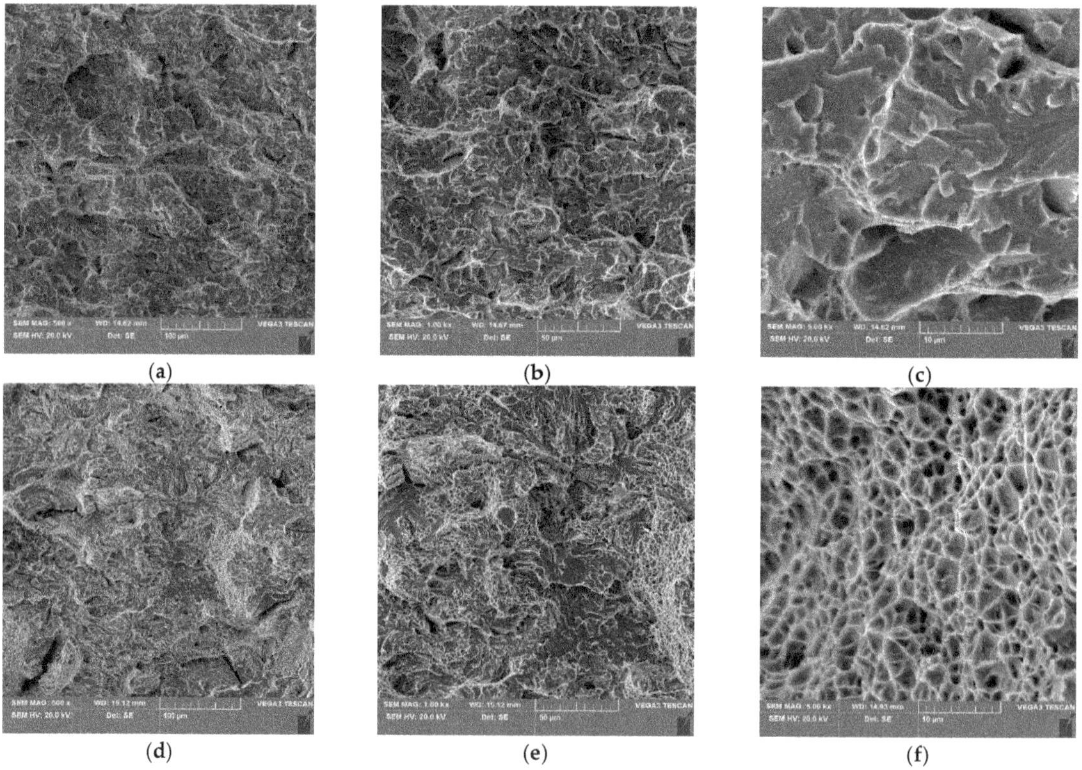

Figure 11. SEM image of the fracture surface of the specimens after impact testing; **(a–c)** VWT specimens, **(d–f)** VHT specimens.

On the fracture surface of the specimens with a notch in the weld (VWT specimens), the interaction between brittle and ductile cracking mechanisms (mixed fracture) was found, with a dominant effect of brittle cracking.

For notched specimens in the HAZ, ductile cracking is the dominant mechanism for propagation of the main crack.

4. Conclusions

This paper presents the results of qualification studies on butt-welded joints made using a hybrid method (laser + MAG) made of S1300QL grade steel sheets.

- Qualification of the test joints was performed based on non-destructive testing and destructive testing.
- The welded joints were characterised by an even, smooth face with no visible spalling. The root was characterised by its correct formation.
- The visual and radiographic examination did not reveal the presence of surface defects.
- The metallographic examination showed that the weld is characterised by a martensitic structure. Columnar crystals are arranged from the fusion line to the weld axis. The heat-affected zone (HAZ) contains minor precipitates of martensite.
- The static tensile test showed that the produced joints were characterised by high strength but did not meet the accepted requirements. These joints failed to meet the minimum tensile strength requirements of the base material, i.e., Rm $\geq$ 1400 MPa.
- A maximum bending angle of 70° was found in the bending test, which did not meet the requirements stipulating that for S1300QL steel the required bending angle should be 90°.
- The impact tests showed that the condition for the required minimum value of breaking work was met for notched specimens in the HAZ. In the case of the specimens with a notch cut into the weld, the minimum breaking work was not achieved, which was the basis for the unsatisfactory impact test results.
- The analysis of the test results shows that obtaining a high-quality joint using S1300QL steel that meets the requirements of the standards is difficult and requires further research. The problem of achieving a proper welded joint is the formation of hard and brittle hardening products during welding, as well as the presence of spherical calcium aluminate particles ($CaO\text{-}Al_2O_3$), which have a detrimental effect on the strength properties of the material.

Author Contributions: Conceptualization, B.K. and M.U.; Funding acquisition, M.M.; Investigation, B.K., M.U. and M.R.; Methodology, M.R.; Resources, M.U.; Supervision, M.M.; Writing—original draft, B.K.; Writing—review & editing, M.M. All authors have read and agreed to the published version of the manuscript.

Funding: This research received no external funding.

Institutional Review Board Statement: Not applicable.

Informed Consent Statement: Not applicable.

Data Availability Statement: Not applicable.

Conflicts of Interest: The authors declare no conflict of interest.

References

1. Ozgowicz, W.; Kalinowska-Ozgowicz, E. Investigations on the impact strength of constructional high-strength Weldox steel at lowered temperature. *Arch. Mater. Sci. Eng.* **2008**, *32*, 89–94.
2. Grigorenko, G.M.; Kostin, W.A.; Poznjakow, W.D.; Żdanow, S.L.; Żukow, W.W. Features of structural transformation in metals HAZ high-strength steel WELDOX 1300 performed by MAG welding. *Weld. Technol. Rev.* **2015**, *87*, 42–46.
3. High Strength and Ultra-High Strength Heavy Plates-Weight Savings Combined with Excellent Weldability, VoestAlpine Grobblech GmbH. 2012. Available online: https://www.yumpu.com/en/document/view/12574930/high-strength-and-ultra-high-strength-heavy-plates-voestalpine (accessed on 10 May 2022).

4. PN-EN 10025-6; Hot Rolled Products of Structural Steels Part 6: Technical Delivery Conditions for Flat Products of High Yield Strength Structural Steels in the Quenched and Tempered Condition. Available online: https://standards.iteh.ai/catalog/standards/cen/8ecdc9d2-f50a-4bf8-8503-eb9fd8e0e276/en-10025-6-2004 (accessed on 12 May 2022).
5. Berg, J.; Stranghöner, N. Fatigue behaviour of high frequency hammer peened ultra-high strength steels. *Int. J. Fatigue* **2016**, *82*, 35–48. [CrossRef]
6. Kah, P.; Pirinen, M.; Suoranta, R.; Martikainen, J. Welding of Ultra High Strength Steels. *Adv. Mater. Res.* **2014**, *849*, 357–365. [CrossRef]
7. Tümer, M.; Schneider-Bröskamp, C.; Enzinger, N. Fusion welding of ultra-high strength structural steels—A review. *J. Manuf. Process.* **2022**, *82*, 203–229. [CrossRef]
8. SSAB Oxelösund. *Offering List of the Original Firm SSAB Oxelösund;* Company catalog in Printed Form, Collective Work; SSAB Oxelösund: Uxelsund, Sweden, 2007.
9. Tyssen Krupp Steel. *Promotional Materials Tyssen Krupp Steel;* Promotional materials of ThyssenKrupp Steel, printed paper catalog, Collective Work; TyssenKrupp Steel: Essen, Germany, 2019.
10. Tuz, L.; Rakoczy, Ł.; Pancikiewicz, K. *Wpływ Podgrzewania Wstępnego na Twardość i Mikrostrukturę Niestopowych Stali Konstrukcyjnych o Wysokiej Wytrzymałości;* Wydawnictwo Naukowe Tygiel: Lublin, Poland, 2017; pp. 7–19. ISBN 978-83-65598-97-4.
11. Tuz, L.; Sulikowski, K. Evaluation of weldability of non-alloy high strenght heat-treated steel. *Przegląd Spaw.* **2018**, *90*, 9–13.
12. Tuz, L.; Pancikiewicz, K.; Sulikowski, K.; Rakoczy, Ł. Evaluation of laser beam butt welded joints of high-strength steel. *Przegląd Spaw.* **2017**, *89*, 90–97. [CrossRef]
13. Gogou, E. Use of High Strength Steel Grades for Economical Bridge Design. Master's Thesis, Delft University of Technology, Delft, The Netherlands, 2012.
14. Tasak, E.; Ziewiec, A. *Spawalność Materiałów Konstrukcyjnych Tom 1. Spawalność Stali;* Wydawnictwo JAK: Kraków, Poland, 2009.
15. Węglowski, M.S.; Zeman, M. Prevention of cold cracking in ultra-high strength steel Weldox 1300. *Arch. Civ. Mech. Eng.* **2014**, *14*, 417–424. [CrossRef]
16. Szczucka-Lasota, B.; Węgrzyn, T.; Piwnik, J.; Jurek, A. Welding of a mixed joint in the structure of mobile platforms made S690QL and S960 MC steel. *Apar. Badaw. Dydakt.* **2020**, *1*, 23–26.
17. Shenghai, Z.; Yifu, S.; Huijuan, Q. The technology and welding joint properties of hybrid laser-TIG welding on thick plate. *Opt. Laser Technol.* **2013**, *48*, 381–388. [CrossRef]
18. Macchietto, C. *Hybrid Laser Arc Welding (HLAW);* Valmont Industries: Omaha, NE, USA, 2011.
19. Denney, P. Hybrid Laser Arc Welding Has Its Time Finally Arrived? Available online: http://www.lincolnelectric.com/assets/US/EN/literature/mc1129.pdf (accessed on 1 June 2022).
20. Urbańczyk, M.; Adamiec, J. Hybrid welding (laser-electric arc MAG) of high yield Point steel S960QL. *Materials* **2021**, *14*, 5447. [CrossRef] [PubMed]
21. Turichin, G.; Klimova-Korsmik, O.; Skylar, M.; Zhitenev, A.; Kurakin, A.; Pozdnyakov, A. Laser-arc hybrid welding perspective ultra-high strength steels: Influence of the chemical composition of weld metal on microstructure and mechanical properties. *Procedia CIRP* **2018**, *74*, 752–756. [CrossRef]
22. Mazar Atabaki, M.; Ma, J.; Yang, G.; Kovacevic, R. Hybrid laser/arc welding of advanced high strength steel in different butt joint configurations. *Mater Des.* **2014**, *64*, 573–587. [CrossRef]
23. Banasik, M.; Urbańczyk, M. Spawanie metodą hybrydową laser + MAG różnych rodzajów złączy. *Biul. Inst. Spaw.* **2017**, *61*, 20–25.
24. Nielsen, S.E. High power laser hybrid welding—Challenges and perspectives. *Phys. Procedia* **2015**, *78*, 24–34. [CrossRef]
25. Sajek, A. Application of FEM simulation method in area of the dynamics of cooling AHSS steel with a complex hybrid welding process. *Weld. World* **2019**, *63*, 1065–1073. [CrossRef]
26. Łomzik, M. Mechanical properties of joints made in steel S1300QL using various welding methods. *Biul. Inst. Spaw. Res.* **2020**, *3*, 9–22. [CrossRef]
27. Zimmer, P. Zur Bewertung der Kaltrisssicherheit von Schweissverbindungen aus hochfesten Feinkornbaustählen. *Phys. Procedia* **2007**, *78*, 24–34.
28. Gaspar, M.; Balogh, A. Gmaw experiments for advanced (Q + T) high strength steels. *Prod. Process. Syst.* **2013**, *6*, 9–24.
29. Gáspár, M.; Sisodia, R.P.S. Weldability analysis of Q + T and TMCP high strength steels by physical simulation. In Proceedings of the IIW 2017 International Conference: Green Welding Technologies for Effective and Reliable Manufacturing, Shanghai, China, 29–30 June 2017; Schaupp, T., Ernst, W., Spindler, H., Kannengiesser, T., Eds.; pp. 166–170.
30. Thomas, S.; Wolfgang, E.; Helmut, S.; Thomas, K. Hydrogen-assisted cracking of GMA welded 960 MPa grade high-strength steels. *Int. J. Hydrogen Energy* **2020**, *45*, 20080–20093. [CrossRef]
31. *ISO 15614-1:2017;* Specification and Qualification of Welding Procedures for Metallic Materials-Welding Procedure Test—Part 1: Arc and Gas Welding of Steels and Arc Welding of Nickel and Nickel Alloys. ISO: Geneva, Switzerland, 2017.
32. *ISO/TR 15608:2017;* Welding—Guidelines for a Metallic Materials Grouping System. Przepisy Klasyfikacji Budowy Statków Morskich. Materiały i Spawanie. Część IX. Polski Rejestr Statków: Gdańsk, Poland, 2021.

materials

Article

Rebuilding of Turbocharger Shafts by Hardfacing

Bogdan Kupiec , Zenon Opiekun * and Andrzej Dec

Department of Casting and Welding, Faculty of Mechanical Engineering and Aeronautics, Rzeszow University of Technology, Al. Powstańców Warszawy 12, 35-959 Rzeszów, Poland
* Correspondence: zopiekun@prz.edu.pl

Abstract: This paper presents the results of structural tests and hardness measurements of rebuilding coatings manually applied by the gas tungsten arc welding (GTAW) method on damaged surfaces of steel shafts of turbochargers of automotive engines. Single- and double-layer coatings were applied in an argon atmosphere by fusing a 1.2 mm diameter wire with a Fluxofil M58 flux core, using a current of 35A and an arc voltage of 8–9 V. The hardfacing resulted in coatings with a martensitic–bainitic structure with fine dispersed carbides rich in M23C6-type chromium. The hardness of the coatings on the rebuilt shafts averaged from about 740HV5 for the single-layer coating to about 770HV5 for the double-layer coating and was two times higher than the hardness of the tempered shafts without coatings.

Keywords: GTAW hardfacing; flux-cored wire; turbocharger shaft

Citation: Kupiec, B.; Opiekun, Z.; Dec, A. Rebuilding of Turbocharger Shafts by Hardfacing. *Materials* **2022**, *15*, 5761. https://doi.org/10.3390/ma15165761

Academic Editors: Marcin Adamiak, Artur Czupryński, Antonín Kříž and Tünde Anna Kovács

Received: 25 July 2022
Accepted: 15 August 2022
Published: 20 August 2022

1. Introduction

Common devices used to increase the specific power of internal combustion engines are turbochargers. The air turbochargers installed as standard in car engines have a design consisting of three main parts: the compressor, where the air is compressed by a skewed-blade rotor, the centrifugal (radial-axial) turbine, where the exhaust gas is expanded, and the turbocharger shaft (Figure 1) [1–3].

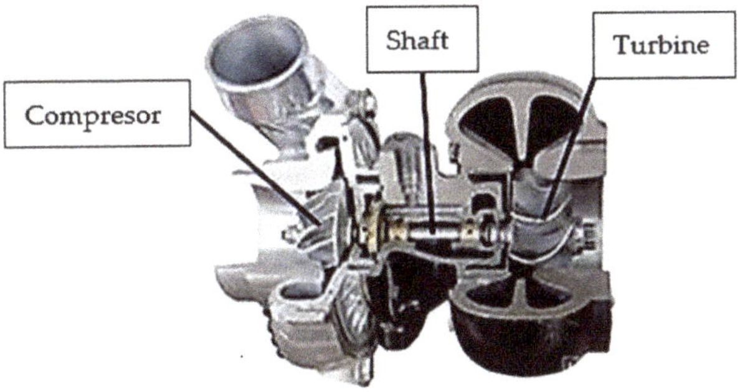

Figure 1. Components of a turbocharger.

The construction of a turbocharger for a motor vehicle is not very complicated, yet the operating conditions require that its components are carefully manufactured, continuously and properly lubricated, and also cooled.

Operational damage to turbochargers is caused by excessively high rotor speeds, which significantly increase the centrifugal forces acting on their components, excessively high temperatures of the gases supplied to the turbine, which can induce dimensional change processes caused by creep, and the deposition of thermolysis products of the superheated oil used to lubricate the shaft [4–6].

The gases flowing through the turbocharger should not contain solid impurities that could cause damage to its moving components, especially the shafts [7]. Turbocharger shafts are made from low-alloy structural steel for tempering [8], as well as corrosion-resistant chrome-nickel steel [9,10]. The shafts are permanently connected to the turbine rotor by welding. The temperature of the exhaust gas flowing into the turbocharger should not exceed about 800 °C for a compression-ignition engine and about 1000 °C for a spark-ignition engine. Higher exhaust gas temperatures can cause damage to the turbocharger through accelerated corrosion, deformation of its body, and thermal fatigue of the shafts [11].

The journals of turbocharger shafts wear abrasively in the areas where the components work together (at the kinetic nodes of the shaft—plain bearing) [12]. The loose particles of hard abrasive material [13,14], during the relative movement of the friction surfaces, cause damage and destruction to the shaft journals by micro-cutting, scratching, furrowing, abrading, and tearing off the protrusions of bumps (Figure 2).

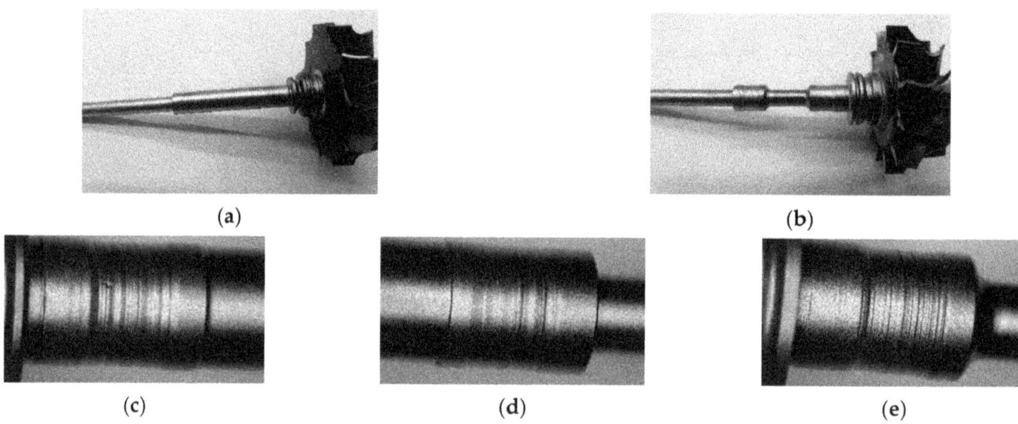

Figure 2. Views of the shafts with turbocharger turbine rotors. Renault Laguna 1.9 DCI, Garret 708,639 model (**a**); Peugeot 3071.6HDI, Garret 7534200375J6 model (**b**). Worn shaft journal surfaces (examples) (**c**–**e**).

Damage to the kinematic nodes of turbochargers causes characteristic vibro-acoustic symptoms and the reduced performance of the internal combustion engine [15–20]. With signs of wear and tear, the shafts with turbine rotors are replaced in their entirety, even after the turbocharger has been in operation for a short time.

The aim of this article is to present the possibility of rebuilding worn top layers of turbocharger shafts by the arc welding of wear-resistant and hard coatings. The turbocharger shaft was regenerated using the GTAW method.

The GTAW method is one of the welding methods for producing welded joints and overlaying coatings, mainly of special steels and non-ferrous metal alloys. The GTAW method is one of the cleanest welding methods in terms of metallurgy. Welded joints and padding welds made with this method are characterized by very high purity and are free from defects. The applied shielding and plasma-forming gases, such as argon and helium or their mixtures, favor a very good shielding of the surfacing area with various coating materials. In the case of surfacing coatings on steel elements with wire with a powder core, the GTAW method gives one of the best coatings. Argon of purity 4.6 (99.996%) as an inert gas of a higher density (45% higher density than air) protects very well, preventing oxidation during surfacing [21,22].

2. Research Material and Methodology

Two shafts with turbine rotors taken from the turbochargers of the Peugeot 307-1.6 HDI and Renault Laguna 1.9 DCI cars were adopted for rebuilding by hardfacing. The top layers of the journals of these shafts were severely damaged by their abrasive wear in the areas of interaction: journal-plain bearing. The thickness of the unevenly worn top layers of the journals averaged 0.2–0.3 mm. Compared to the journal diameters of new shafts of an, approximately, 7.8 mm diameter, the loss in thickness represents 4–8% of the diameter. The unevenly worn top layers of the shaft journals mainly showed deep scratches and furrowing effects caused by displaced loose particles of hard abrasive (Figure 3).

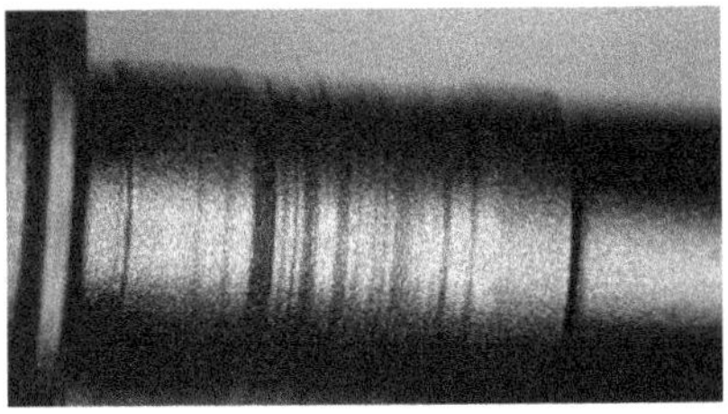

(a)

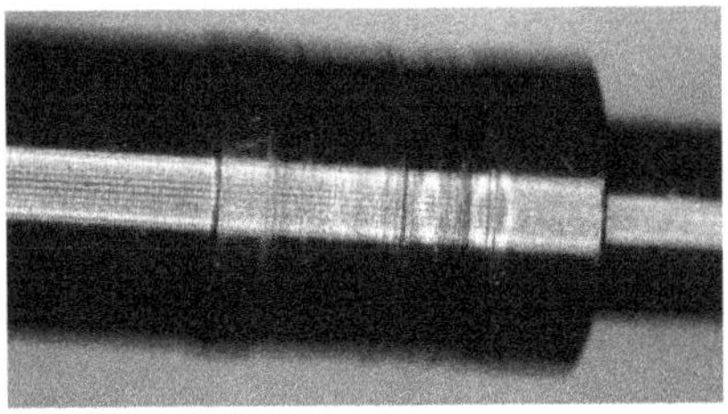

(b)

Figure 3. Worn top layers of turbocharger shaft journals: Renault Laguna 1.9 DCI, Garret 708,639 model (**a**), Peugeot 3071.6HDI, Garret 7534200375J6 model (**b**).

The rebuilding of the top layers of the shaft journals was carried out by hardfacing them with the GTAW method using argon as a shielding and plasma-forming gas at a flow rate of approximately 8 L/min, using a wire with a diameter of 1.2 mm and a flux core containing approximately 0.6% C, 1.9% Mn, 0.7% Si, 5.4% Cr, and the remainder being Fe. The hardfacing of the coatings was carried out manually using a current of about 35A and an arc voltage of 8–9 V. In order to cover the entire surface of the shaft journal, beads approximately 2.5 mm wide were laid lengthwise so that they overlapped each other by approximately 0.5 of their width (W), according to the diagram shown in Figure 4.

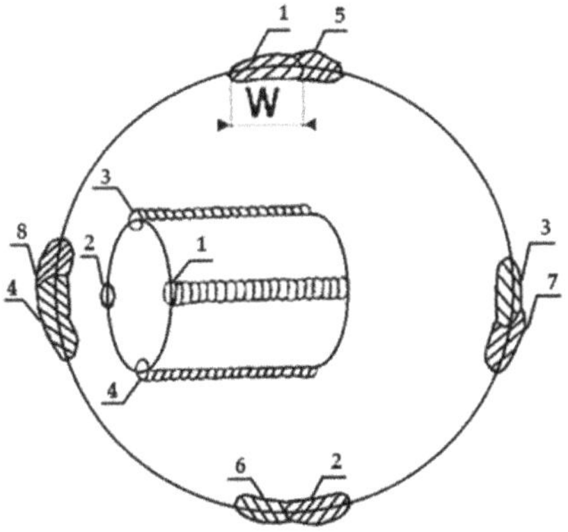

Figure 4. Laying sequence of the individual beads of the hardfacing coating (diagram). The numbers 1–4 in the internal diagram show the order in which the first stitches are laid on the damaged turbocharger shaft. The numbers 1–8 on the outer circle show the execution of the next stitches.

The systematic arrangement of beads around the periphery of the shaft journal ensures that axiality is maintained despite the small diameter of the shaft. Similar to the single-layer coating, a two-layer coating was made on the shaft journal by longitudinally hardfacing the second layer of the coating, respecting the width of the beads and the order in which they were laid. The course of the hardfacing of the turbocharger shaft journals is shown in Figure 5.

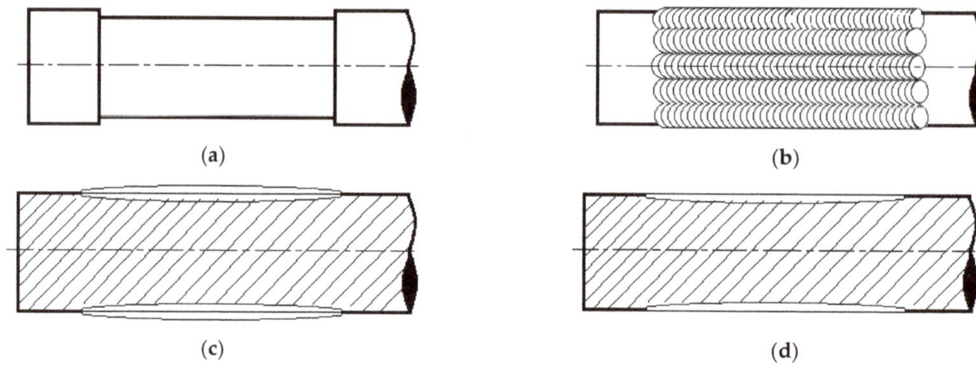

Figure 5. Course of rebuilding by hardfacing of shaft journals. Sandblasted surface of used shaft (**a**), bead pattern after coating (**b**,**c**), shaft after finish grinding (**d**) (diagram).

Metallographic examination of the shafts was carried out on metallographic specimens, on their cross-sections transverse to the direction of hardfacing of the coatings. The metallographic specimens were made by mechanical polishing and etching of the cross-sections of the shafts embedded in an electrically conductive mass. The macrostructure was observed using a Makrolite optical microscope, and the microstructure using a Neophot 2 metallographic light microscope and a Tescan scanning electron microscope with an Oxford EDS chemical composition microanalysis attachment. Shaft hardness and coating hardness were measured using a ZHV10 hardness tester, Zwick/Roell.

The chemical composition of the steel from which the turbocharger shafts were made was determined by spectral analysis using a Bruker Q4 Tasmann spectrometer.

3. Findings and Discussion

3.1. Metallographic Examination

The macro-structures of the shaft journal cross-sections after hardfacing of the single-layer and double-layer coating, as well as the cross-section of the new shaft (without hardfacing), are shown in Figure 6.

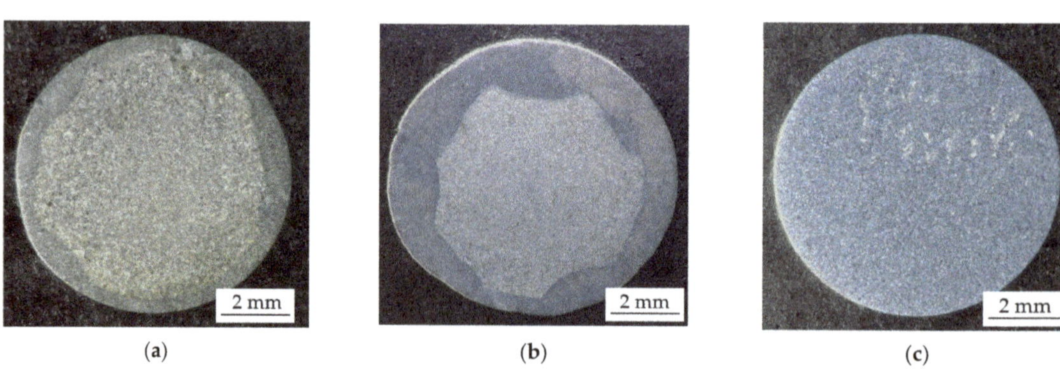

Figure 6. Cross-sections of turbocharger shaft journals. Single-layer coating (**a**), double-layer coating (**b**), new shaft journal (**c**). The specimens were etched with Kalling's reagent; 3× magnification.

The effect of the arc manual longitudinal hardfacing of the journals is uneven coating thicknesses, with the coating made by double-layer hardfacing having the greater average thickness. In both cases of single- and double-layer hardfacing, complete coverage of the journal surfaces with flux-cored wire hardfacing coatings was achieved (after finishing machining).

The microstructures of the shaft journals observed using a light metallographic microscope are shown in Figures 7–9.

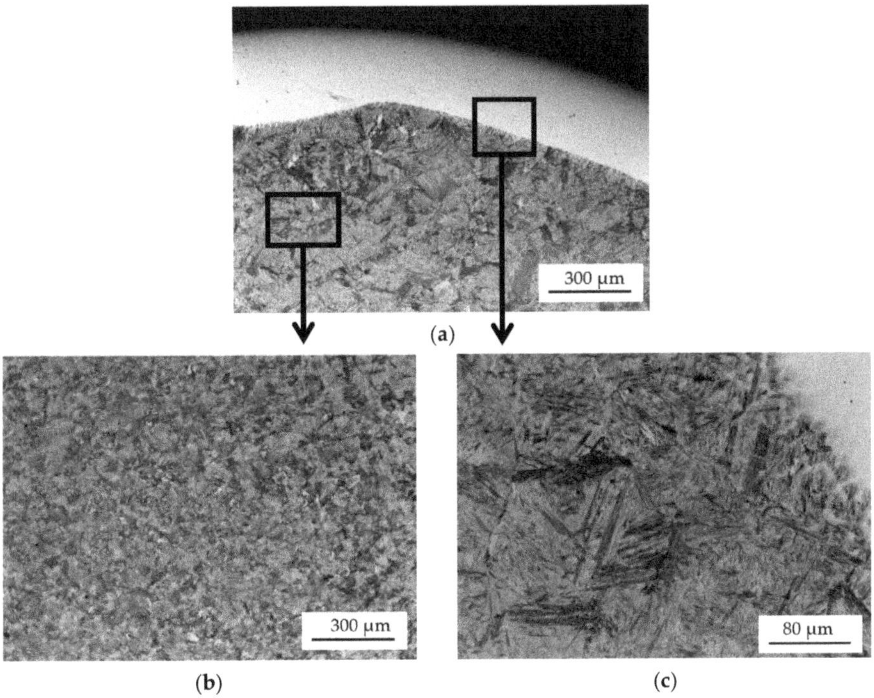

Figure 7. Microstructure of a turbocharger shaft journal cross-section after single-layer hardfacing: bright padding weld coating (**a**), shaft journal core (**b**), transition zone: coating—shaft journal material (c). The specimen was etched with 4% nital.

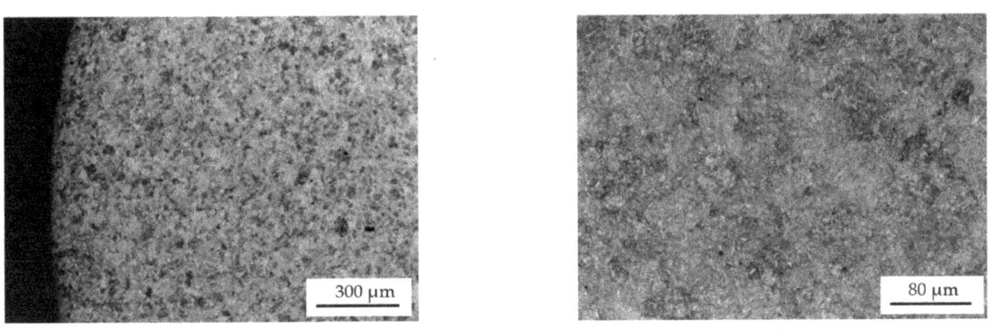

Figure 8. Microstructure of the new shaft journal (without coating). The specimen was etched with Kalling's reagent (**a**,**b**).

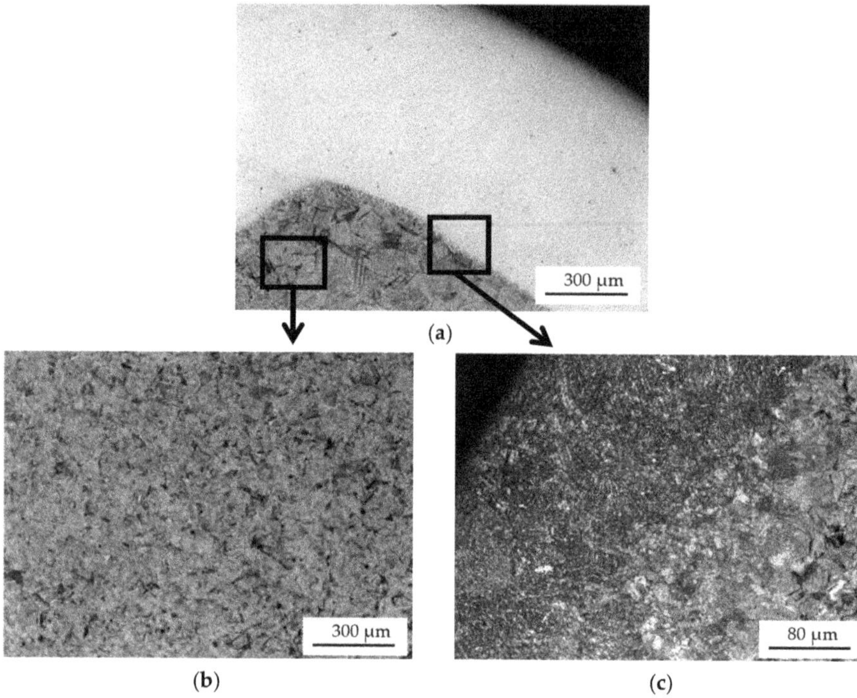

Figure 9. Microstructure of a cross-section of a turbocharger shaft journal after two-layer hardfacing: bright padding weld coating (etched with 4% nital) (**a**), microstructure of the shaft journal core (**b**), transition zone: coating—material (etched with Kalling's reagent) (**c**).

The microstructure of the two coatings arc-applied with a flux-cored wire is similar. Their microstructure, revealed by etching with Kalling's reagent (Figure 9c), is fine-grained. In contrast, the sub-coating zone and core of the single-layer hardfaced shaft is coarse-grained. The 100–150 μm grains are coarse-grained tempered martensite. In the case of the double-layer coating, by reheating the core of the shaft journal during the application of the second coating, its structure was crystallized and it has a fine-grained structure similar to that of the new shaft journal (Figures 8a and 9b).

The results of the microstructural examination of the turbocharger shaft journals using a scanning electron microscope (SEM) with an X-ray microanalysis (EDS) attachment are shown in Figures 10–12. Metallographic specimens for these studies were etched more strongly with Kalling's reagent.

The coatings that are arc hardfaced with a flux-cored wire have a similar microstructure (Figures 10c and 12b). The structure of the coatings is a fine-acicular mixture of martensite and lower bainite with dispersed carbides. The microstructure of the core of a single-layer hardfaced shaft is coarse-grained (Figure 10d) with coarse-plate martensite and lower bainite. The core of the double-layer hardfaced shaft is fine-grained.

For the turbocharger shafts, a low-alloy structural steel for tempering was used with mark 42CrMo4 (acc. to PN-EN 10083-1, 1999) and the following chemical composition; 0.44% C, 0.94% Cr, 0.17% Mo, 0.17% Si, 0.73% Mn, and the remainder being Fe. The structure of the tempered shafts is fine-grained and is composed of highly tempered martensite (Figure 11).

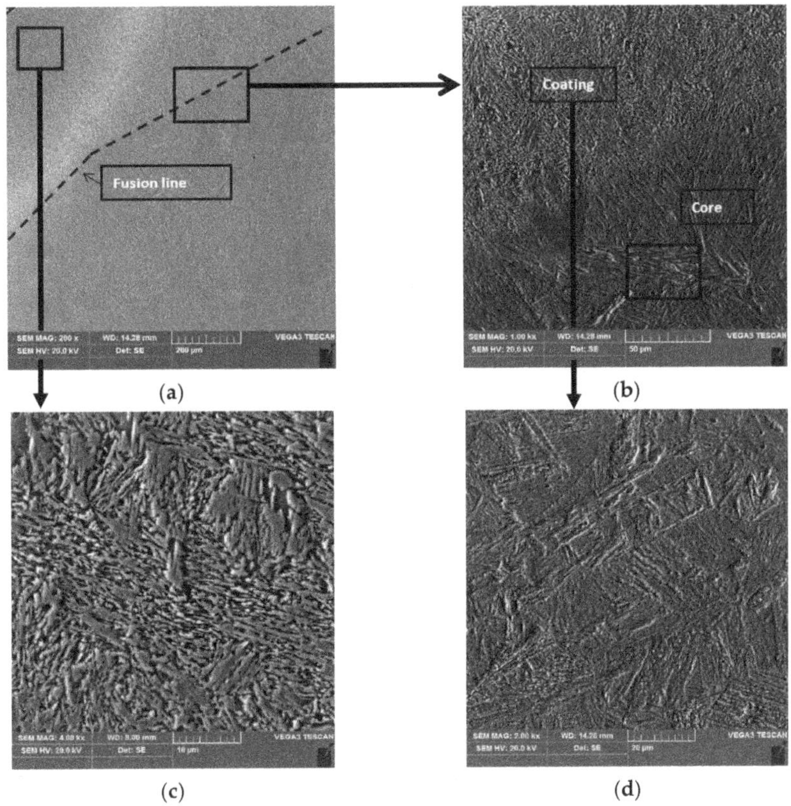

Figure 10. Microstructure of a turbocharger shaft journal cross-section after single-layer hardfacing. Transition zone microstructure, coating-core (**a**), detail from (**a**), coating-core (**b**), coating microstructure (**c**), core microstructure (**d**).

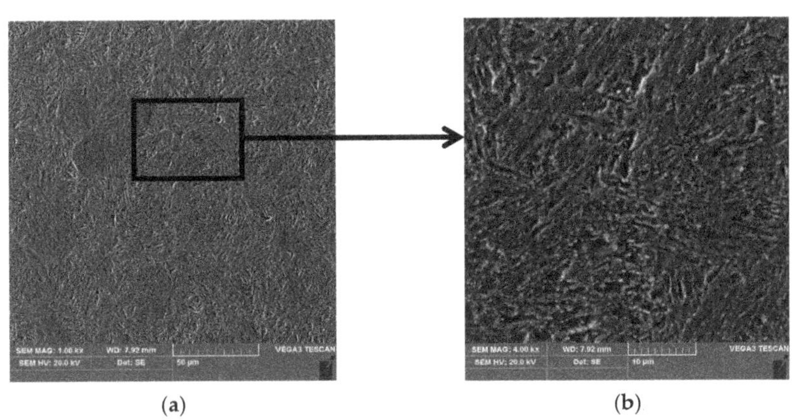

Figure 11. Microstructure of the new shaft journal (without hardfacing) (**a**), detail from (**a**) visible fine grains of highly tempered martensite (**b**).

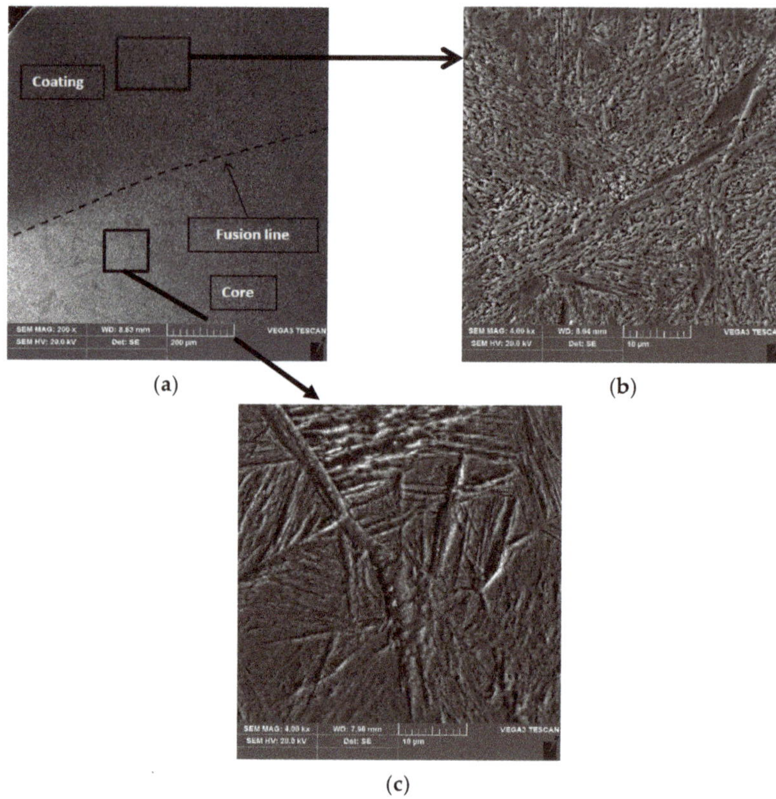

Figure 12. Microstructure of a turbocharger shaft journal cross-section after double-layer hardfacing. Transition zone microstructure, coating-core (**a**), coating microstructure (**b**), core microstructure (**c**).

3.2. X-ray Microanalysis

The place of performance of the spot X-ray microanalysis of the coating is shown in Figure 13a. The results are summarized in Table 1. Fine carbide separations, dispersively distributed between the lamellar–acicular martensite, were analyzed.

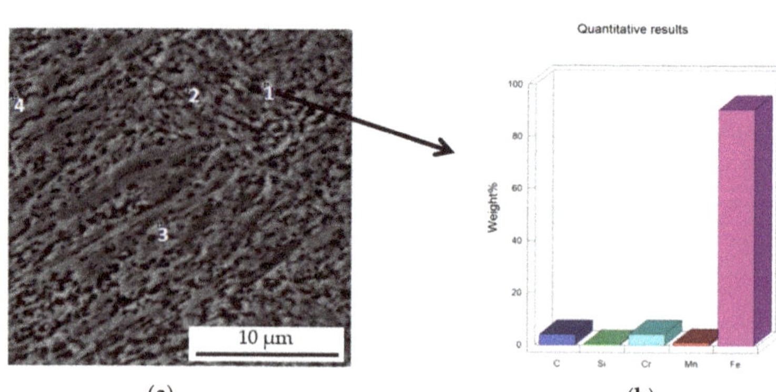

Figure 13. Microstructure of the flux-cored arc-hardfaced coating on the turbocharger shaft journal with micro-areas (points1–4) (**a**), bar chart of elemental content in micro-area 1 (**b**).

Table 1. The result of the micro analysis of the chemical composition from points 1–4. The compactness of the elements % by weight.

No. of the Micro-Area	C	Si	Cr	Mn	Fe
1.	4.01 (16.15) *	-	4.18 (3.90) *	0.87 (0.77) *	balance
2.	4.01 (16.15) *	0.52 (0.89) *	3.86 (3.59) *	1.05 (0.93) *	balance
3.	5.28 (20.47) *	0.50 (0.83) *	3.82 (3.42) *	1.17 (0.99) *	balance
4.	4.48 (17.87) *	-	4.23 (3.90) *	1.11 (0.97) *	balance

* The compactness of the elements in atomic %.

These tests confirmed that the fine spherical dispersionally distributed phases in the martensitic matrix of the hardfaced coating are iron- and chromium-containing carbides of the $M_{23}C_6$-type coatings. These carbides are hard, therefore, the hardfaced coating should have a higher abrasion resistance than the original uncoated rollers.

3.3. Hardness Measurements

Measurements of the hardness of the cores and journal coating of the turbocharger shafts were carried out using a Vickers indenter hardness tester under a 5N load. Five measurements were taken for each coating and core of the hardfaced shafts and the new shaft. Table 2 summarizes the results of these measurements.

Table 2. HV5 hardness values of coatings and journal cores of turbocharger shafts.

	Single-Layer Coating	Journal Core with Single-Layer Coating	Double-Layer Coating	Journal Core with Double-Layer Coating	New Shaft
Hardness, HV5	742, 749, 740, 745, 746	576, 575, 602, 580, 581	761, 772, 770, 773, 778	494, 423, 445, 450, 460	380, 384, 382, 380, 378
On average, HV5	745	580	770	455	380

The hardnesses of the coatings range from approx. 740HV5 to approx. 780HV5, and their value depends on the measurement location. The thinner coatings hardfaced in a single layer that have a higher proportion of core material components have a lower hardness. In the double-layer coatings, especially in their wider areas, hardness values near the journal surface are highest. In these areas, X-ray analysis of the chemical composition revealed alloying element contents similar to those found in the flux-cored wire with which the shafts were hardfaced. The hardnesses of the journal cores of the hardfaced shafts have different values and are significantly greater than the hardness of the tempered new shaft. The hardness of the journal core of the shaft hardfaced with a double-layer by double structural transformation is significantly lower than that of the journal of the shaft hardfaced with a single-layer.

4. Conclusions

- Worn surface layers of steel journals of turbocharger shafts of automotive engines (e.g., Peugeot 307-1.6 HDI and Laguna 1.9 DCI) can be subjected to the process of their rebuilding by arc hardfacing of hard coatings.

- Coatings applied to the journals of the remanufactured shafts are characterized by a fine-grained martensitic structure with fine spherical dispersive hard carbides containing $M_{23}C_6$-type chromium arranged in the matrix.
- As a result of surfacing with wire with powder cores with an increased content of 0.6% C and 5.4% Cr, coatings were obtained with a hardness twice as high as the hardness of the turbo-compressor shaft journal cores.
- Coating structure, with a hardness of almost twice that of the highly tempered martensite structure of new shafts, should, in turn, provide better abrasive wear resistance for these important structural components of turbochargers.

Author Contributions: Investigation, B.K. and Z.O.; Methodology, B.K., Z.O. and A.D.; Resources, A.D.; Supervision, Z.O.; Writing—original draft, Z.O.; Writing—review & editing, B.K. All authors have read and agreed to the published version of the manuscript.

Funding: This research received no external funding.

Institutional Review Board Statement: Not applicable.

Informed Consent Statement: Not applicable.

Data Availability Statement: Not applicable.

Conflicts of Interest: The authors declare no conflict of interest.

References

1. Seifoori, S.; Parrany, A.M.; Khodayari, M. A high-cycle fatigue failure analysis for the turbocharger shaft of BELAZ 75131 mining dump truck. *Eng. Fail. Anal.* **2020**, *116*, 104752. [CrossRef]
2. Idzior, M.; Bieliński, M.; Borowczyk, T.; Daszkiewicz, P.; Stobnicki, P. A simulation study of turbocharger intake leakage on its efficiency. *Combust. Engines* **2012**, *51*, 64–71. [CrossRef]
3. Hadryś, D.; Bąkowski, H.; Stanik, Z.; Kubik, A. Analysis of shaft wear in turbocharges of automotive vehicles. *Transp. Probl.* **2019**, *14*, 85–95. [CrossRef]
4. Dudziński, W.; Haimann, K.; Lachowicz, M. Struktura i własności materiałowe łożysk ślizgowych turbosprężarek współpracujących z silnikami spalinowymi. *Tribol. Tarcie Zużycie Smarowanie* **2002**, *4*, 1133–1141.
5. Danilecki, K. Trends in the development of turbocharging systems in automotive vehicles. *Siln. Spalinowe* **2008**, *47*, 61–76. [CrossRef]
6. Schweizer, B. Oil whirl, oil whip and whirl/whip synchronization occurring in rotor systems with full-floating ring bearings. *J. Nonlinear Dyn.* **2009**, *328*, 156–190. [CrossRef]
7. Idzior, M.; Bieliński, M.; Borowczyk, T.; Karpiuk, W. Analiza wpływu warunków eksploatacji na stan techniczny turbosprężarek doładowanych silników spalinowych. *Logistyka* **2011**, *3*, 1129–1139.
8. Hermawan, M.R.; Sonawanb, H. Case Study: Failure analysis of induced draft fan after serious inspection by overlay welding. *Eng. Fail. Anal.* **2020**, *118*, 104855. [CrossRef]
9. Shilun, Z.; Xiong, Z.; Gang, Z. Comprehensive welding technology for type 304 stainless steel rotating shaft. *Procedia Eng.* **2011**, *24*, 840–844. [CrossRef]
10. Zangeneh, S.; Ketabchi, M.; Kalaki, A. Fracture failure analysis of AISI 304L stainless steel shaft. *Eng. Fail. Anal.* **2014**, *36*, 155–165. [CrossRef]
11. Burkinshaw, M.; Blacker, D. The high temperature tribological performance of turbocharger wastegate materials. In *11th International Conference on Turbochargers and Turbocharging: 13–14 May 2014*; Institution of Machanical Engineers: London, UK, 2014; pp. 289–298.
12. Romagnoli, A.; Manivannan, A.; Rajoo, S.; Chiong, M.S.; Feneley, A.; Pesiridis, A.; Martinez-Botas, R.F. A review of heat transfer in turbochargers. *Renew. Sustain. Energy Rev.* **2017**, *79*, 1442–1460. [CrossRef]
13. Deligant, M.; Podevin, P.; Descombes, G. Experimental identification of turbocharger mechanical friction loss. *Energy* **2012**, *39*, 388–394. [CrossRef]
14. Chuepeng, S.; Saipom, S. Lubricant thermo-viscosity effects on turbocharger performance at low engine load. *Appl. Therm. Eng.* **2018**, *139*, 334–340. [CrossRef]
15. Tian, L.; Wang, W.J.; Peng, Z.J. Effects of bearing outer clearance on the dynamic behaviours of the full floating ring bearing supported turbocharger. *Mech. Syst. Signal Process.* **2012**, *31*, 155–175. [CrossRef]
16. Zheng, X.Q.; Zhang, Y.J.; Yang, M.Y. Research and development on transonic compressor of high pressure ratio turbocharger for vehicle internal combustion engines. *Sci. China Technol. Sci.* **2010**, *53*, 1817–1823. [CrossRef]
17. Wojciechowski, H.; Żółtowski, B. Exploitation of Car Turbochargers-Selected Problems, 8th International Conference Diagnostics of Machines and Vehicles. *MATEC Web Conf.* **2019**, *302*, 01024. [CrossRef]

18. Kaszkowiak, J.; Borowski, S.; Dulcet, E.; Zastempowski, M. Damage Analysis of Turbochargers. *Logistyka* **2015**, *4*, 1885–1892.
19. Idzior, M.; Wichtowska, K.; Kinal, G. The investigation of conditions of work and their influence on work of shafts of turbochangers combustion engines. *Logistyka* **2014**, *3*, 2395–2403.
20. Yang, K.; Fletcher, K.A.; Styer, J.P.; Lam, W.Y.; Guinther, G.H. Engine Oil Components Effects on Turbocharger Protection and the Relevance of the TEOST 33C Test for Gasoline Turbocharger Deposit Protection. *SAE Int. J. Fuels Lubr.* **2017**, *10*, 815–821. [CrossRef]
21. Bridigum, T. *Welding: Everything You Need to Know*; Motorbooks International: Osceola, WI, USA, 2022; 160p.
22. Singh, R. *Applied Welding Engineering, Processes, Codes and Standards*; Butterworth-Heinemann: Oxford, UK, 2020.

materials

MDPI

Article

Optimization of Bead Geometry during Tungsten Inert Gas Welding Using Grey Relational and Finite Element Analysis

Muhammad Hanif [1], Abdul Hakim Shah [2], Imran Shah [3] and Jabir Mumtaz [4,*]

[1] Schools of Mechanical Science and Engineering, Huazhong University of Science and Technology, Wuhan 430074, China; engr.hanif94@yahoo.com
[2] Department of Physics and Chemistry, Khushal Khan Khattak University Karak, Karak 27400, Pakistan; dr.abdulhakim@kkkuk.edu.cn
[3] Department of Aerospace Engineering, College of Aeronautical Engineering, National University of Sciences and Technology, Risalpur 24090, Pakistan; imranshahswabi@gmail.com
[4] College of Mechanical and Electrical Engineering, Wenzhou University, Wenzhou 325035, China
* Correspondence: jabirmumtaz@live.com

Abstract: Mild steel welded products are widely used for their excellent ductility. Tungsten inert gas (TIG) welding is a high-quality, pollution-free welding process suitable for a base part thickness greater than 3 mm. Fabricating mild steel products with an optimized welding process, material properties, and parameters is important to achieve better weld quality and minimum stresses/distortion. This study uses the finite element method to analyze the temperature and thermal stress fields during TIG welding for optimum bead geometry. The bead geometry was optimized using grey relational analysis by considering the flow rate, welding current, and gap distance. The welding current was the most influential factor affecting the performance measures, followed by the gas flow rate. The effect of welding parameters, such as welding voltage, efficiency, and speed on the temperature field and thermal stress were also numerically investigated. The maximum temperature and thermal stress induced in the weld part were 2083.63 °C and 424 MPa, respectively, for the given heat flux of 0.62×10^6 W/m^2. Results showed that the temperature increases with the voltage and efficiency of the weld joint but decreases with an increase in welding speed.

Keywords: TIG welding; finite element method; grey relational analysis; thermal stress; mild steel

Citation: Hanif, M.; Shah, A.H.; Shah, I.; Mumtaz, J. Optimization of Bead Geometry during Tungsten Inert Gas Welding Using Grey Relational and Finite Element Analysis. *Materials* **2023**, *16*, 3732. https://doi.org/10.3390/ma16103732

Academic Editors: Marcin Adamiak, Artur Czupryński, Antonín Kříž and Tünde Anna Kovács

Received: 7 April 2023
Revised: 8 May 2023
Accepted: 9 May 2023
Published: 15 May 2023

1. Introduction

Mild steel is a type of low-carbon steel that is malleable and ductile, making it ideal for applications that require a low strength-to-weight ratio, such as cages, frames, and fencing. According to Silva et al. (2018), mild steel is preferred over other steels because it endures high stresses and results in longer service life [1]. Due to its malleability, it can be quickly shaped, drilled, welded, and cut. Welding is the fabrication and repair of metal products and an essential process in every industry. It is a simple, cost-effective, and dependable method for joining metals. In welding processes, weld joint quality is usually affected by the weld material's attributes, mechanical properties, and the highly concentrated localized zone. Saha et al. (2017) stated that the weldability of a material depends upon the properties changing during the welding process and the strength of the centralized weld zone [2]. Datta et al. (2008) noted that the weld functionality features also depend on weld joint geometry, which is directly influenced by the input parameters [3].

Several other welding processes have been proposed to join the specific types and dimensions of work parts: laser welding, TIG welding, electron beam welding, etc. [4]. Cary et al. noted that TIG welding is extensively used in industries because of high weld quality, protection of the weld pool by using an inert gas shield, its versatility, repairability, and flexibility to adapt. Research showed that an optimal welding speed could produce a joint with high mechanical strength, excellent microstructure, and good morphology [5]. Eshwar et al. (2014) found that the input parameter has a substantial role in determining the weld quality, which primarily includes bead width and height, and penetrating the area, thermal, and residual stresses, resulting in minimizing the life of a weld joint [6]. Researchers have tried to work on different welding processes regarding weld bead geometry, temperature, and stress fields. For instance, Dhas et al. (2007) have proposed an adaptive neuro-fuzzy inference system (ANFIS) for determining the optimum input parameters for bead width [7]. A multi-objective optimization of process parameters was carried out to find the optimal bead geometry using the Taguchi method [8]. Fernando et al. (2021) investigated the impact of metal transfer modes on the symmetry of bead geometry. They measured bead geometry using a 3D scanner and found that the metal transfer mode affected the bead geometry symmetry. It provides insight into the effect of metal transfer modes on the quality of WAAM aluminum parts [9]. Improvements in weld quality have been made by combining grey relational analysis with the Taguchi method. Chandrasekhar et al. (2015) have developed an artificial neural network (ANN) to estimate the weld bead width and depth of penetration by using IR images during TIG welding of 6 mm thick 316 LN stainless steel by changing the values of the welding current [10]. Chokkalingham et al. (2012) have also developed a similar model for better estimation of bead width and depth of penetration [11]. Kolahan et al. (2010) established a simulated annealing methodology to analyze and improve the geometry of the weld joint (bead width, bead height, and penetration) in gas metal arc welding [12]. Zhang et al. (2017) have developed an online system for the quality of aluminum alloy welding, which automatically evaluates the quality parameters in real-time using online arc sound, voltage, and spectrum signals [13]. The microstructure, hardness, and corrosion behavior of gas tungsten arc welding (GTAW) Inconel 625 super alloy over A517 carbon steel was also investigated using ERNiCrMo3 filler metal [14]. Bandhu et al. (2022) have experimentally studied the low alloy steel after heat treatment [15]. It was concluded that regulated metal deposition weldments have better mechanical properties than gas welding for the same steel alloy. Jiang et al. (2017) have addressed the quality issues (defects) in the welding process [16]. They proposed a novel approach for classifying weld defects, such as porosity, slag inclusion, lack of penetration, etc., based on the Dumpster–Shafer evidence theory.

In recent years, the main problem faced by researchers is optimizing the factors that affect the welding joint [17] and the geometry of the welding area to predict the temperature distribution, minimum residual stresses, and distortion [18]. Understanding the thermal flow in welding is theoretically and experimentally important. The impact of stress relief heat treatment on the microstructure and mechanical properties of dissimilar GTAW weld joints between Inconel 625 and A106 carbon steel was also investigated [19]. The results provide insights into the effects of heat treatment on the properties of such weld joints. Sepe et al. (2014) demonstrated the residual stresses and temperature distribution in metal arc welding using FEM analysis, finding higher thermal stress in heat-affected and fusion zones [20]. The stresses decrease with the increase in temperature and pre-heating and post-heating methods. Katherasan et al. (2014) performed a simulation study for bead geometry and demonstrated the effects of process parameters such as feed rate, voltage, speed, and torch angle on bead width, reinforcement, and penetration depth using particle swarm optimization (PSO) for flux-cored arc welding [21]. The effect of different process parameters such as voltage, welding current, and wire feed rate on the bead geometry and microstructure of WAAM parts were studied and the results showed that the process parameters significantly impacted the bead geometry and microstructure. It provides insights into optimizing WAAM process parameters for improved part quality and can help guide the WAAM process design [22]. El-Sayed et al. (2017) also predicted the residual stresses and temperature-induced in friction stir welding using finite element analysis [23].

A new calibration coefficient was computed for aluminum plates subjected to uniaxial load. An appropriate coefficient and residual stress compensation value could result in in-situ thermal stresses with expected accuracy within acceptable ranges of the industry's specification limits [24]. Kulkarni et al. (2019) demonstrated TIG welding joining dissimilar metals for stainless steel (P91 steel and AISI 3161) with Inconel 800 and Inconel 600 [25]. It was observed that an Inconel 600 interlayer and a fully austenitic structure could be achieved without any mechanical loss. Based on the above literature, it has been realized that very little or no work has been reported to study the thermal and structural analysis and variations in temperature with the TIG welding parameters for the optimized process.

This research primarily focuses on analyzing multi-objective optimization of the TIG welding process for an optimum factor selection affecting bead geometry, temperature, and thermal stresses. Taguchi-based grey relational analysis for multiple objectives was employed to determine significant factors affecting the performance measures. The temperature variations over time and thermal stresses for the optimum geometry were studied. ANSYS APDL was used to generate the model, and finite element analysis (FEA) was employed to predict the responses. Heat flux was also determined by conducting the experiments with the selected levels of factors, which was further used in FEA in the V-shaped butt joint. As per the required dimensions, a 3D solid element with DOF, temperature (SOLID-70), and a 2D solid element with four nodes (PLANE-55) was used. A transient analysis followed by the coupled field analysis was performed to predict temperature variations and thermal stresses. The variation of temperature by welding parameters was also investigated.

2. Materials and Methods

TIG welding was achieved on the mild steel (MS) plates, and its chemical composition is presented in Table 1 [26]. A CNC milling machine was used to prepare test samples of dimensions $80 \times 20 \times 3\ mm^3$ for the V-shaped butt joint. The samples were appropriately clamped at the ends during welding. In experiments, argon gas was used because of its inert behavior, protecting the molten metal from impurities in the atmosphere and being heavier than air. The joining of metals was performed by melting the work partly with an arc fixed between the electrode (tungsten) diameter of 3.2 mm and the base part. A high-frequency spark delivered a conductive pathway through the air.

Table 1. Mild steel composition.

C (%)	Si (%)	Mn (%)	S (%)	P (%)
0.05–0.25	<0.35	0.30–0.90	<0.05	<0.05

Due to the spark produced, the surface of the base work partly melted, and a molten pool was formed. Figure 1 shows a schematic diagram and setup of TIG welding [27].

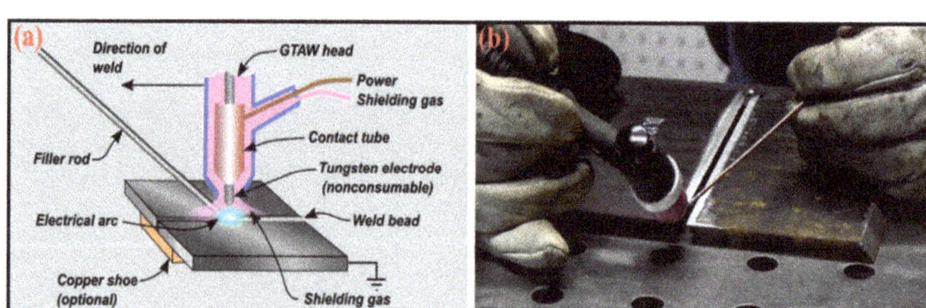

Figure 1. TIG Welding mechanism. (**a**) illustration of TIG welding process (**b**) experimental setup.

Based on an extensive literature review, the selection of factors and their levels have been provided in Table 2. The bead geometry was optimized for further analysis of temperature and thermal stresses, as shown in Figure 2.

Table 2. Factors and their levels.

Input Parameter	Units	Levels		
		Low	Middle	High
A: Gas Flow Rate	L/min	10	12	14
B: Welding Current	A	60	80	100
C: Gap Distance	mm	1.5	2.0	2.5

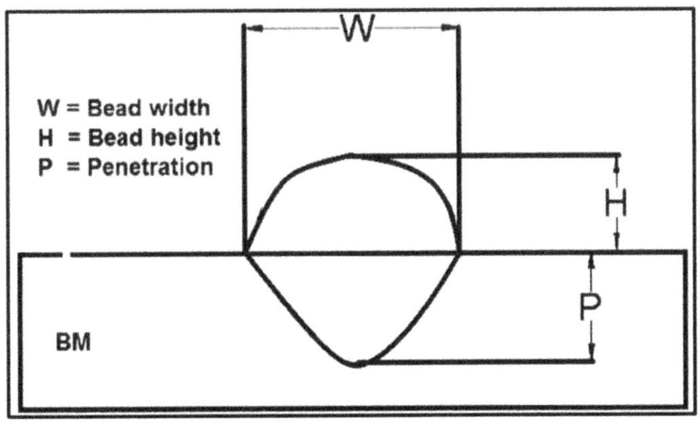

Figure 2. Bead geometry for V-shape butt joint.

The bead geometry was divided into bead height, width, and penetration, which were measured using linear measuring instruments.

3. Experimental Design and Analysis

Taguchi design is an efficient, straightforward approach and saves more time by minimizing the number of experiments, as stated by Choudhury et al. [28]. Hence it is used in the present study. Table 3 shows the results of output responses concerning different sets of input factors. Nine experiments were performed by Taguchi orthogonal Array ($L_9 3^3$(OA)). A grey relational analysis (GRA) was employed to find the optimal setting for a final geometrical model for further investigation by FEA. GRA is a multi-objective optimization technique in which various responses can be optimized simultaneously. Three input factors, i.e., rate of flow (F), welding current (I), and gap distance (G), were selected to study their influence on bead height and width (H and W) and area of penetration (P) of the weld. Grey's relational coefficient was determined to find the correspondence in actual and theoretical results.

Two criteria correspond to the output, i.e., the smaller- and larger-the-better cases. In the present study, weld geometry comprises bead height (H) and bead width (W) that corresponds to the lower case that can be represented by Equation (1):

$$x_i(k) = \frac{\max y_i(k) - y_i(k)}{\max y_i(k) - \min y_i(k)} \quad (1)$$

And similarly, the response variable, penetration (P), follows the larger-the-better case, as shown in Equation (2):

$$x_i(k) = \frac{y_i(k) - \min y_i(k)}{\max y_i(k) - \min y_i(k)} \quad (2)$$

In the above expressions (Equations (1) and (2)), $x_i(k)$ represents the grey relation generated value, where $\min y_i(k)$ and $\max y_i(k)$ shows the lowest and highest values for all kth responses, respectively, whereas $x_0(k)$ represents the initial (ideal) sequence for output responses (k = 1–9). The final grey grades show the DOF between sequences $x_0(k)$ and $x_i(k)$. The grey relational coefficient, $\delta_i(k)$ was calculated using Equation (3):

$$\delta_i(k) = \frac{\Delta_{\min} + \Psi\Delta_{\max}}{\Delta_{0i}(k) + \Psi\Delta_{\max}} \tag{3}$$

In Equation (3), $\Delta_{0i}(k)$ shows the deviation sequence of the reference, $x^*_{0(k)}$ and $x^*_{i(k)}$, i.e., the ideal and comparability sequences. The symbol Ψ represents the identification coefficient, and its value ranges from 0–1. Usually, it takes a value of 0.5 to ensure equal importance of all factors. The deviation sequence $\Delta_{0i}(k)$ is determined using Equation (4):

$$\Delta_{0i}(k) = |x_0{}^*(k) - x_i{}^*(k)| \tag{4}$$

$$y_i = \frac{1}{n}\sum_{k=1}^{n} \delta_i(k) \tag{5}$$

whereas in Equation (5), y_i and n show the final grey grades and response variables for corresponding experiments, respectively. Higher values of grey grades show better agreement between the ideal and given reference, i.e., $x_0(k)$ and $x_i(k)$. The reference sequence shows the better sequence; thus, the high value of grade corresponds to combinations of input parameters closer to the optimum for each experiment using $L_9 3^3$(OA). Plots of the main effect and grand mean of grey relational grades are essential for efficient analysis.

Table 3. Design matrix.

Input Factors				Output Factors		
				Smaller-the-Better		Larger-the-Better
Run	**F: Gas Flow Rate**	**I: Welding Current**	**G: Gap Distance**	**H: Bead Height (mm)**	**W: Bead Width (mm)**	**P: Penetration (mm)**
1	10	60	1.5	0.136	5.535	1.344
2	10	80	2.0	0.093	4.456	1.029
3	10	100	2.5	0.267	6.324	1.204
4	12	60	2.0	0.183	6.829	1.045
5	12	80	2.5	0.129	5.232	1.079
6	12	100	1.5	0.263	7.304	1.234
7	14	60	2.5	0.129	4.690	1.127
8	14	80	1.5	0.143	5.640	1.006
9	14	100	2.0	0.188	6.339	1.137

4. Grey Relational Analysis

Responses are measured after performing experiments according to the design plan, as provided in Table 3. After experimentation and data collection, the sequences $x_i(k)$ and $x^*_{i(k)}$ are calculated using Equations (2) and (3) by keeping both cases in mind, i.e., smaller and larger-the-better cases, respectively. Similarly, the deviation sequence $\Delta_{0i}(k)$ was determined using Equation (4) and, Δ_{max} and Δ_{min} are the higher and lower values in deviation sequences in each response, respectively, hence $\Delta_{max} = 1$ and $\Delta_{min} = 0$. As discussed earlier, the Ψ value is 0.5, which means all factors have equal importance. In Equation (3), $\delta_i(k)$ i.e., the grey relational coefficient for each run was determined, as shown in Table 4. Grey grades, y_i were determined by taking the average of grey coefficients using Equation (5). Table 4 shows that the second experiment gives the highest relational grade, which means it agrees with the theoretical results.

Table 4. Ranking of Grey Relational Grades in the optimization process.

Run No.	Grey-Coefficient			Grey Grades	Ranking
	Bead Height $\delta_i(1)$	Bead Width $\delta_i(2)$	Penetration $\delta_i(3)$		
1	0.669	0.569	1.000	0.746	2
2	1.000	1.000	0.349	0.783	1
3	0.333	0.433	0.547	0.438	7
4	0.492	0.375	0.361	0.409	9
5	0.707	0.647	0.389	0.581	4
6	0.339	0.333	0.606	0.426	8
7	0.707	0.859	0.438	0.668	3
8	0.635	0.546	0.333	0.505	5
9	0.478	0.431	0.449	0.453	6

It has been concluded that multiple performance characteristics have been transformed into optimized grey grades. As it is an orthogonal experimental design, it is easy to determine the main effects of input parameters with different grey grades. For instance, the average grey grade value for gas flow rate (F) at the first, second, and third levels were computed by taking the mean value of grey grades from the 1st to 3rd, 4th to 6th, and 7th to 9th experiments, as presented in Table 5.

Table 5. Mean grey grades.

Factors	Grey Relational Grades			Main Effect	Rank
	Level 1	Level 2	Level 3		
Gas Flow Rate	0.656 *	0.472	0.542	0.183	2
Welding current	0.608	0.623 *	0.439	0.184	1
Gap Distance	0.559	0.548	0.562 *	0.014	3

* Optimum levels (Mean grey relational grades $(y_m) = 0.5566$).

Similarly, the mean grey grades can be determined for the other factors, namely welding current and gap distance. The results for each factor at each level have been summarized in Table 5. As discussed earlier, product quality will be nearer to the optimum as the value of grey grades increases. Therefore, for adequate performance measures, higher grey grades are desirable. The optimal combination of the factors for optimum bead geometry is F1, A2, and G3, as presented in Table 5. The signal-to-noise ratios for each factorial combination relative to optimization use Equation (6), as provided in Table 6.

$$S/N = -10\log\left[\frac{1}{n}\sum_{i=1}^{n}\frac{1}{{y_i}^2}\right] \quad (6)$$

Table 6. Mean responses.

Factors	Mean S/N Response Values			Difference	Rank
	Level 1	Level 2	Level 3		
Gas Flow Rate	−3.949	−6.629	−5.442	2.680	2
Welding Current	−4.603	−4.258	−7.159	2.901	1
Gap Distance	−5.299	−5.590	−5.131	0.458	3

Note: The total mean OF s/n ratio is −5.345.

In the above equation, n is the number of experiments and y_i shows the measured output response. Mean S/N ratios for the three factors at each level are computed and presented in Table 6. MINITAB 16 software was used to plot the parameters' main effects, as shown in Figure 3. This figure illustrates the optimal condition of factors, i.e., F1, A2, and G3 (i.e., gas flow rate (F) = 10 L/min, welding current (I) = 80 A, and gap distance (G) = 2.5 mm), which is the same as obtained by grey relational analysis.

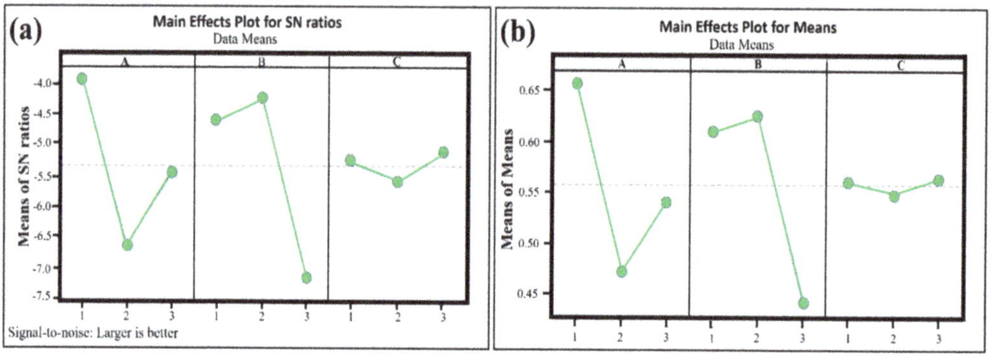

Figure 3. Results for verification of grey relational grades by signal-to-noise (SN) ratio (**a**) main effect plot for S/N ratio of process parameters (**b**) mean effect plot for grey relational grade. A: Gas Flow Rate, B: Welding Current, C: Gap Distance.

5. Finite Element Analysis (FEA)

The primary purpose of the finite element analysis method is to determine the solutions to different engineering problems with complex geometries to reduce the calculations and saves time. Experimentation can be used to find unknown factors and optimal combinations, but it requires additional testing and samples to study each behavior. This will has a high cost and time is needed to prepare each model and procure the equipment. The FEA technique predicts the approximate solutions to save time and cost within an acceptable range. Depending upon the accuracy level of the model, the optimized structure or shape is integrated into the finite element analysis. Different boundary conditions must be satisfied according to the structure and field variables. The steps are generally categorized into three main stages: (a) pre-processing, (b) solution, and (c) post-processing. Pre-processing includes the geometric and material properties of the elements, e.g., dimensions, element types, meshing the model, loadings, and boundary constraints. In the solution, the model is built to find the unknown variables. In post-processing, results are evaluated and interpreted.

5.1. Implementation of FEA (Material, Modeling, and Properties)

The present study analyzed the solid model using the finite element method (ANSYS APDL 2020R1) of the butt welded joint (mild steel). The model comprised two mild steel plates with dimensions 80 mm × 20 mm × 3 mm, welded at the ends using a V-shaped butt joint. The welded parts are modeled and analyzed to observe the temperature distribution and stresses induced during welding. The study was conducted by taking a single pass. Figure 4a,b shows the solid and meshed geometric models. The temperature at each node was adapted in the welding environment for the element types, PLANE 55 (2D), containing four nodes, and SOLID70 (3D) with a single DOF. Meshing was performed in four areas (mm) as A1 (0.0012), A2 (0.0025), A3 (0.005), and A4 (0.0065) for better estimation. Transient thermal analysis was first performed by providing heat flux concerning time to predict the temperature change. Then, stresses and distortion were obtained by coupling the thermal and static structural analyses. The mesh size used in both steps was the same. It was assumed that the base metals and welded regions have the same thermal properties. There is no penetration, and overfilled welds are considered for simplicity.

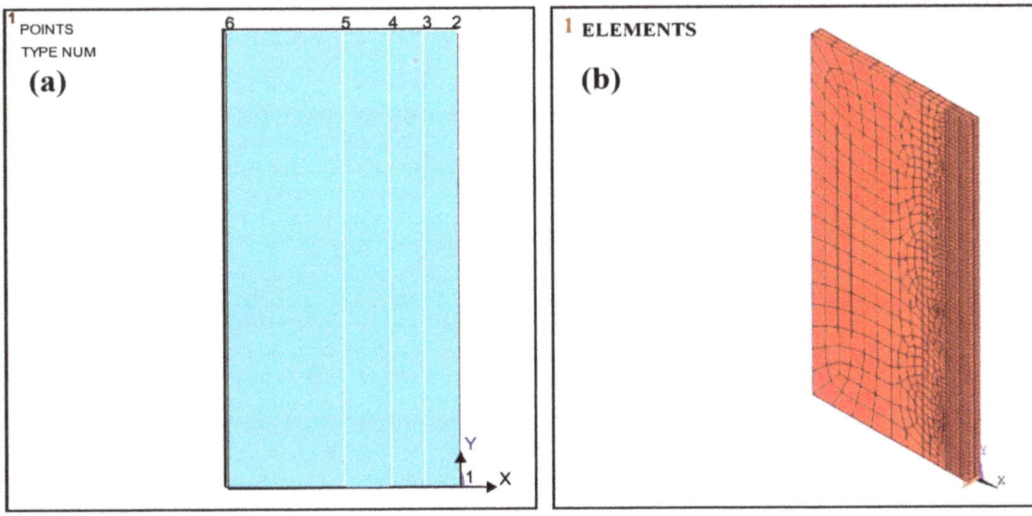

Figure 4. (**a**) Solid model. (**b**) Meshed model using ANSYS Mechanical APDL. 1. first pick point on welding area, 2. second pick point on welding area, 3. third pick point on welding area, 4. fourth pick point on welding area, 5. fifth pick point on welding area, 6. sixth pick point on welding area.

The temperature-dependent properties, such as Poisson's ratio, Young's modulus, conductivity, density, and coefficient of thermal expansion, were characterized through differential scanning calorimetry and dynamic mechanical analysis, as illustrated in Figure 5a–f. The melting temperature of mild steel is 1450 °C.

5.2. Thermal Loading

The heat flux varies with time after the thermal load is applied during thermal analysis. Figure 6 shows that the load inputs are provided in three steps (t1–t3), and the total time is up to 1100 s. For the first ten seconds, a ramp input is given, followed by a step input up to 100 s, and then again, a ramp input is given up to the end. When welding begins, the heat steadily increases until it reaches a particular value. For a short period, it remains stable (no change) and then decreases steadily until the temperature of the welded plates reaches room temperature.

The input heat was determined by the product of arc efficiency, provided voltage, and welding current, which are 0.7 V, 15 V, and 80 A, respectively, in the present study. The maximum heat flux determined is 0.62×10^6 W/m^2. The node solutions for temperature in this step were regarded as inputs into the stress analysis. This is because the residual stresses produced in heating and cooling periods result from temperature fields.

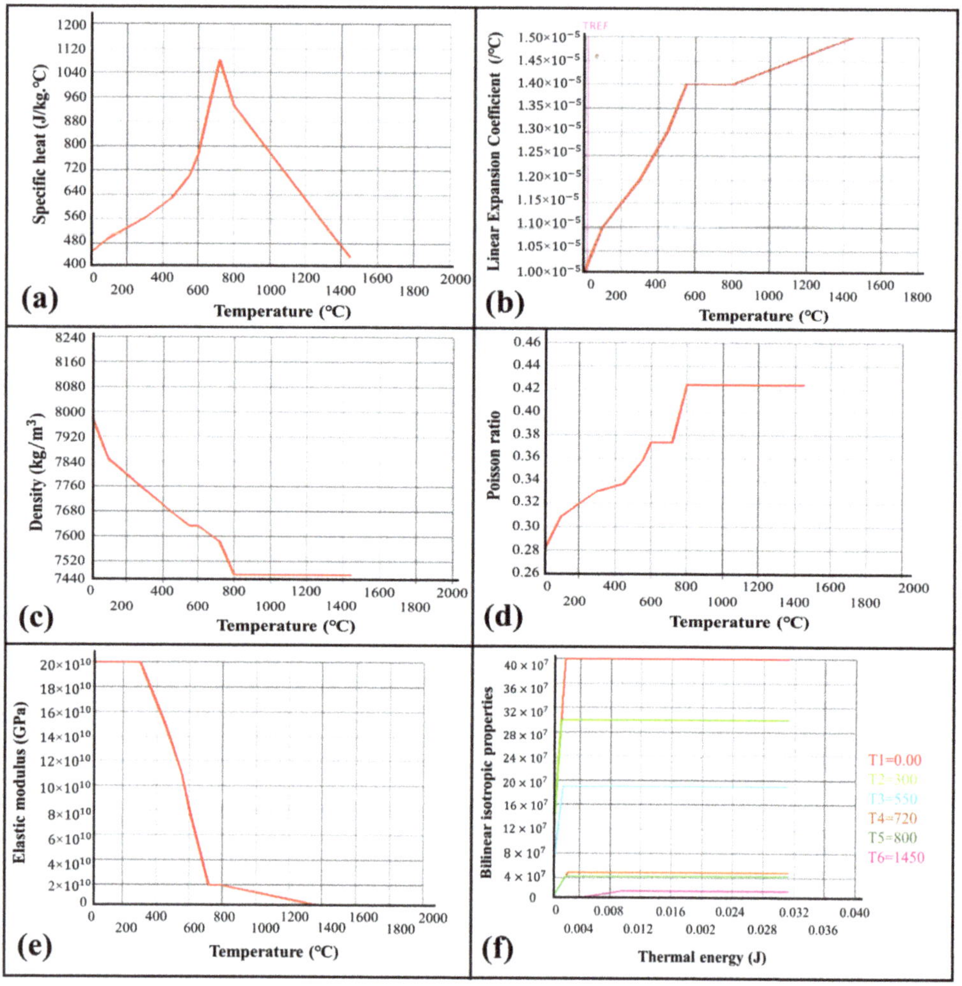

Figure 5. Temperature-dependent thermal properties of mild steel considered for the analysis. (**a**) Specific heat. (**b**) Linear expansion coefficient. (**c**) Density. (**d**) Poisson's ratio. (**e**) Elastic modulus coefficient. (**f**) Bilinear isotropic properties.

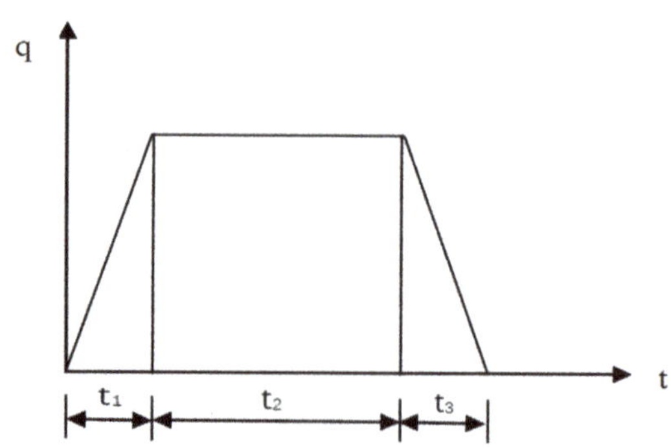

Figure 6. Load inputs function.

5.3. Boundary Conditions

The boundary conditions should be set for the thermal and static analysis, which can be applied to the plates for different parameters. The boundary conditions for other parameters used in this study are the following.

5.4. Thermal Analysis

The convection heat was applied to the surface area of the plates, which are meshed in identical elements in the symmetric model, as illustrated in Figure 7. The Gaussian heat source was applied on surface areas A5 and A10.

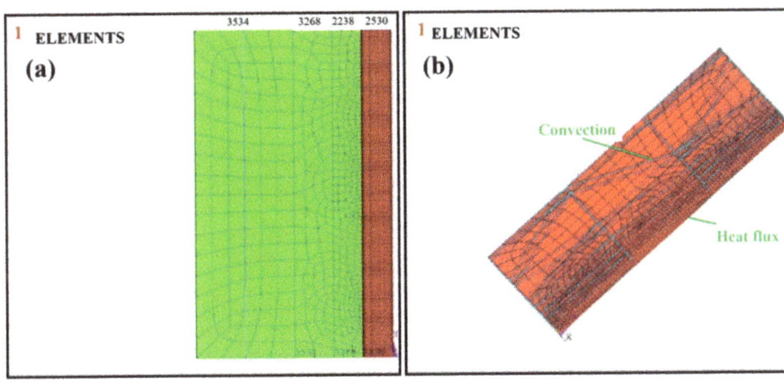

Figure 7. Boundary conditions applied to the thermal analysis. (**a**) number of elements in the constrained area (**b**) application of convection and heat flux to the plate area in symmetric model.

The welded region where the plates are in contact is supposed to be insulated, i.e., no heat flow across this region. The value of the coefficient of convection is 20 W/m^2 °C. The uniform temperature, 220 °C, was given to other remaining areas.

6. Static Analysis

After the thermal analysis of temperature distribution, the model is inserted into the static analysis, so new conditions should be applied. During static analysis, the degree of freedom should be zero to constrain the model in all directions.

7. Heat Transfer and Mechanical Analysis

Heat transfer and mechanical analysis is the first step in the FEA while running the TIG welding simulation using ANSYS. The following equation can describe the formulation in FEA for the elements:

$$[C(T)]\{\hat{T}\} + [K(T)]\{T\} = \{Q(T)\} \tag{7}$$

The heat conduction equation must be integrated with regard to time for this study. The Crank–Nicolson/Euler theta integration method is used to solve these system equations. This element type can conduct heat in three dimensions. The heat conduction versus time equation must primarily be integrated into this kind of analysis. The heat is provided by the electrode as the heat flux and is considered input for heat transfer from the electrode to the plates based on setting factors and arc efficiency. The heat flux density at any point from the heat source center can be computed using the following equation:

$$Q = Q_m e^{\frac{-3r^2}{R^2}} \tag{8}$$

$$r^2 = (X - 0)^2 + (Y - V \times Time)^2 \tag{9}$$

Since, at the y-axis, the value of x is zero, Equation (8) can be written as:

$$Q = Q_m e^{\frac{-3(X^2+(Y-VT)^2)}{R^2}} \tag{10}$$

where Q_m is the maximum heat at the center of the heating source, sub-degree R is the effective radius of the heating source, and r is the distance between the center of the arc heating source and point A. The heat source model estimates temperature with finite element analysis in TIG welding. For stress analysis, heat transfer analysis was carried out to compute the temperature at the nodes as the time function, and then the structural analysis was performed using the temperatures attained from the heat analysis. The input values of the selected factors and constant factors used in the simulation are given in Table 7.

Table 7. Input parameters for simulation.

Weldment Length (L) (mm)	Weldment Width (W) (mm)	Weldment Height (H) (mm)	Welding Voltage (U), Volts	Welding Current (I), Amp	Welding Speed (V), mm/sec	Welding Thermal Efficiency (YITA)	Effective Heat arc Rad© (R), (mm)
0.1	4.456	0.093	15	80	0.01	0.7	0.003

The constant voltage of 15 V was supplied to each experiment, and the selected value for welding current achieved from the grey relational analysis was 80 A. The value of total heat flux at the welded region was calculated using the following equation:

$$q = \frac{Q}{A} \tag{11}$$

where Q is heating flux density, and A is the area of the welded region, which is the same for all the experiments. By computing the value of Q using the equation and the area of the welded triangular region ($Q = 840$ W and $A = 4.5 \times 300 \times 10^{-6}$ m^2), the total heat flux was determined ($q = 0.62 \times 10^6$ W/m^2). The FEA for the thermal and structured analysis of TIG welding using ANSYS Mechanical APDL works according to the following Equations:

$$K\left(\frac{\partial^2 T}{\partial x^2} + \frac{\partial^2 T}{\partial y^2} + \frac{\partial^2 T}{\partial z^2}\right) - q + h(T - T\dot{\alpha}) = 0 \tag{12}$$

$$[C]\{T\} = \{f\} \tag{13}$$

$$[K^*]\{q\} = \{F\} \tag{14}$$

$$[K]^* = \frac{AE}{L}\begin{pmatrix} 1 & -1 \\ -1 & 1 \end{pmatrix} \tag{15}$$

$$F = \int B^t D\varepsilon_0 \delta\theta \tag{16}$$

The notations used in the above equations "SDPBS. http://sdpbs.math.uwm.edu/about.php (accessed on 22 December 2022)" are explained in Table 8. The total time to arrive at the solution for temperature was 1100 s, illustrated in Figure 8, and the number of sub-steps in each iteration was 50. The x-axis and y-axis are labeled as cumulative iteration numbers and absolute convergence norms.

Table 8. Notations and their explanation.

Symbol	Meaning	Symbol	Meaning
K	Thermal conductivity	A	Welding area
T	Initial temperature	E	Young's modulus
X	Plate width	F	Force function in structural
Y	Plate length	L	Length of material
Z	Plate thickness	[B]	Strain displacement matrix
q	Heat flux (Displacement function)	[D]	Elasticity matrix
h	Coefficient of heat transfer	f	Force function in thermal
Tà	Exposed temperature	[*K*]	Stiffness matrix
C	Shape matrix		

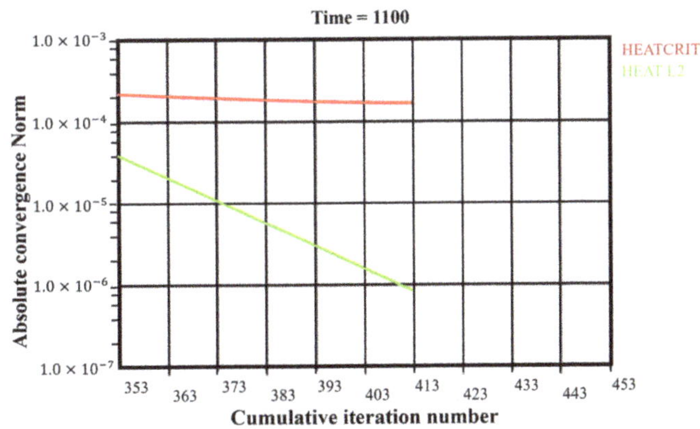

Figure 8. Convergence curve for thermal analysis.

The absolute convergence norm uses normalized values to quickly solve the variables using the Newton–Raphson method. A solution within the range means the function value should be higher than the value of the vector norm of the function for each sub-step. It is observed that the HEAT CRIT values are higher than the HEAT L2 (vector norm) at every sub-step, so it can be concluded that the solution is within tolerance and is correctly converged.

8. Results and Discussion

8.1. Regression Model for Grey Relational Grades and Corresponding Response Variable

The quadratic regression model was employed to analyze the grey relational grades for optimizing the corresponding bead height, width, and penetration. The regression model for an overall grey grade is expressed by Equation (18):

$$\begin{aligned} GRG &= 0.5917 - 0.0607F - 0.0007I - 0.0107G \\ &-0.0247FI + 0.1673FG - 0.0077IG + 0.1907F^2 \\ &-0.1833I^2 + G^2 \end{aligned} \tag{17}$$

The mathematical models predict the response variables (bead height, bead width, and penetration) for the given input parameters to the grey relational grades given by the Equations (19)–(21). A two-factor interaction model was suggested to predict the actual bead height, bead width, and penetration:

$$\begin{aligned} H &= -0.0857 + 0.0573F + 0.01401I - 0.7404G \\ &-0.0015FI + 0.0334FG + 0.0034IG \\ &-0.1833I^2 + G^2 \end{aligned} \tag{18}$$

$$W = +1.8961 + 0.3428F + 0.04298I - 0.1970G \\ -0.0021FI - 0.0716FG + 0.00211IG \tag{19}$$

$$P = +5.8028 - 0.1879F + 0.01389I - 4.2446G \\ -0.00328FI + 0.02321FG + 0.01587IG \tag{20}$$

In the above equations, the terms GRG, H, W, and P show the grey grades, bead height, width, and penetration, respectively. The parameters F, I, and G are the gas flow rate, welding current, and gas distance, respectively. Statistical optimum value for grey relational grade obtained for the welding current range is 61 to 83 (A), and the gas flow rate is up to 1.79 (L/min). According to the optimum results achieved from grey relational analysis, simulation of the TIG welding has been performed on ANSYS APDL 2020R1 using finite element analysis to analyze the temperature field and thermal stresses induced. Autodesk Inventor carried out the welding and then imported it to the ANSYS workbench for further research. The following results have been achieved for temperature and stress analysis. First, the temperature variations over time and thermal stresses for the optimum geometry are studied. ANSYS APDL was used to generate the model, and finite element analysis (FEA) was employed to predict the responses. Second, a transient analysis followed by the coupled field analysis was performed to predict temperature variations and thermal stresses. Finally, the voltage, efficiency, and welding variation on the temperature field are presented.

8.2. Temperature Analysis

Based on the simulation and analysis of the optimized weld bead geometry of the model, the temperature field data from the thermal analysis are loaded into the stress analysis. Figure 9a shows the start of the TIG welding to move the torch at the first subsequent step.

The minimum temperature is equal to the uniform temperature (22 °C), and the maximum temperature provided to melt the boundary for welding both the plates is 1310.99 °C for V = 0.01 mm/s at the first point of a welded region, which is less than the melting temperature (1450 °C). The maximum temperature observed in the welded area at the last pick is 2083.63 °C which exceeds the melting temperature of mild steel (1450 °C), as shown in Figure 9b. This depicts the temperature at the weld region subject to the boundary conditions. The maximum temperature should be at the flux area. It also shows that the temperature drops quickly from the boundary to the central region and is distributed through the transverse and longitudinal axis, as provided in Figure 9c. The thermal area also changes as the heat source moves along the path and the pool that melts moves with the heat source. The maximum temperature drops to 24.26 °C and becomes nearly equal to the uniform/room temperature, showing that the plates quickly cool down after welding.

This occurs in the cooling cycle, which shows a slight decrease in temperature after the thermal phase until it reaches room temperature. Figure 9d describes the temperature variation with the time away from the heat zone. This curve is achieved by selecting the nearest node to the welded region. It depicts the temperature change with time and distance. When a certain amount of heat is applied, the temperature gradually increases to the peak value during the first phase, then drops to room temperature at a constant rate, called a cooling cycle. It is clear from the graph that the peak temperature recorded for the given time of 1100 s is 2083.63 °C on the y-axis. For a small range of the x-axis, i.e., 0–20 s, it can be observed that the peak temperature is reached very quickly, but as the torch moves further, the temperature is distributed uniformly till the end of the pass. Then the cooling cycle takes place to achieve its room temperature.

The results of heating are quite similar to Liu et al. (2022) [29] and Nie et al. (2020) [30] for aluminum, which showed the reliability of our conclusions.

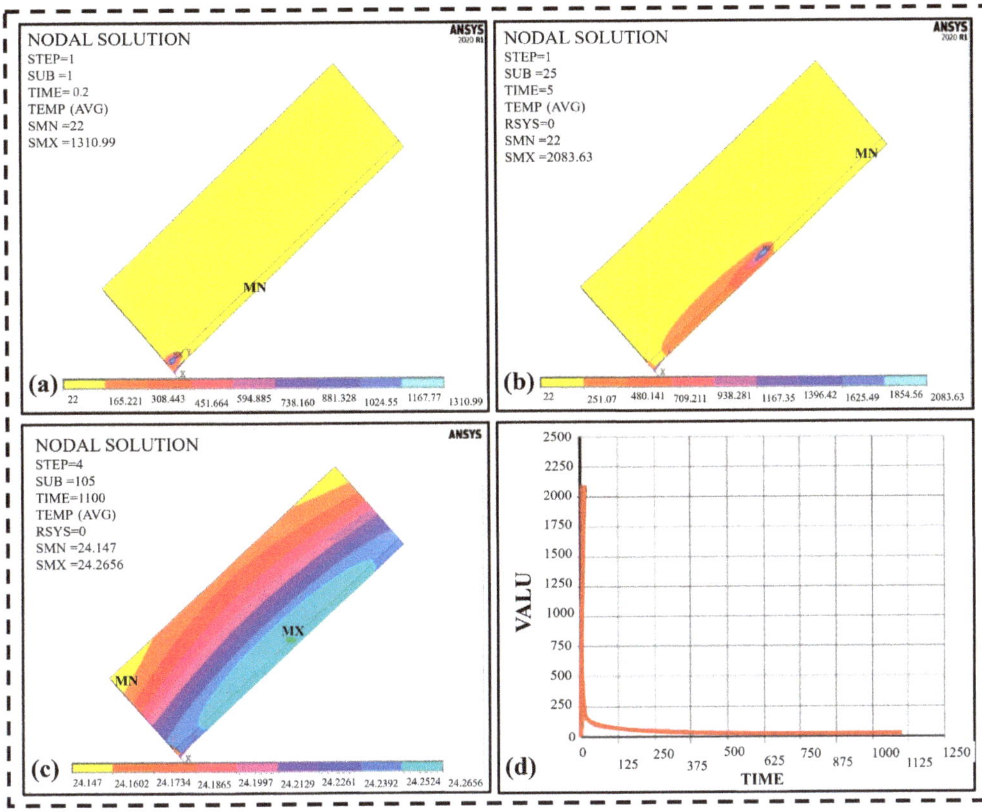

Figure 9. Temperature distribution at time t at subsequent steps: (**a**) 1st pick (beginning of weld) (**b**) 25th pick (**c**) after the weld pass (cooling) (**d**) Variation of temperature with the time for 1100 s (one complete pass).

8.3. Stress Analysis

Prediction of thermal stresses is essential to minimize the failure within welded plates of mild steel, which causes the failure of the welds. Thermal stresses result from thermal analysis in the subsequent process, as shown in Figure 10a–c. At the beginning of the welding process, the tensile thermal stresses are developed but then turn into compressive thermal stresses upon solidification (cooling). When the heat source moves away, the parts are cooled down, and the atoms are compressed, resulting in nodes being compressed. Figure 10a shows the beginning of the welding process when the heat source starts to melt the material, which indicates that the maximum thermal stress recorded is 416 MPa at the third step. As the heat source moves further (STEP-10), the thermal stress increases and reaches 424 MPa, as presented in Figure 10b. Finally, when the first pass was completed (STEP-205, TIME-1100s), thermal stresses decreased and dropped to 388 MPa, as presented in Figure 10c. These changes are associated with temperature variations, which correctly describe the process [30].

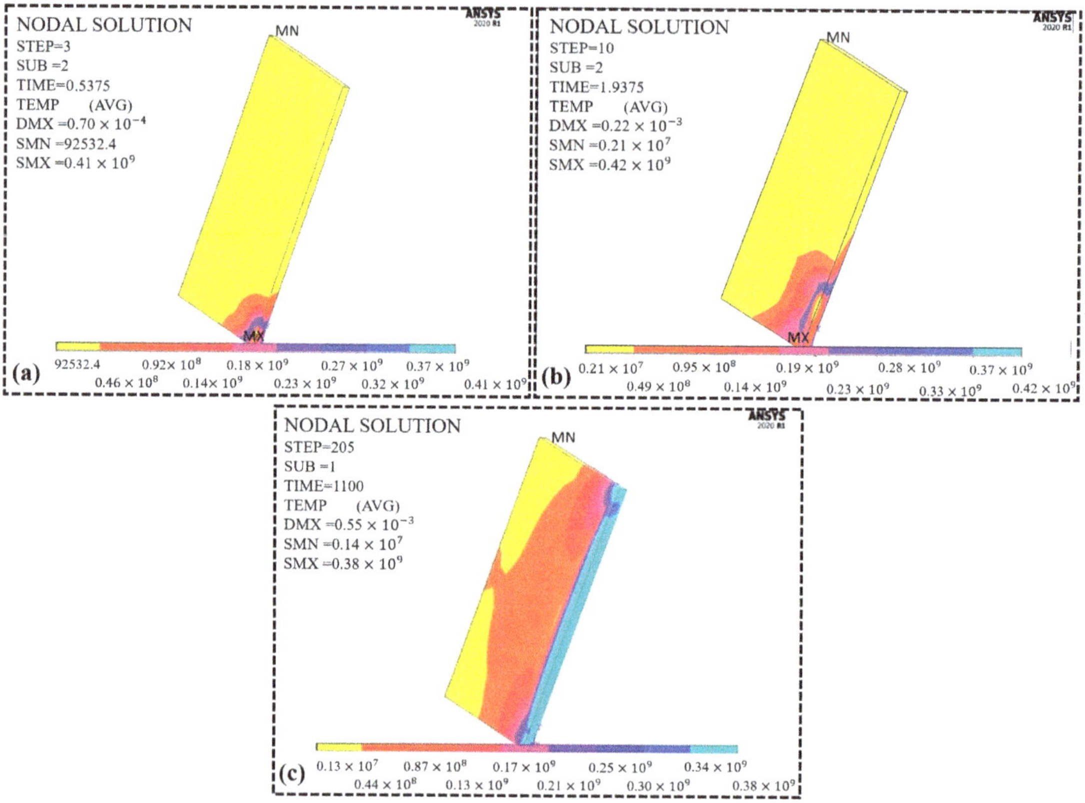

Figure 10. Thermal stress distribution (**a**) 1st pick (**b**) last pick (complete pass) (**c**) after the weld pass (cooling).

The thermal stresses increase due to undergoing early heating and cooling mechanisms, which results in thermal gradients and associated thermal stresses. Thermal gradients decrease as the temperature increases, which results in higher thermal stresses. As the process repeats in TIG welding, thermal stress may be produced inside the parts. Heat treatment can minimize this issue before other operations, such as surface cleaning, finishing, welding, etc. The thermal stress will affect the welding quality of the mild steel parts due to deflection. The pre-heating mechanism will increase the material's ductility, which improves the chances of relieving the thermal stresses. It will produce deformation in different directions to minimize thermal stresses, i.e., the x-axis, y-axis, and z-axis. Figure 11a shows the deformation in the x-direction along the path of the heat source. This is due to the thermal/residual stresses, mainly resulting from temperature variations. The maximum deformation recorded along the direction of the heat source is 0.440×10^{-4} m (0.044 mm).

Similarly, the maximum deformation along the y-axis is 0.0047 mm, which is the transverse direction to the direction of the heat source, as shown in Figure 11b. The maximum deformation is demonstrated at the welding start due to the maximum stress in this region. As the heat source moves linearly along the x-axis, less deformation occurs as there is no movement along the y-axis. The maximum stress is distributed uniformly in the central region, shown by the red area. Figure 11c shows the deformation along the z-axis indicated by the plates' thickness. The maximum deformation recorded along the thickness of the plates is 0.556 mm. The resultant deformation is the vector sum of all the components, as shown in Figure 11d. The maximum resultant deformation recorded in the plates is 0.558 mm. The main reason for induced deformation is the thermal stresses resulting from temperature variations. The more uniform the stress is distributed, the less deformation will be induced in the plates.

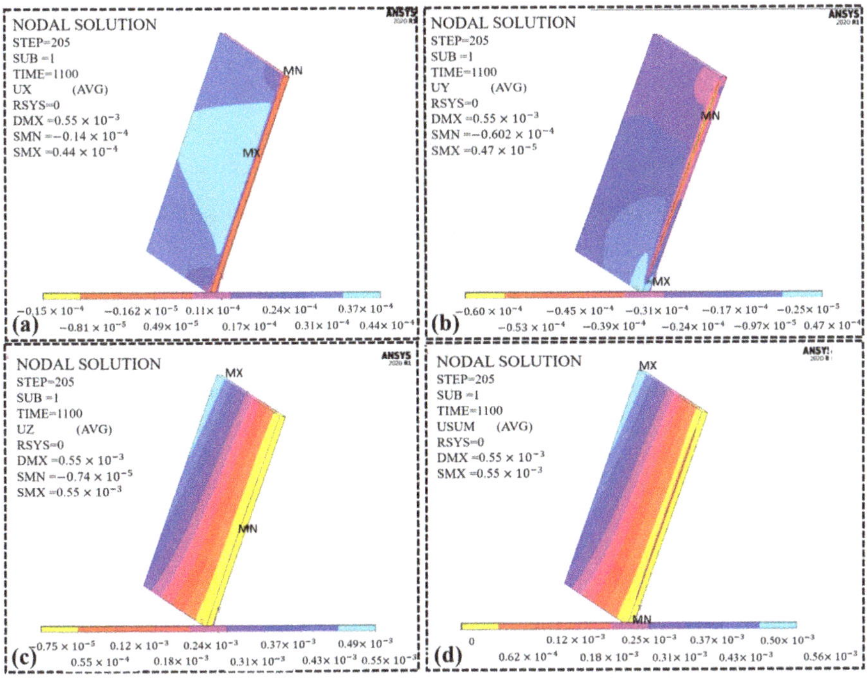

Figure 11. Deformation along with the three-axis (**a**) along the x-axis in the welding direction, (**b**) along the y-axis transverse to the welding direction, (**c**) along the z-axis through the thickness of the plates, and (**d**) vector sum of all the components.

8.4. Effect of Welding Parameters on Temperature Field

After finite element analysis of TIG welding of an optimized bead geometry, it is necessary to determine the effect of parameters on temperature to minimize the thermal stresses in welding. In the present analysis, the parameters considered for investigation vary in voltage, efficiency, and welding speed. The thermal stresses are produced due to temperature variations caused by the process parameters, so the temperature needs to be investigated. The results are summarized in Table 9.

Table 9. Variations in temperature with the welding parameters.

Variables		Welding Current, A	Welding Area, mm^2	Heat Flux, W/m^2	Other Parameters		Temperature
Voltage, U (volts)	12	80	1350	0.48×10^6	E = 0.7	V = 0.01	1362.22
	13	80	1350	0.52×10^6	E = 0.7	V = 0.01	1771.05
	14	80	1350	0.57×10^6	E = 0.7	V = 0.01	1926.80
	15	80	1350	0.62×10^6	E = 0.7	V = 0.01	2083.63
Efficiency, E (%)	0.6	80	1350	0.53×10^6	U = 15	V = 0.01	1750.01
	0.7	80	1350	0.62×10^6	U = 15	V = 0.01	2083.63
	0.8	80	1350	0.71×10^6	U = 15	V = 0,01	4001.78
Welding Speed, V (mm/sec)	0.01	80	1350	0.62×10^6	U = 15	E = 0.7	2083.63
	0.02	80	1350	0.62×10^6	U = 15	E = 0.7	1329.57
	0.03	80	1350	0.62×10^6	U = 15	E = 0.7	243.78

8.5. Effect of Variations in Voltage on Temperature

The change in voltage will lead to temperature variation, resulting in thermal stresses. The values taken for voltage are 12 V, 13 V, 14 V, and 15 V, and keep other parameters constant, are shown in Figure 12a–d. It can be observed from the figure that as the voltage increases from 12 V–15 V, the temperature also increases from 1362.22 °C to 2083.63 °C. The peak temperatures for voltages of 12, 13, 14, and 15 volts are 1362.22 °C, 1771.05 °C, 1926.80 °C, and 2083.63 °C, respectively. Therefore, it is concluded that the temperature increases with the voltage increase. The results were also verified by comparing with Ghazvinloo et al. [31], which showed that voltage significantly impacts the temperature, which will further affect fatigue life, energy, and bead penetration.

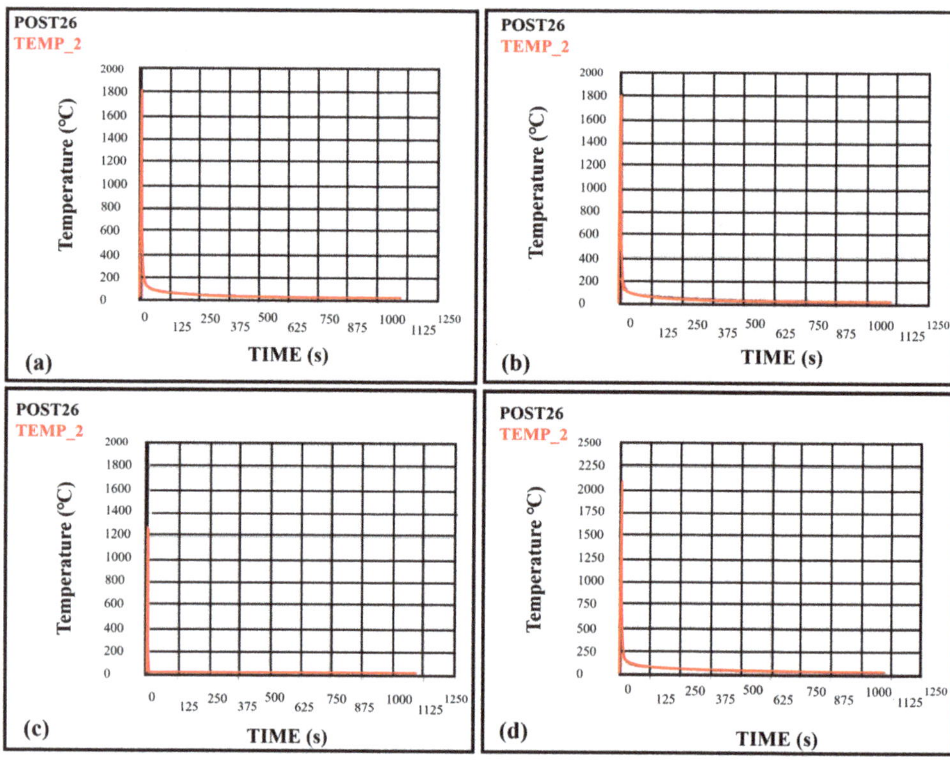

Figure 12. Temperature variation with time at the weld zone (**a**) for 12 Volts (**b**) for 13 Volts (**c**) for 14 Volts (**d**) for 15 Volts.

8.6. Effect of Variations in Efficiency on Temperature

The efficiency variation will also change temperature, further leading to thermal stresses. The efficiencies taken in this study are 0.60, 0.70, and 0.80, and other parameters are kept constant, as shown in Figure 13a–c. It can be determined from the figure that the temperatures at the weld zone for efficiencies of 0.60, 0.70, and 0.80 are 1750.01 °C, 2083.63 °C, and 4001.78 °C, respectively. Therefore, it is concluded that as the efficiency of the weld joint increases, the peak temperature also increases [32]. Arc efficiency measures how effectively the electric arc is used during TIG welding. It is calculated by comparing the amount of heat generated by the arc to the amount of electrical energy consumed. To calculate the arc efficiency (E) during TIG welding, arc voltage (V), arc current (I), welding speed (v), and electrode efficiency ($E_{electrode}$). The formula for efficiency is:

$$E = \left[\frac{H_i}{E_i}\right] \times 100\% \quad (21)$$

$$\text{where, } \left\{ \begin{array}{l} H_i = V \times I \times E_{electrode} \times 60 \\ E_i = V \times I \times 60 \end{array} \right\} \quad (22)$$

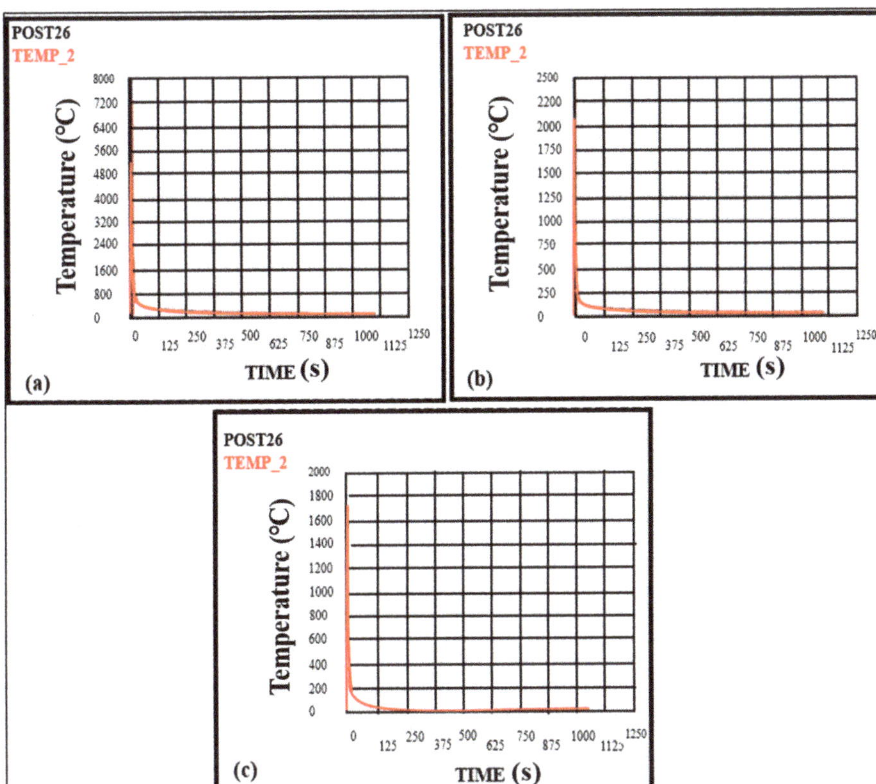

Figure 13. Temperature variation with time at the weld zone (**a**) for 0.6 Efficiency (**b**) for 0.7 Efficiency (**c**) for 0.8 Efficiency.

This is a factor that takes into account how efficiently the electrode is transferring welding current to the workpiece. The electrode efficiency for TIG welding is typically around 70%. Note that voltage, welding current, and welding speed values should be measured during the welding process. The electrode efficiency value is an estimate based on the type of electrode being used.

8.7. Effect of Variation in Welding Speed on Temperature

The change in welding speed also affects the temperature field, which will cause thermal stresses and anisotropy in the weld pool. The temperature variation is also studied for the welding speed of 0.01, 0.02, and 0.03 mm/sec, as shown in Figure 14a–c. The temperatures obtained for the welding speed of 0.01, 0.02, and 0.03 mm/sec are 2083.63 °C, 1329.57 °C, and 243.78 °C, respectively. It is concluded that the temperature decreases with the increase in welding speed and reaches its lowest value. In a way, the temperature field can be optimized by varying the welding parameters. Welding speed also significantly influences the microstructure and mechanical properties of the joining due to the variation of temperature field, which is also stated by Cevik et al. (2021) [33].

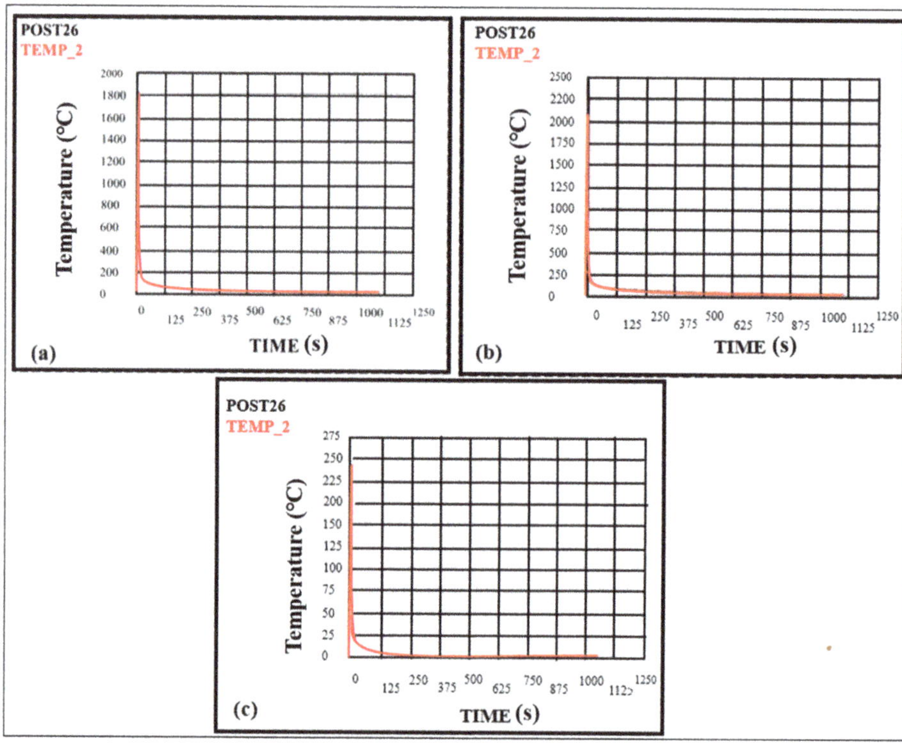

Figure 14. Temperature variation with time at the weld zone (**a**) for v = 0.01 mm/s (**b**) for v = 0.02 mm/s (**c**) for v = 0.03 mm/s.

Confirmation Test for Optimization

After selecting optimal parameters, it is necessary to predict and verify the responses of the welding process with an optimum combination of process parameters. Firstly, the developed models for all the responses were validated through six additional runs. The values for input models (other than the selected design) lie in the defined ranges taken for additional confirmatory runs. The final calculations and results are summarized in Table 10. The percent error for each predicted and experimental response for comparison was determined by Equation (21) [34]. A percent error of less than 5% will be in reasonable agreement to predict the bead height, bead width, and penetration with reasonable accuracy. Therefore, it can be observed from the table that all the predicted and experimental results are in good agreement and validate the developed models. The predictable error in grey relational grade ($\hat{y}$) is calculated by optimum levels of input process parameters using Equation (22) [35] as:

$$Error(\%) = \left| \frac{Experimental - Predicted}{Predicted} \right| \times 100 \quad (23)$$

$$\hat{y} = y_m + \sum (\overline{y} - y_m) \quad (24)$$

Table 10. Comparison of Experimental and Predicted Results.

Runs	Input Variables					Response Variables		
	Gas Flow Rate	Welding Current	Gap Distance			Bead Height	Bead Width	Penetration
1	13	90	1.8	Experimental		0.187	6.090	1.125
				Predicted	GRD = 0.514	0.191	6.230	1.138
				Error		2.094	2.241	1.142
2	13	90	2.3	Experimental		0.197	5.930	1.251
				Predicted	GRD = 0.583	0.193	5.790	1.239
				Error		2.073	2.418	0.968
3	11	70	2.3	Experimental		0.08	5.217	0.890
				Predicted	GRD = 0.550	0.0791	5.303	0.850
				Error		1.138	1.622	4.705
4	11	70	1.8	Experimental		0.141	5.568	1.090
				Predicted	GRD = 0.639	0.145	5.692	1.140
				Error		2.759	2.178	4.385
5	13	70	1.8	Experimental		0.195	5.870	1.168
				Predicted	GRD = 0.556	0.191	5.810	1.137
				Error		2.094	1.033	2.726
6	11	90	2.3	Experimental		0.178	5.989	1.197
				Predicted	GRD = 0.524	0.173	5.836	1.141
				Error		2.890	2.622	4.907

In the above expression, y_m represents the total mean value of the grey relational grade, $\overline{y_i}$ shows the grade value at an optimum level, and $\hat{y}$ represents the predicted value of the grey relational grade. Table 11 compares the predicted values with the actual values of bead geometry using optimal welding parameter combinations.

Table 11. Confirmation Test Results.

Process Parameters Level	Initial Factor Setting	Optimal Factor Setting	
		Predicted Optimal Factors Setting	Experimental Optimal Factors Setting
	F1, A1, G1	F1, A2, G3	F1, A2, G3
Bead height (BH)	0.136		0.106
Bead width (W)	5.535		5.728
Penetration(P)	1.344		1.938
S/N ratio for overall GRD	−3.17	−2.65	−1.78
Overall Grey GRD	0.69	0.74	0.83

Improvement in S/N ratio 1.39, Improvement in grey relational grade 0.14.

In this research, performance characteristics have been selected with the parameters of bead geometry and given a weightage equal to all responses. The results show that using the optimum setting of process parameters (F1, A2, and G3) caused higher penetration and less bead height and bead width.

9. Conclusions

In this study, we utilized the Taguchi-grey relational analysis approach to optimize the bead geometry in TIG welding. We also conducted numerical simulations using finite element analysis to analyze temperature and stress variations. Our results showed that the welding current was the most influential factor affecting the performance measures, followed by the flow rate. We found that the optimal values to optimize the bead geometry were a heat flux of 0.62×10^6 W/m^2, a gas flow rate of 10 L/min, a welding current of 80 A, and a gap distance of 2.5 mm. The maximum deformations induced by the thermal stress along the x-axis, y-axis, and z-axis were 0.044 mm, 0.0047 mm, and 0.556 mm, respectively.

Less deformation occurred along the y-direction as there was no motion of the heat source. By changing the voltage, efficiency, and welding speed, we observed that the temperature field increased with voltage and efficiency but decreased by increasing the welding speed of the torch. Our models were validated through additional experiments.

Overall, this study demonstrates the potential of using Taguchi-grey relational and finite element analyses to optimize the bead geometry in TIG welding. By achieving uniform temperature and stress distribution, we can improve the mechanical properties and quality of the weld while saving resources and reducing costs. In future studies, we can consider heat dissipation or losses and multiple passes for even more precise modeling.

Author Contributions: Conceptualization, M.H. and A.H.S.; Methodology, M.H. and A.H.S.; Software, I.S. and J.M.; Writing—original draft, M.H. and A.H.S.; Writing—review & editing, I.S.; Supervision, J.M.; Funding acquisition, I.S. and J.M. All authors have read and agreed to the published version of the manuscript.

Funding: International and Hong Kong, Macao and Taiwan high end talent exchange funding of Guangdong Province (No. 2022A0505020007), Research and Academic Exchange on Intelligent Balance Optimization Methods for Assembly Lines.

Institutional Review Board Statement: Not applicable.

Informed Consent Statement: Not applicable.

Data Availability Statement: The data presented in this study is available on request from the corresponding author.

Conflicts of Interest: The authors declare no conflict of interest.

References

1. e Silva, R.H.G.; dos Santos Paes, L.E.; Okuyama, M.P.; de Sousa, G.L.; Viviani, A.B.; Cirino, L.M.; Schwedersky, M.B. TIG welding process with dynamic feeding: A characterization approach. *Int. J. Adv. Manuf. Technol.* **2018**, *96*, 4467–4475. [CrossRef]
2. Saha, D.; Sushma, S.; Jalageri, M. Parametric optimization of Tig welding for M1020 using Taguchi-grey relation based design method. In Proceedings of the International Conference on Emerging Trends in Science, Engineering and Management, Hyderabad, India, 17–18 March 2017.
3. Datta, S.; Bandyopadhyay, A.; Pal, P.K. Grey-based Taguchi method for optimization of bead geometry in submerged arc bead-on-plate welding. *Int. J. Adv. Manuf. Technol.* **2008**, *39*, 1136–1143. [CrossRef]
4. Amiri, E.; Ostovan, F.; Toozandehjani, M.; Shafiei, E.; Mohamed, I.F. Study and selection of most appropriate filler rod for GTAW of S32750 super duplex steel joints: A comprehensive study on microstructural, mechanical and corrosion properties. *Mater. Chem. Phys.* **2021**, *270*, 124839. [CrossRef]
5. Cui, S.; Pang, S.; Pang, D.; Zhang, Z. Influence of Welding Speeds on the Morphology, Mechanical Properties, and Microstructure of 2205 DSS Welded Joint by K-TIG Welding. *Materials* **2021**, *14*, 3426. [CrossRef]
6. Eshwar, D.; Kumar, A. Taguchi based mechanical property optimization of as weld Al-65032 alloy using TIG welding. *J. Mech. Civ. Eng.* **2014**, *11*, 56–62. [CrossRef]
7. Dhas, J.; Kumanan, S. ANFIS for prediction of weld bead width in a submerged arc welding process. *J. Sci. Ind. Res.* **2007**, *66*, 335–338.
8. Greyjevo, O.; Metodo, A.I.T. Optimization of weld bead geometry in TIG welding process using grey relation analysis and Taguchi method. *Mater. Tehnol.* **2009**, *43*, 143–149.
9. Veiga, F.; Suárez, A.; Aldalur, E.; Bhujangrao, T. Effect of the Metal Transfer Mode on the Symmetry of Bead Geometry in WAAM Aluminum. *Symmetry* **2021**, *13*, 1245. [CrossRef]
10. Chandrasekhar, N.; Vasudevan, M.; Bhaduri, A.; Jayakumar, T. Intelligent modeling for estimating weld bead width and depth of penetration from infra-red thermal images of the weld pool. *J. Intell. Manuf.* **2015**, *26*, 59–71. [CrossRef]
11. Chokkalingham, S.; Chandrasekhar, N.; Vasudevan, M. Predicting the depth of penetration and weld bead width from the infra red thermal image of the weld pool using artificial neural network modeling. *J. Intell. Manuf.* **2012**, *23*, 1995–2001. [CrossRef]
12. Kolahan, F.; Heidari, M. A new approach for predicting and optimizing weld bead geometry in GMAW. *Int. J. Mech. Syst. Sci. Eng.* **2010**, *2*, 138–142.
13. Zhang, Z.; Chen, S. Real-time seam penetration identification in arc welding based on fusion of sound, voltage and spectrum signals. *J. Intell. Manuf.* **2017**, *28*, 207–218. [CrossRef]
14. Ostovan, F.; Hasanzadeh, E.; Toozandehjani, M.; Shafiei, E.; Jamaluddin, K.R.; Amrin, A. Microstructure, Hardness and Corrosion Behavior of Gas Tungsten Arc Welding Clad Inconel 625 Super Alloy over A517 Carbon Steel Using ERNiCrMo3 Filler Metal. *J. Mater. Eng. Perform.* **2020**, *29*, 6919–6930. [CrossRef]

15. Bandhu, D.; Vora, J.J.; Das, S.; Thakur, A.; Kumari, S.; Abhishek, K.; Sastry, M.N. Experimental study on application of gas metal arc welding based regulated metal deposition technique for low alloy steel. *Mater. Manuf. Process.* **2022**, *37*, 1727–1745. [CrossRef]
16. Maurya, A.K.; Pandey, C.; Chhibber, R. Influence of heat input on weld integrity of weldments of two dissimilar steels. *Mater. Manuf. Process.* **2023**, *38*, 379–400. [CrossRef]
17. Assefa, A.T.; Ahmed, G.M.S.; Alamri, S.; Edacherian, A.; Jiru, M.G.; Pandey, V.; Hossain, N. Experimental Investigation and Parametric Optimization of the Tungsten Inert Gas Welding Process Parameters of Dissimilar Metals. *Materials* **2022**, *15*, 4426. [CrossRef]
18. Huang, H.; Yin, X.; Feng, Z.; Ma, N. Finite Element Analysis and In-Situ Measurement of Out-of-Plane Distortion in Thin Plate TIG Welding. *Materials* **2019**, *12*, 141. [CrossRef]
19. Seddighi, S.; Ostovan, F.; Shafiei, E.; Toozandehjani, M. A study on the effect of stress relief heat treatment on the microstructure and mechanical properties of dissimilar GTAW weld joints of Inconel 625 and A106 carbon steel. *Mater. Res. Express* **2019**, *6*, 086582. [CrossRef]
20. Sepe, R.; Armentani, E.; Lamanna, G.; Caputo, F. Evaluation by FEM of the influence of the preheating and post-heating treatments on residual stresses in welding. In *Key Engineering Materials*; Trans Tech Publications, Ltd.: Stafa-Zurich, Switzerland, 2015; pp. 93–96.
21. Katherasan, D.; Elias, J.V.; Sathiya, P.; Haq, A.N. Simulation and parameter optimization of flux cored arc welding using artificial neural network and particle swarm optimization algorithm. *J. Intell. Manuf.* **2014**, *25*, 67–76. [CrossRef]
22. Dinovitzer, M.; Chen, X.; Laliberte, J.; Huang, X.; Frei, H. Effect of wire and arc additive manufacturing (WAAM) process parameters on bead geometry and microstructure. *Addit. Manuf.* **2019**, *26*, 138–146. [CrossRef]
23. El-Sayed, M.; Shash, A.; Abd-Rabou, M. Finite element modeling of aluminum alloy AA5083-O friction stir welding process. *J. Mater. Process. Technol.* **2018**, *252*, 13–24. [CrossRef]
24. Li, Y.; Li, Y.; Ma, X.; Zhang, X.; Fu, D.; Yan, Q. Study on Welding Deformation and Optimization of Fixture Scheme for Thin-Walled Flame Cylinder. *Materials* **2022**, *15*, 6418. [CrossRef] [PubMed]
25. Kulkarni, A.; Dwivedi, D.K.; Vasudevan, M. Dissimilar metal welding of P91 steel-AISI 316L SS with Incoloy 800 and Inconel 600 interlayers by using activated TIG welding process and its effect on the microstructure and mechanical properties. *J. Mater. Process. Technol.* **2019**, *274*, 116280. [CrossRef]
26. Qin, G.; Meng, X.; Fu, B. High speed tandem gas tungsten arc welding process of thin stainless steel plate. *J. Mater. Process. Technol.* **2015**, *220*, 58–64. [CrossRef]
27. Kumar, R.; Mevada, N.R.; Rathore, S.; Agarwal, N.; Rajput, V.; Barad, A.S. Experimental investigation and optimization of TIG welding parameters on aluminum 6061 alloy using firefly algorithm. In Proceedings of the IOP Conference Series: Materials Science and Engineering, Telangana, India, 3–4 July 2017; p. 012153.
28. Choudhury, N.; Bandyopadhyay, A.; Rudrapati, R. Design optimization of process parameters for TIG welding based on Taguchi method. *Int. J. Curr. Eng. Technol.* **2014**, *2*, 12–16.
29. Liu, G.; Xiong, J.; Li, D.; Zhang, G. Effect of heat input on auxiliary wire GMA-AM TiAl alloys. *Mater. Manuf. Process.* **2022**, 1–10. [CrossRef]
30. Nie, L.; Wu, Y.; Gong, H. Prediction of temperature and residual stress distributions in friction stir welding of aluminum alloy. *Int. J. Adv. Manuf. Technol.* **2020**, *106*, 3301–3310. [CrossRef]
31. Ghazvinloo, H.; Honarbakhsh-Raouf, A.; Shadfar, N. Effect of arc voltage, welding current and welding speed on fatigue life, impact energy and bead penetration of AA6061 joints produced by robotic MIG welding. *Indian J. Sci. Technol.* **2010**, *3*, 156–162. [CrossRef]
32. Singh, B.; Singhal, P.; Saxena, K.K.; Saxena, R.K. Influences of latent heat on temperature field, weld bead dimensions and melting efficiency during welding simulation. *Met. Mater. Int.* **2021**, *27*, 2848–2866. [CrossRef]
33. Çevik, B.; Ozcatalbas, Y.; Gülenç, B. Effect of welding speed on the mechanical properties and weld defects of 7075 Al alloy joined by FSW. *Kov. Mater* **2016**, *54*, 241–247.
34. Hanif, M.; Zhang, L.; Ahmad, W. Analysis of a novel sustainability framework based on economic and environmental aspects in graphene-based dielectric electric discharge machining. *Int. J. Adv. Manuf. Technol.* **2022**, *119*, 6287–6306. [CrossRef]
35. Hanif, M.; Wasim, A.; Shah, A.H.; Noor, S.; Sajid, M.; Mujtaba, N. Optimization of process parameters using graphene-based dielectric in electric discharge machining of AISI D2 steel. *Int. J. Adv. Manuf. Technol.* **2019**, *103*, 3735–3749. [CrossRef]

materials

MDPI

Article

Arc Characteristics of Ultrasonic-Magnetic Coaxial Hybrid GTAW

Wenbo Du [1,*], Wenlong Li [1,2], Yue Li [3] and Chao Chen [2]

[1] National Key Laboratory for Remanufacturing, Army Academy of Armored Forces, Beijing 100072, China
[2] College of Mechanical and Electrical Engineering, Northeast Forestry University, Harbin 150040, China
[3] Beijing Institute of Radio Measurement, Beijing 100039, China
* Correspondence: dwbneu@163.com; Tel.: +86-188-4318-9190; Fax: +86-0451-82190397

Abstract: Ultrasonic-magnetic field coaxial hybrid GTAW(U-M-GTAW) is a new non-melting electrode welding method proposed by combining ultrasonic assisted GTAW(U-GTAW) and magnetic assisted GTAW(M-GTAW) on the regulation characteristics of the GTAW arc. U-M-GTAW introduces ultrasonic and magnetic field effects into GTAW to improve arc characteristics. The orthogonal experiment was designed to investigate the degree of influence of different process parameters on the arc. The degree of influence of ultrasonic power P, radiator height H, magnetic field current C_W, welding current C_W and tungsten electrode height H_T on ΔL_1 (degree of arc root diameter change), ΔL_2 (degree of maximum diameter change) and ΔS (degree of area change) were analyzed. In the parameter range, P has the greatest degree of influence on ΔL_1 and ΔL_2. As all process parameters increase, L_1 shows a tendency to decrease, indicating an increase in the compression of the arc root. ΔL_2 with the increase in P and C_W shows a trend of first decreasing and then increasing. ΔL_2 with the increase in H decreases, indicating that the acoustic radiation force increases, the arc energy increases, and the dark region decreases. The magnetic field current increases, the bottom of the arc expands, and the height of the tungsten electrode increases, the arc dispersion and thus the difference between the dark and luminous regions at the bottom increases, resulting in ΔL_2 with the increase in C_M and H_T increases. C_W has the greatest degree of influence on ΔS. ΔS decreases and then increases as P and H increase, which indicates that the force on acoustic radiation increases and then decreases in the range. An increase in the magnetic field current increases the rotation of the arc, leading to an increase in the arc area. An increase in welding current leads to an increase in arc energy, expansion of the arc morphology, and an increase in ΔS. The tungsten electrode height increases, the arc diverges, the dark region increases, the luminous area decreases, and ΔS increases. Finally, combined with the analysis of ultrasonic field and magnetic field theory, changes in process parameters will affect the force of the arc and thus the arc morphology. The U-M-GTAW arc under the action of acoustic radiation force, the plasma flow is shifted in the direction of the arc axis, and the arc contraction, under the action of magnetic field force to generate circumferential current, the arc undergoes periodic rotation, which improves GTAW arc characteristics.

Keywords: ultrasonic; magnetic; GTAW; arc characteristics; orthogonal experiments

check for updates

Citation: Du, W.; Li, W.; Li, Y.; Chen, C. Arc Characteristics of Ultrasonic-Magnetic Coaxial Hybrid GTAW. *Materials* **2022**, *15*, 8130. https://doi.org/10.3390/ma15228130

Academic Editors: Artur Czupryński and Shinichi Tashiro

Received: 28 August 2022
Accepted: 26 October 2022
Published: 16 November 2022

1. Introduction

Gas tungsten arc welding (GTAW) is suitable for high-quality welding, with the advantages of stable arc, easy control of the heat source, excellent forming, and high quality of the weld bead, but its shortcomings such as shallow welding depth and low productivity limit the development of the method. To improve the welding productivity and expand the application range of GTAW, domestic and international scholars have carried out a lot of work to regulate the arc and improve the energy density of the arc, including plasma arc welding (PAW) [1], activating flux TIG (A-TIG) [2], multi-tungsten electrode TIG (A-TIG) [2], and keyhole TIG (K-TIG) [3].

Meanwhile, some scholars proposed a multi-field hybrid GTAW method to regulate the arc, such as U-GTAW [4] and M-GTAW [5]. Composite welding can effectively improve the energy utilization of the heat source, increase the welding speed, increase the weld penetration, and improve the quality of the weld bead, which is the future trend of GTAW process development. Ultrasound refers to the vibration frequency of more than 20 kHz sound waves, ultrasound is a carrier of energy transfer and energy form, when its intensity exceeds a certain value, it can make some physical, chemical, and biological properties or state of the sound-transmitting medium change through its interaction with the sound-transmitting medium, or to speed up the process of such changes, which is called power ultrasound [6]. U-GTAW is the introduction of power ultrasound with the welding arc as the medium to realize the compound of ultrasound energy and welding arc. When the ultrasound passes through the arc plasma, as the power ultrasound is applied energy, it will inevitably have a certain effect on the arc and the molten pool. Sun et al. [4,7–10] were the first to compound ultrasonic to arc. They found that after applying ultrasonic, the arc was compressed so that the pressure and temperature in its central region increased significantly, and the stiffness of the arc increased. At the same time, the pressure distribution of the arc obeyed a Gaussian distribution. Xie et al. [11] found that the ultrasonic composite arc under the joint action of the internal and external acoustic fields, the degree of plasma constraint is significantly increased, the brightness of the arc is enhanced, and the range of the high-temperature region of the arc column is extended to the anode, and the intermediate particles appear to agglomerate and jitter up and down at a certain frequency. Cai et al. [12] studied periodic ultrasonic assisted TIG welding of 2219 aluminum alloy and found that the coarse precipitation phase tended to be uniformly dispersed in the weld zone under ultrasonic vibration. Compared to TIG, the hardness of the weld zone increased by 8.43% and the tensile strength increased by 29.02%.

The research on the effect of magnetic fields on the welding process originated in the 1960s and was initially used for grain refinement [13]. Later, various scholars found that the magnetic field can change the arc morphology, influence the metal transfer process, control the heat input and molten pool behavior, improve the weld formation, joint organization, mechanical properties and physical and chemical properties, reduce welding defects and improve the quality of weld formation [14–16]. Fan et al. [17] modeled the arc and the melt pool and found that the maximum arc velocity as well as the melt pool velocity could be increased by the applied axial magnetic field. Liu et al. [18] found in axial magnetic field-assisted CMT welding of pure copper and aluminum alloys that the magnetic field reduced the average heat input and the thickness of the Al_2Cu intermetallic compound layer formed at the interface from 50 μm to 5 μm. The joint strength increased with increasing magnetic field. For a 21 mT and 10 Hz magnetic field, the maximum loading force reached 1.67 kN, resulting in a 33% increase in joint strength. Li et al. [19] applied an axial magnetic field to the brazing of AA6061 aluminum alloy with HSLA350 steel, which found that the Lorentz force caused the aluminum matrix and the second term to be refined, and the intermetallic compounds at the weld interface were significantly reduced making the hardness of the joint increased by 14%. Anderson et al. [20] found that when TIG welding 304 stainless steel the applied magnetic field increased the weld area by 100%, narrowed the width of the fusion zone by 25% and increased the depth of fusion by 3.0 mm. Moreover, the external magnetic field reduces the percentage of δ ferrite, improves the filling of the weld metal to the melt pool, and reduces the primary and secondary dendrite spacing.

The previous research has confirmed the feasibility of U-M-GTAW. Compared with the normal welding arc, the composite arc has certain ultrasonic and magnetic properties that are a feature of the system, and experiments have found that the composite arc is more concentrated under the action of ultrasonic and magnetic fields while generating periodic rotation [21,22]. However, the relationship between different process parameters and the arc and the mechanism of action is not clear, thus, it is necessary to study the degree of influence of different process parameters on the arc characteristics of U-M-GTAW.

2. Experimental Materials and Methods

The experiment was completed by U-M-GTAW system established based on the self-designed composite welding torch, whose structure is shown in Figure 1. The maximum output power of the ultrasonic power supply was 2400 W, and the vibration frequency was 20 kHz. The Master TIG 335AC/DC welding power supply was used to provide a stable welding process. The maximum output current of the magnetic field power supply was 3 A, the AC pulse frequency is 40 Hz, and the number of solenoid coil turns was 140. The protection flow was 25 L/min (99.99%Ar). The shooting frame rate of the high-speed camera was 2000 fps, and the exposure time was 20 μs. The tungsten electrode diameter was 3 mm, and the tip angle was 30°. The tungsten electrode tip loss was continuously observed during the experiment to ensure that the tungsten electrode angle remained unchanged, and the welding anode was a water-cooled copper block.

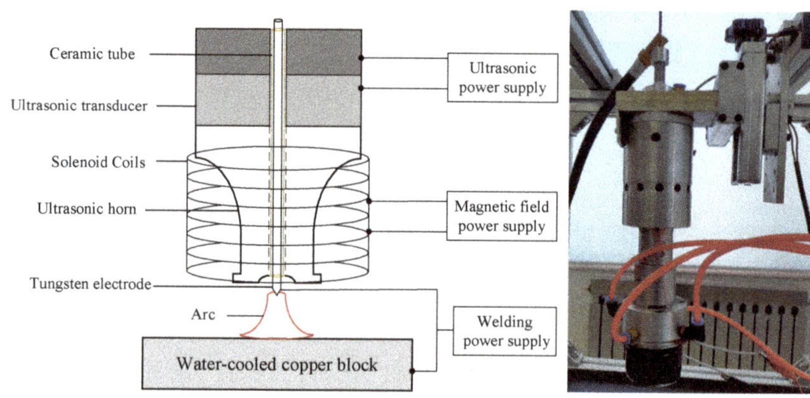

Figure 1. U-M-GTAW system.

The core component of the U-M-GTAW was the composite torch; which consisted of a modified common welding torch; ultrasonic transmitting equipment consisted of an ultrasonic transducer and an ultrasonic horn; and a solenoid coil. The exploded diagram of the composite welding torch is shown in Figure 2. Power ultrasound was generated by an ultrasonic transducer that converted the electrical signal output from the ultrasonic power supply into mechanical vibration, which was amplified by an ultrasonic horn and finally transmitted to the welding arc to achieve the impact of ultrasound on the welding process. The ceramic nozzle passed through the hollow through-hole of the ultrasonic transducer and the horn; and the tungsten electrode passed through the ceramic nozzle insulated from the ultrasonic emission equipment. The solenoid coils were rigidly connected to the ultrasonic horn; so that the ultrasonic emission equipment; tungsten electrode and solenoid coils were in the same axis by mechanical coupling. This structure could ensure the continuous transmission of ultrasonic energy and avoid the phenomenon of arc bias blowing during ultrasonic application; and also made the designed composite welding torch compact.

Arc measurement parameters are shown in Figure 3: the dark region and luminous region are expressed in S_1 and S_2, respectively, the diameter of the dark region and luminous region at the root of the arc are d_1 and d_2, the diameters of the dark region and luminous region at the maximum diameter were D_1 and D_2. Using the first set of data as a baseline, the arc variation degrees at different locations were expressed as ΔL_1 (Degree of arc root diameter change), ΔL_2 (Degree of maximum diameter change), and ΔS (Degree of area change), as shown in Equations (1)–(3):

$$\Delta L_1 = \frac{d_1 - d_2}{D_{2-1}} \times 100\% \quad (1)$$

$$\Delta L_2 = \frac{D_1 - D_2}{D_{2-1}} \times 100\% \tag{2}$$

$$\Delta S = \left| \frac{S_1 - S_2}{S_{2-1}} \times 100\% \right| \tag{3}$$

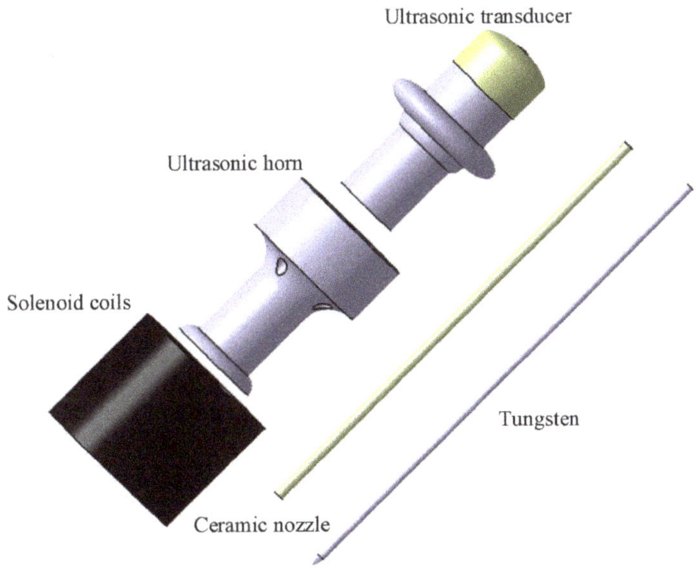

Figure 2. The exploded diagram of the composite welding torch.

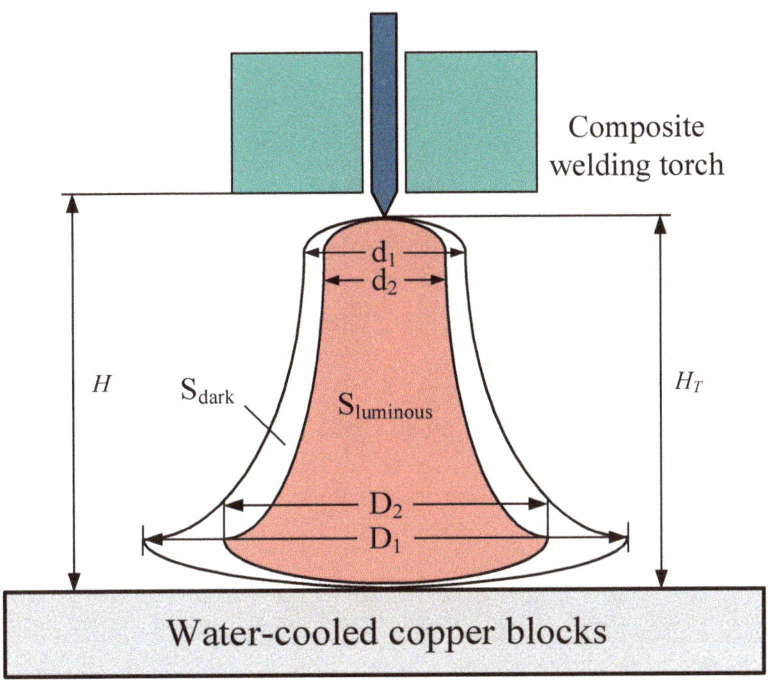

Figure 3. The arc parameter diagram.

In order to better distinguish the boundary of the arc parameters, the arc image was processed as follows. Figure 4a shows the original image, the pixels below a certain threshold are removed from the image, as shown in Figure 4b, and then the brightness of the arc image is output as one color for the part above a certain value and another color for the part below a certain value, as shown in Figure 4c.

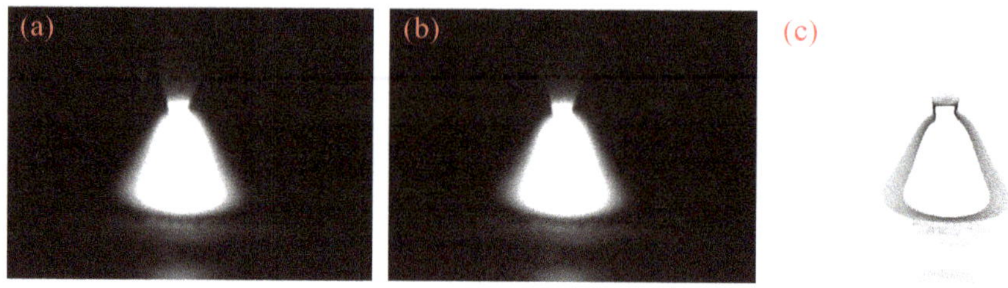

Figure 4. Arc image processing process. (**a**) original image, (**b**) threshold processing image, and (**c**) extracted image.

Orthogonal experiments were designed to better investigate the degree of influence of different factor variations on the arc morphology of U-M-GTAW, choosing ultrasonic power (P), radiator height (H), magnetic field current (C_M), welding current (C_W), and tungsten electrode height (H_T) as the five factors, and three optimal parameters were selected as levels for each factor based on the results of previous single-factor experiments.

3. Results

The orthogonal experimental table and results are shown in Table 1, and the arc images are shown in Figure 5.

Table 1. The orthogonal experimental table and results.

NO.	P/(W)	H/(mm)	C_M/(A)	C_W/(A)	H_T/(mm)	ΔL_1/(%)	ΔL_2/(%)	ΔS/(%)
1	900	15	1	40	2	156	121	108.56
2	900	17	2	60	3	100	194	3.89
3	900	19	3	80	4	56	192	508.12
4	1200	15	1	60	3	43	105	172.20
5	1200	17	2	80	4	40	90	316.72
6	1200	19	3	40	2	34	67	59.63
7	1500	15	2	40	4	47	184	124.78
8	1500	17	3	60	2	41	73	107.18
9	1500	19	1	80	3	42	71	208.22
10	900	15	3	80	3	76	260	352.89
11	900	17	1	40	4	64	224	9.79
12	900	19	2	60	2	67	124	123.07
13	1200	15	2	80	2	48	79	154.88
14	1200	17	3	40	3	44	120	60.31
15	1200	19	1	60	4	46	105	169.93
16	1500	15	3	60	4	54	102	230.29
17	1500	17	1	80	2	39	81	132.09
18	1500	19	2	40	3	45	103	149.28

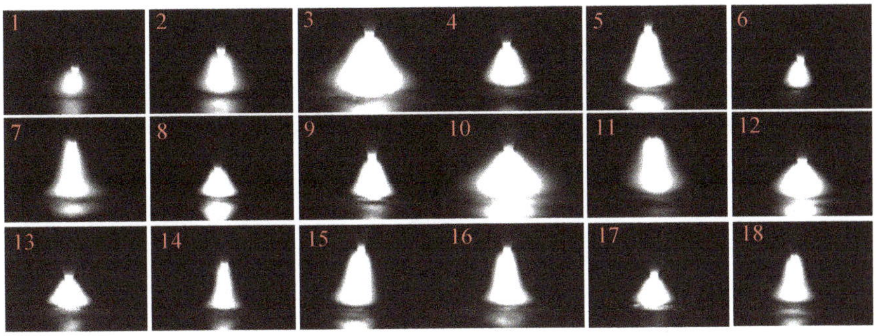

Figure 5. The arc images. (Note: the numbers 1,2,3, ...18 is expressed the images of arc shape in Table 1).

The results of the analysis of ΔL_1 for different process parameters are shown in Table 2 and the trend diagram of the influence of process parameters on ΔL_1 is shown in Figure 6.

Table 2. Results of the orthogonal experimental analysis of ΔL_1.

Mean	P/(W)	H/(mm)	C_M/(A)	C_W/(A)	H_T/(mm)
Level 1/(%)	86.500	70.667	65.000	65.000	64.167
Level 2/(%)	42.500	54.667	54.833	58.500	58.333
Level 3/(%)	44.667	48.333	50.833	50.167	51.167
Range/(%)	44.000	22.334	14.167	14.833	13.000

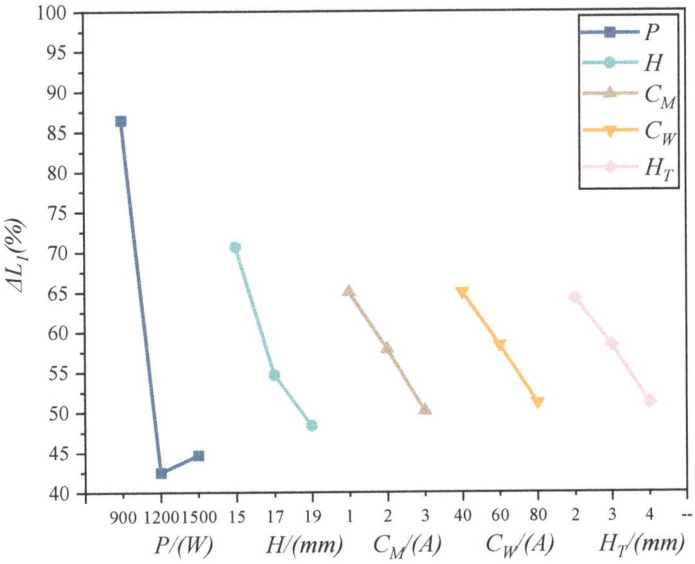

Figure 6. Trend diagram of ΔL_1 with the variation of process parameters.

As seen in Table 2, the range of P is much larger than that of other factors, indicating that P has the greatest degree of influence on ΔL_1, and the degree of influence on ΔL_1 is $P > H > C_W > C_M > H_T$. From Figure 5, ΔL_1 decreases and then slightly increases with increasing P, which indicates that the contraction of the arc root is better when P is at 1200 W and decreases with increasing H, C_M, C_W, and H_T in the above parameter range. As H increases, the intensity of the standing wave ultrasonic field increases, the acoustic radiation force increases, the degree of arc contraction increases, and ΔL_1 decreases. With the increase in C_M, the compression effect of electromagnetic force on the root of the arc increases, and ΔL_1 decreases. With the increase in C_W, the arc energy increases, the dark bright area decreases, and ΔL_1 decreases. With the increase in HT, the arc length increases, the radial diameter of the arc decreases, and ΔL_1 decreases.

The results of the analysis of ΔL_2 for different process parameters are shown in Table 3 and the trend diagram of the influence of process parameters on ΔL_2 is shown in Figure 7.

Table 3. Results of the orthogonal experimental analysis of ΔL_2.

Mean	P/(W)	H/(mm)	C_M/(A)	C_W/(A)	H_T/(mm)
Level 1/(%)	185.833	141.833	117.833	136.500	90.833
Level 2/(%)	94.333	130.333	129.000	117.167	142.167
Level 3/(%)	102.333	110.333	135.667	128.833	149.500
Range/(%)	91.500	31.500	17.834	19.333	58.667

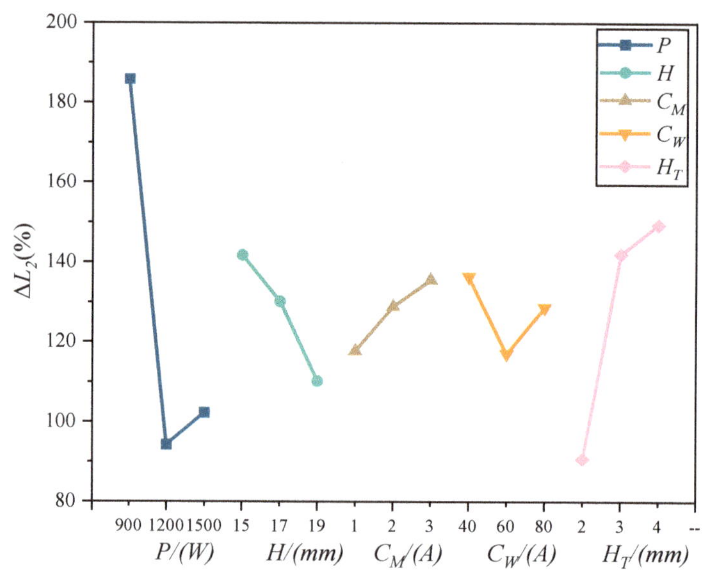

Figure 7. Trend diagram of ΔL_2 with the variation of process parameters.

Figure 3 the degree of influence on ΔL_2 is $P > H_T > H > C_W > C_M$. In Figure 6, ΔL_2 tends to decrease and then increase with the increase in P and C_W, which indicates that the compression of ΔL_2 is better when P = 1200 W and C_W = 60 A. With the increase in C_M and H_T ΔL_2 tends to increase, which indicates that the amplitude of arc rotation increases and the arc bottom expands. The decreasing trend with increasing H indicates that the acoustic radiation force increases and the arc contracts.

The results of the analysis of ΔS for different process parameters are shown in Table 4 and the trend diagram of the influence of process parameters on ΔS is shown in Figure 8.

Table 4. Results of orthogonal experimental analysis of ΔS.

Mean	P/(W)	H/(mm)	C_M/(A)	C_W/(A)	H_T/(mm)
Level 1/(%)	184.387	190.600	133.465	85.392	114.235
Level 2/(%)	155.612	104.997	145.437	134.427	157.798
Level 3/(%)	158.640	203.042	239.737	278.820	226.605
Range/(%)	28.775	98.045	86.272	193.428	112.370

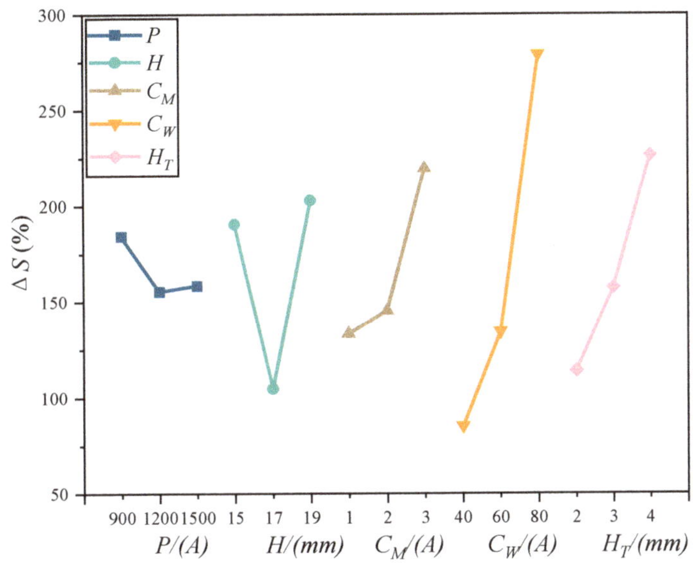

Figure 8. Trend diagram of ΔS with the variation of process parameters.

From Table 4, the degree of influence on ΔS is $C_W > H_T > H > C_M > P$. In Figure 7, ΔS increases with the increase in C_M, C_W, and H_T, which is caused by the increase in C_M, the increase in arc rotation amplitude so that the increase in arc area, the increase in arc energy with the increase in C_W, the decrease in dark region, the increase in luminous region, and the increase in H_T, the arc elongation, the expansion of luminous region downward, and the increase in area. ΔS increases. With the increase in P and H, ΔS first shows a trend of decreasing and then increasing, which means that the arc compression effect is better when $P = 1200$ W and $H = 17$ mm.

4. Discussions

Plasma is a collective term for charged and neutral particles in arc space, including cations, electrons, molecules and atoms. The motion behavior of the particles in the arc space is very complex, there is a violent collision between the particles, to achieve the exchange of arc heat, transfer process, and the regular movement of electrons can achieve the current conduction, and through the collision between the particles will transfer energy to the neutral particles and ionization, is the arc itself to create charged particles to maintain its conductivity of the most important way.

The mechanical force of arc mainly includes electromagnetic contraction force and plasma flow force. When the current flows through the conductor, the entire current can be seen as several parallel current lines, these current lines will produce mutual attraction between the conductor cross-section tends to shrink. It will produce cross-sectional contraction, a phenomenon called electromagnetic contraction effect, and the resulting force is called electromagnetic contraction force. The plasma flow force in the arc is formed by the high-speed flow of high-temperature particles. The action of the arc static pressure causes a certain pressure difference between the arc cathode and the workpiece, resulting in a continuous flow, i.e., plasma flow. The additional pressure formed when it reaches the surface of the workpiece is called the plasma flow force. Among the arc forces, the plasma flow force is the main one and plays a decisive role, which can account for more than 80% of the total mechanical force of the arc. For the arc in the plasma flow at any position in a particle force analysis, as shown in Figure 9a.

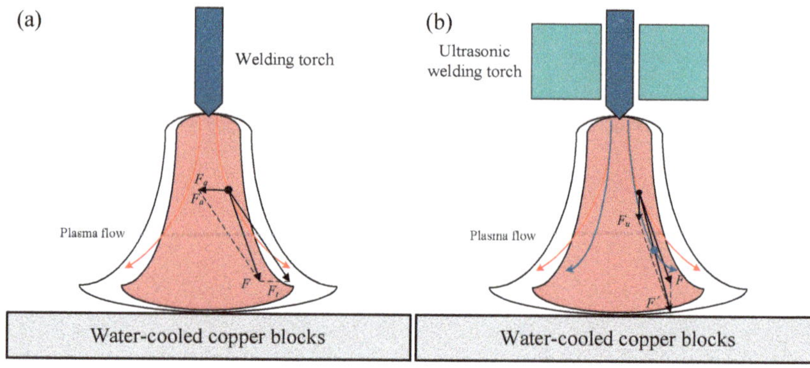

Figure 9. Arc force models. (**a**) traditional welding torch, (**b**) Ultrasonic welding torch.

F_t is plasma flow force, F_a is electromagnetic contraction force, whose magnitude is determined by Equations (4) and (5), respectively. Visible, electromagnetic contraction force and plasma flow force are both numerical quantities related to the arc current, arc distribution form, and form.

$$F_a = KI^2 \lg(R_b/R_a) \tag{4}$$

where R_b is the radius of the bottom surface of the arc; R_a is the radius of the root surface of the arc; K is the coefficient, $K = \mu/4\pi$ (μ is the magnetic permeability of the medium).

$$F_t = (K_1/2)I^2 \tag{5}$$

where K_1 is the parameter related to the form of arc pressure distribution; I is the welding current.

The conventional GTAW arc is subjected to a combined electromagnetic force F_a and plasma flow force F_t, resulting in resultant force F. The resultant force F deflects farther from the electrode axis, resulting in an extension of the arc morphology. When the current increases, F_a and F_t increase, the combined force F increases, so the arc morphology increases, the degree of change in area increases. The arc heat production increases, the luminous area increases, so the degree of change in the diameter tends to decrease.

The acoustic radiation force F_u generated by ultrasound is a single direction force along the direction of ultrasound propagation, i.e., along the axis of the arc, as shown in Figure 9b.

The acoustic radiation force F_u together with the resultant force F of the plasma flow force and the electromagnetic contraction force produce a new resultant force F', which is more biased in the direction of the axis of the arc. The intervention of the acoustic radiation force F_u changes the original flow direction of the plasma flow, forcing the arc particles to be displaced along the direction of the resultant force F', and the particle position moves off axis. Because the whole arc is in the ultrasonic field effect, all the particles in the arc are subjected to acoustic radiation force and forced displacement, the particles from the cathode to the anode trajectory shortened, which is manifested as the overall compression of the arc. The U-GTAW arc is strongly compressed in the radial direction due to the constraining effect of the additional acoustic radiation force, and the arc temperature, energy density, and plasma flow rate are significantly increased. In addition, the magnitude of the acoustic radiation force is influenced by the phase difference between the incident and reflected waves, and H is the most important factor affecting the phase difference. At some specific values of H, the incident ultrasonic waves and the reflected waves peak simultaneously, and the medium-fixed acoustic field is the resonant field with the maximum energy density. At some other specific values of H, the incident ultrasonic waves and the reflected waves cancel, which means that the amplitude of the standing waves is zero and the ultrasonic field has the lowest energy density. Therefore, the change in H affects the magnitude of the acoustic radiation force and thus the degree of change in the arc morphology.

Axial magnetic field distribution in the arc is the middle of the magnetic field density is higher, the surrounding magnetic field density is lower, and the greater the frequency of the magnetic field, the faster the speed of magnetic field change, the more uneven the distribution of magnetic lines of force inside and outside the arc, and charged particles in the axial magnetic field is the trajectory of the spiral curve, both radial motion, and axial motion [23]. When an axial magnetic field is applied in addition to the applied ultrasonic field, the plasma is then subjected to magnetic forces in addition to F_a, F_t, and F_L.

The distance from any charged particle in the arc to the axis of the arc is r, and the current I at this point can be decomposed into the axial components I_{xy} and I_z, the angle between the current I and the axial component I_z is θ, $I_z = I\cos\theta$, and $I_{xy} = I\sin\theta$ so that the Ampere's force F_A per unit arc column at the point on the arc is:

$$F_A = BI_{xy} = BI\sin\theta \tag{6}$$

For the unit arc column length, it is subject to the aerodynamic drag is:

$$F_D = \frac{1}{2}\rho v^2 AC_D \tag{7}$$

where F_D is aerodynamic drag; C_D is aerodynamic drag coefficient; v is the speed of arc column motion; ρ is the gas density; A is the windward area per unit length of arc column.

When the unit particle reaches a steady state, i.e., when a circumferential current is generated within the arc, the arc rotates. The greater the magnetic field current, the greater the F_A, the greater the circumferential current of the arc in the steady state, and therefore the greater the rotation of the arc, the greater the ΔS. It is found that in the arc length is long enough or the particle motion is slow in the arc, the Lorentz force F_M on the particle in the upper middle section is greater than the centrifugal force $F_M{}'$ of the current particle motion, the radius of particle motion gradually decreases, the arc shows contraction. In the middle and lower sections of the particle force is the opposite arc shows expansion. Therefore, ΔL_1 and ΔL_2 show a decreasing and increasing trend, respectively [5,24].

At this point, the combined force with the magnetic field forces acting on the U-M-GTAW arc, as shown in Figure 10, resulting in the arc both contraction and periodic rotation of the arc characteristics. When the process parameters change, the force of the arc changes and affects the arc morphology.

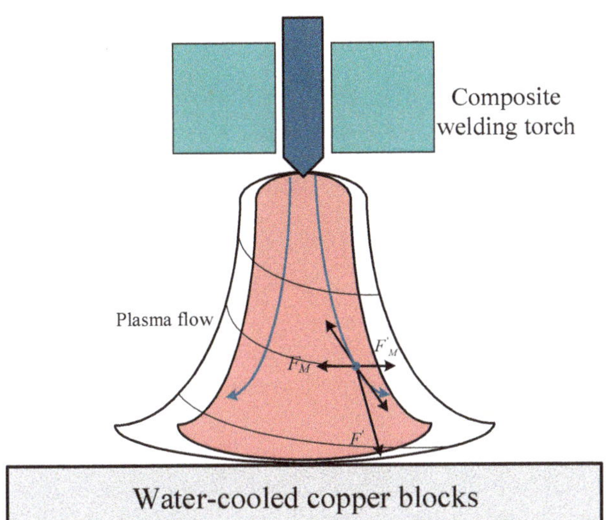

Figure 10. U-M-GTAW arc force mode.

5. Conclusions

1. The U-M-GTAW welding system was designed, and the experimental results showed that the U-M-GTAW arc rotates periodically while compressing under the action of the applied ultrasonic field and the axial magnetic field, which combines the effects of ultrasonic GTAW and magnetically controlled GTAW on the arc.
2. The orthogonal experiment was designed to investigate the effect of different process parameters on the degree of change in arc morphology, and the degree of effect on ΔL_1 was $P > H > C_W > C_M > H$, the degree of effect on ΔL_2 was $P > H_T > H > C_W > C_M$, and the degree of effect on ΔS is $C_W > H_T > H > C_M > P$.
3. Within the above parameters, ΔL_1 decreases and then slightly increases with increasing P, and decreases with increasing H, C_M, C_W, and H_T. ΔL_2 tends to decrease and then increase with increasing P and C_W, increases with increasing C_M and H_T, and decreases with increasing H. ΔS increases with increasing C_M, C_W, and H_T. With the increase in P and H, ΔS tends to increase first and then decrease.
4. U-M-GTAW under the action of acoustic radiation force, the arc is strongly compressed in the radial direction, and the temperature, energy density, and plasma flow rate of the arc are significantly increased. Periodic rotation is generated under the action of the magnetic field force. When the process parameters change, the force of the arc changes and affects the arc morphology.

Author Contributions: Funding acquisition, W.D.; investigation, W.D.; Methodology, Y.L.; Validation, W.D.; Visualization, C.C.; Data curation, C.C.; Writing—original draft, W.L. All authors have read and agreed to the published version of the manuscript.

Funding: This work was financially supported by the Fundamental Research Funds for the Central Universities through Grant No. 2572021BF02.

Institutional Review Board Statement: Not applicable.

Informed Consent Statement: Not applicable.

Data Availability Statement: Not applicable.

Conflicts of Interest: No conflict of interest exists in the submission of this manuscript, and the manuscript is approved by all authors for publication.

References

1. Zhang, C.; Wu, C.S.; Tian, S.S. Effect of ultrasonic vibration on current density and keyholing capability of plasma arc. *Sci. Technol. Weld. Join.* **2020**, *25*, 422–430. [CrossRef]
2. Huang, B.; Yang, J.; Yin, W.; Chen, P.; Zhu, Y.; Li, J. Research Progress and Prospect of A-TIG Welding. *Mater. Rev.* **2016**, *30*, 76–80.
3. Liu, Z.M.; Chen, S.Y.; Yuan, X.; Zuo, A.; Zhang, T.; Luo, Z. Magnetic-enhanced keyhole TIG welding process. *Int. J. Adv. Manuf. Technol.* **2018**, *99*, 275–285. [CrossRef]
4. Sun, Q.; Lin, S.; Yang, C.; Fan, Y. The arc characteristic of ultrasonic assisted TIG welding. *China Weld.* **2008**, *17*, 52–57.
5. Chang, Y.; Yang, X.; Li, D.; Li, D. Arc shapes of TIG welding in a longitudinal magnetic field. *Trans. China Weld. Inst.* **2010**, *31*, 49–52.
6. Zhao, J.; Dai, G.; Han, J.; Huang, X. Research Progress of Application of Power Ultrasound in Metal Wwelding. *Hot Work. Technol.* **2012**, *41*, 144–147+150.
7. Sun, Q.J.; Lin, S.B.; Yang, C.L.; Zhao, G.Q. Penetration increase of AISI 304 using ultrasonic assisted tungsten inert gas welding. *Sci. Technol. Weld. Join.* **2009**, *14*, 765–767. [CrossRef]
8. Yangyang, F.A.N.; Qingjie, S.U.N.; Chunli, Y.; Lin, S.B. TIG welding of the stainless steel 304 based on the ultrasonic vibration. *Trans. China Weld. Inst.* **2009**, *30*, 91–94.
9. Fan, C.; Yang, C.; Lin, S.; Zhang, Y. Arc Characteristics of Ultrasonic Wave-Assisted GMAW. *Weld. J.* **2013**, *92*, 375S–380S.
10. Sun, Q. Characteristic of Arc Pressure in Ultrasonic-TIG Hybrid Welding. *J. Mech. Eng.* **2011**, *47*, 53–57. [CrossRef]
11. Xie, W.; Fan, C.; Yang, C.; Lin, S.; Zhang, Y. Characteristics of acoustic-controlled arc in ultrasonic wave-assisted arc. *Acta Phys. Sin.* **2015**, *64*, 421–429.
12. Cai, X.; Lin, S.; Wang, X.; Yang, C.; Fan, C. Characteristics of Periodic Ultrasonic Assisted TIG Welding for 2219 Aluminum Alloys. *Materials* **2019**, *12*, 4081. [CrossRef] [PubMed]
13. Brown, D.J.W.J. The Effect of Electromagnetic Stirring and Mechanical Vibration. *Weld. J.* **1962**, *41*, 241s–250s.

14. Shoichi, M.; Yukio, M.; Koki, T.; Yasushi, T.; Yukinori, M.; Yusuke, M. Study on the application for electromagnetic controlled molten pool welding process in overhead and flat position welding. *Sci. Technol. Weld. Join.* **2013**, *18*, 38–44. [CrossRef]
15. Chen, J.; Wei, Y.; Zhan, X.; Pan, P. Weld profile, microstructure, and mechanical property of laser-welded butt joints of 5A06 Al alloy with static magnetic field support. *Int. J. Adv. Manuf. Technol.* **2017**, *92*, 1677–1686. [CrossRef]
16. Li, M.; Xu, J.; Huang, Y.; Rong, Y. Improving Keyhole Stability by External Magnetic Field in Full Penetration Laser Welding. *JOM* **2018**, *70*, 1261–1266.
17. Xiao, L.; Fan, D.; Huang, J. Tungsten cathode-arc plasma-weld pool interaction in the magnetically rotated or deflected gas tungsten arc welding configuration. *J. Manuf. Process.* **2018**, *32*, 127–137. [CrossRef]
18. Liu, Y.B.; Sun, Q.J.; Wang, H.; Zhang, H.M.; Cai, S.J.; Feng, J.C. Effect of the axial external magnetic field on copper/aluminium arc weld joining. *Sci. Technol. Weld. Join.* **2016**, *21*, 460–465. [CrossRef]
19. Li, R.; Yuan, X.; Zhang, H.; Yang, J.; Wu, K.; Li, T.; Wang, G.; Tao, S. Effect of axial magnetic field on TIG welding–brazing of AA6061 aluminum alloy to HSLA350 steel. *J. Mater. Res. Technol.* **2021**, *12*, 882–893. [CrossRef]
20. Queiroz, A.V.D.; Fernandes, M.T.; Silva, L.; Demarque, R.; Xavier, C.R.; De Castro, J.A. Effects of an External Magnetic Field on the Microstructural and Mechanical Properties of the Fusion Zone in TIG Welding. *Metals* **2020**, *10*, 714. [CrossRef]
21. Chen, C.; Li, W.; Du, W.; Du, W.; Liu, J.; Zhang, H. Feasibility analysis of standing wave ultrasonic—Axial magnetic field hybrid for controlling GTAW arc characteristics. *J. Manuf. Process.* **2022**, *80*, 187–195. [CrossRef]
22. Chen, C.; Li, W.; Fan, C.; Du, W.; Zhang, H. Understanding the changing mechanism of arc characteristics in ultrasound-magnetic field coaxial hybrid gas tungsten arc welding. *Plasma Sci. Technol.* **2022**, *24*, 115506. [CrossRef]
23. Chen, S.; Meng, D.; Su, Z.; Jiang, F.; Lu, Y. Effects of longitudinal magnetic field on non-consumable gas shielded arc welding. *Trans. China Weld. Inst.* **2014**, *35*, 5–8.
24. Wu, H.; Chang, Y.L.; Babkin, A.; Lee, B. The behavior of TIG welding arc in a high-frequency axial magnetic field. *Weld. World* **2020**, *65*, 95–104. [CrossRef]

materials

MDPI

Article

Microstructures and Properties Investigation on DP980 Dual-Phase Steel CMT + P Welded Joints

Yan Liu [1,*], Zhaozhen Liu [2], Yongman Chen [2], Chunlin He [1], Ao Liu [2] and Xiaoang Liu [2]

[1] Liaoning Provincial Key Laboratory of Advanced Material Preparation Technology, Shenyang University, Shenyang 110044, China
[2] School of Mechanical Engineering, Shenyang University, Shenyang 110044, China
* Correspondence: liuyan1979@syu.edu.cn; Tel.: +86-024-62269467

Abstract: The increasing demand for the lightweight production of advanced high-strength steel puts forward higher requirements for the quality of welded joint forming. The common CMT welding process has certain limitations and is difficult to meet the needs of lightweight manufacturing. In this study, the CMT + P welding technology was used to weld the DP980 dual-phase steel plate with 1.2 mm thickness. The ER120S-G welding wire was used as the filling material to conduct a 70° angle lap welding experiment. The effects of wire feeding speed (3 m/min~5 m/min) and welding speed (350 mm/min~600 mm/min) on the forming, microstructures, and mechanical properties of DP980 dual-phase steel welded joints were analyzed. The results show that the CMT + P welding process can produce lap weldments with good formability and properties. The welded joints can be divided into the weld zone, the HAZ, and the base metal zone, where the HAZ can be divided into the coarse-grained zone and the softened zone. The role of the elements Ni and Si is to promote the production of martensite and to increase the ferrite strength in welded joints. As the wire feeding speed increases, the grain size of the coarse grain zone in the HAZ increases from 31.90 μm to 50.93 μm; while the welding speed increases, the grain size of the coarse grain zone decreases from 45.48 μm to 35.73 μm. The average microhardness of the weld zone is 420 HV. In contrast, the average microhardness of the softening zone in HAZ is reduced to 250 HV. When the wire feeding speed is 4 m/min and the welding speed is 550 mm/min, the tensile properties of the weldment are optimal, its tensile strength can reach 973 MPa, and the tensile fracture is ductile fracture.

Keywords: DP980 dual-phase steel; CMT + P welding technology; microstructures; mechanical property; softened zone

Citation: Liu, Y.; Liu, Z.; Chen, Y.; He, C.; Liu, A.; Liu, X. Microstructures and Properties Investigation on DP980 Dual-Phase Steel CMT + P Welded Joints. *Materials* **2022**, *15*, 5880. https://doi.org/10.3390/ma15175880

Academic Editors: Adam Grajcar, Marcin Adamiak, Artur Czupryński, Antonín Kříž and Tünde Anna Kovács

Received: 24 July 2022
Accepted: 22 August 2022
Published: 25 August 2022

1. Introduction

To accomplish the goal of green development, energy conservation, and emission reduction, lightweight components are increasingly used in rail transit and other manufacturing fields [1]. As a new generation of advanced high-strength steel, dual-phase steel has higher strength and better toughness than traditional high-strength steel [2–5]. In the field of automobile manufacturing, dual-phase steel is used instead of conventional high-strength steel to manufacture automobile chassis, brackets, and control arms, which not only makes the automobile have a stronger anti-impact ability but also reduces the weight of the automobile by 10–20%. Therefore, dual-phase steel is one of the ideal structural materials in the modern industry [6–8]. Although dual-phase steel has good welding performance, the welded components with good forming effect cannot be produced by using the traditional tungsten inert gas (TIG) and melt inert gas (MIG) welding process. As a new and improved MIG connection technology, cold metal transfer (CMT) has the welding characteristics of low heat input and no spatter after welding. It can solve some welding defects in traditional fusion welding and has a broad application prospect in the field of dual-phase steel sheet welding [9–12]. Therefore, the combination of advanced welding technology and high-strength dual-phase steel can solve the problems of low welding quality and poor forming in practical production. It can further promote the broad application of dual-phase steel in the modern automobile manufacturing industry.

Researchers at home and abroad have conducted some related investigations on the welding process of dual-phase steel over the past few years. Cui et al. [13] used resistance spot welding technology to conduct welding experiments on DP780 dual-phase steel with 1.6 mm thickness and studied the failure behavior of welded joints. Finally, the tensile part broke under 21.08 kN load. Deng et al. [14] conducted welding experiments on DP800 dual-phase steel with 1.2 mm thickness by laser welding. The results showed that the microhardness of the weldments varied from 221 HV to 243 HV. Sezgin et al. [15] used electron beam welding technology to weld DP1000 dual-phase steel with 1.2 mm thickness. They found that the average tensile strength of the weldments was 656 MPa. In addition, Huang et al. [16] studied the arc behavior of the CMT welding in a unique environment, and the results showed that the welding arc stability of the CMT welding technology was good, and the test data was higher than the standard value, which could reach the operation requirements of the unique environment.

According to the short-circuit transition principle of the CMT, the welding process is characterized by continuous low heat input, while it is only suitable for welding thin plates. If the heat input is increased, the area of the softened zone in the heat-affected zone (HAZ) will be increased, which will affect the service performance of materials. However, reducing the heat input is prone to welding defects such as incomplete fusion and incomplete penetration [17–19]. Based on the CMT welding process, the cold metal transfer plus pulse technology (CMT + P) is proposed to improve the narrow control range of the CMT welding heat input [20–22]. Pulse can change the current output by accelerating the flow rate of electrons, which makes up for the deficiency of the typical CMT welding process, that is, the small weld depth and only suitable for thin plate welding. It is easy to obtain a welded joint with a relatively large weld depth-to-width ratio [23–27]. In this study, the DP980 dual-phase steel plate with 1.2 mm thickness was welded by the CMT + P welding technology. The effects of wire feeding speed and welding speed on the forming, microstructures, and mechanical properties of DP980 dual-phase steel welded joints were analyzed to provide experimental data and theoretical support for realizing high-quality and high-efficiency welding of dual-phase steel.

2. Materials and Methods

In this experiment, the research object is the DP980 dual-phase steel for automobile frame. The size of the test plate is 150 mm × 125 mm × 1.2 mm. ER120S-G welding wire with diameter $\Phi = 1.0$ mm is selected. The chemical compositions of both are shown in Table 1. The experimental protective gasses are CO_2 and Ar mixed in a ratio of 1:4.

Table 1. Chemical compositions of DP980 dual-phase steel and ER120S-G welding wire (mass fraction/%).

Materials	C	Si	Mn	P	S	Nb	Ti	Mo	B	N	Als	Cr	Ni
DP980	0.091	0.307	2.464	0.012	0.001	0.016	0.049	0.153	0.002	0.005	0.047	-	-
ER120S-G	0.080	0.770	1.780	0.008	0.002	-	-	0.580	-	-	-	0.340	2.310

The TPS-3200 CMT welder produced by the Fronius company in Austria was employed as welding equipment. DP980 dual-phase steel was welded by overlapping. The overlapping width of the two plates was 11 mm, the included angle between the welding torch and the parent metal of the workpiece was 70°, and the dry elongation of the welding wire was 10 mm. The schematic diagram of the welding process is shown in Figure 1. Before welding, the basic metal surface of DP980 dual-phase steel was polished with sandpaper to remove the surface oxide film. Then, the base metal was clamped firmly on the welding platform surface, and the CMT + P welding mode was used for the welding experiment on the CMT welder. The working principle of pulse current is shown in Figure 2. The gas flow rate was controlled at 25 L/min. The arclength correction coefficient was 0%. In the CMT + P welding mode, the forming, microstructure, and mechanical properties of welded joints are studied by adjusting the wire feeding speed (3 m/min~5 m/min) and welding speed (350 mm/min~600 mm/min).

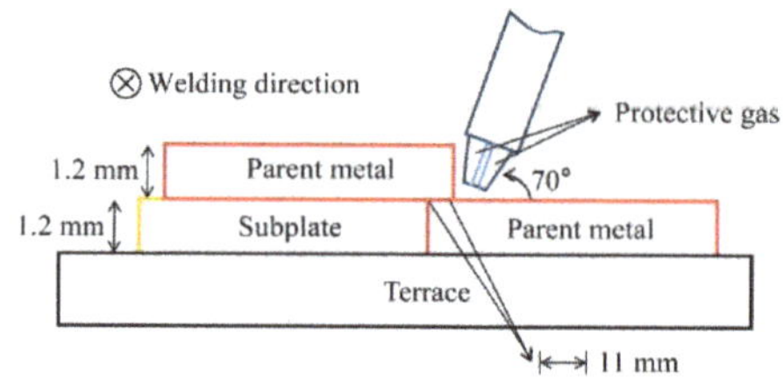

Figure 1. Cross-section view of the CMT + P welding process.

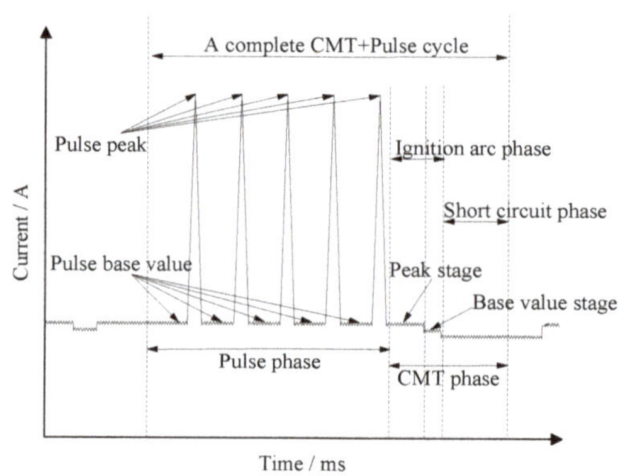

Figure 2. Schematic diagram of the CMT + P welding.

After welding, the metallographic samples of 30 mm× 10 mm× 1.2 mm were prepared by an electric spark CNC wire-cutting machine and were corroded with 4% nitric acid alcohol reagent after grinding and polishing. The energy dispersive spectrometer (EDS) equipped with a scanning electron microscope (SEM) was used to scan the welded joint area to determine the main elements of the welded joint and analyze the effect of alloying elements in the welding process. The resolution of the SEM is 1.0 nm (15 kv), and the acceleration voltage is 0.5~30 kv. The OM morphology of the welded joint HAZ was observed by the metallographic microscope. The grain size of the coarse-grain zone was measured according to GB/T 6394-2017 "Determination of estimating the average grain size of the metal". Secondly, the microhardness distribution of welded joints was measured by a digital Vickers hardness testing machine with a load of 200 g and was applied for 15 s. The welded joint area was dotted every 0.3 mm from the position of the weld to the base metal. Three groups of samples were tested, and the microhardness curves were plotted using the Origin mapping software based on the measurement results. Finally, the tensile properties of welded joints were tested by a microcomputer-controlled electronic universal testing machine. The size of the tensile sample was set according to the GB/T 2651-2008 "Tensile test method on welded joints", as shown in Figure 3. A test plate with an equal thickness was added on both sides of the lapweld, and the tensile test was conducted at a tensile rate of 2 mm/min. The fracture microstructures of the welded joints were observed by the SEM.

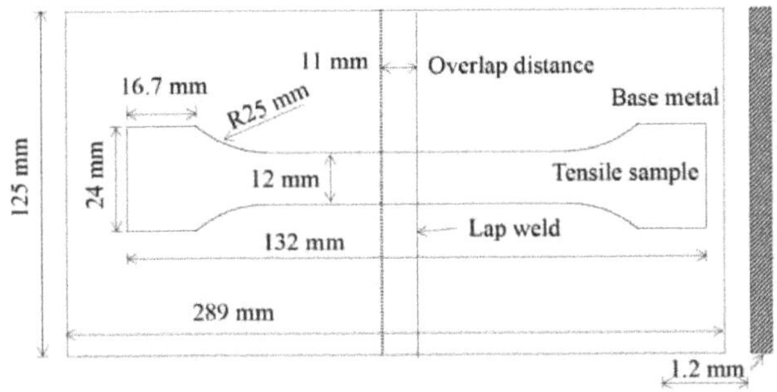

Figure 3. Schematic diagram of tensile sample size.

Welding line energy is defined as the ratio of welding power to welding speed, which reflects the quantity of heat supplied by the welding energy input of each welding piece within the unit distance. With the change of welding parameters, the welding line energy also changes. The shift in welding line energy will lead to changes in microstructures and mechanical properties in HAZ. Therefore, this paper focuses on the effect of the welding process conditions on the welding line energy and how this change affects the mechanical properties of the welded joints. The calculation formula of welding line energy is:

$$Q = UI/V \tag{1}$$

where:

Q—welding line energy (J/cm),
U—welding voltage (V),
I—welding current (A),
V—welding speed (cm/s).

The depth-to-width ratio can be used to evaluate the effect of weld forming. When applying different welding processes to weld the same material, under the specified weld depth, the larger the weld depth-to-width ratio, the better the welding forming effect. If the weld width is too wide, the HAZ area in the welded joints will increase. It influences the comprehensive performance of the welded structure. In addition, an excessively wide weld will reduce the protective effect of the gas and result in the mixing of harmful elements during the welding process. Therefore, it is of great significance to measure the depth-to-width ratio of the weld in this experiment, and its calculation formula is:

$$m = D/W \tag{2}$$

where:

m—the depth-to-width ratio of the weld,
D—weld depth (μm),
W—weld width (μm).

3. Results

3.1. *Typical Macromorphology and Element Distribution of Welded Joints*

The typical macroscopic morphology of the welded joint fabricated by CMT + P technology is shown in Figure 4. It can be gained that the surface forming effect of the welded joint is well. There are no welding defects such as spatter residue, non-fusion, and arc crater. In Figure 4, the welded joints are divided into weld zone, HAZ, and base metal zone. The HAZ is between the base metal and the weld zone. It is symmetrically distributed at the center of the weld zone. The HAZ close to the fusion line is the weakest area between the welding wire and the base metal. Therefore, the performance of the HAZ determines the overall performance of the component. Starting from the wire feeding speed and welding speed, the paper mainly explored the effects of the CMT + P welding process parameters on the welding forming, microstructures, and properties of the HAZ in DP980 dual-phase steel welded joints.

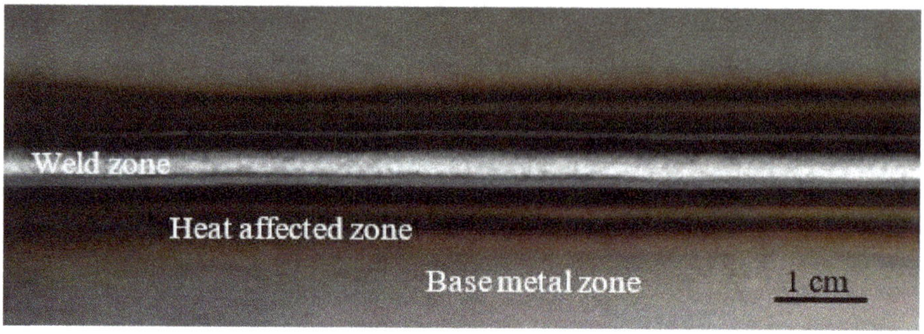

Figure 4. The CMT + P welded joint of DP980 dual-phase steel.

In the welding process, the effect of alloying elements on the properties of the welded joints is noticeable. Different wire components have different effects on the structure of welded joints [28]. The ER120S-G welding wire is selected as the experimental welding wire after careful consideration. First, the selection of welding wire follows the principle of "equal strength matching". The tensile strength of the welding wire is 1043 MPa. Secondly, compared with the base metal, the contents of Si and Ni elements are increased in the welding wire. The Si has a more vital deoxidation ability than the Mn, and it is easily soluble in ferrite. Therefore, it can strengthen the toughness of the ferrite [29,30]. The addition of Ni to steel can improve the hardenability of the dual-phase steel and increase its strength. There are two methods to improve the mechanical properties of welded joints. One is to decrease the relative content of the ferrite by adjusting the welding process parameters. Another is to add alloying elements to enhance the toughness of the microstructures. Finally, the content of Mn in the welding wire is reduced. The Mn content should not exceed 1.8% to improve the strength and toughness of the welded joint. The main element compositions of the welded joints were analyzed by the EDS line scanning, and the results are shown in Figure 5. The element of the Si has an excellent toughening effect and low crack sensitivity in welded joints. The boiling point of the Mn is 1962 °C and the general arc welding temperature can reach about 3000 °C, which is easy to cause the evaporation and burning loss of the Mn [31,32]. The element of the Ni can decrease the brittle transition temperature and increase the hardness and strength of welded joints [33–35]. Moreover, Ni and Fe belong to an infinite solid solution, which can reduce the content of proeutectoid ferrite to improve the strength of welded joints.

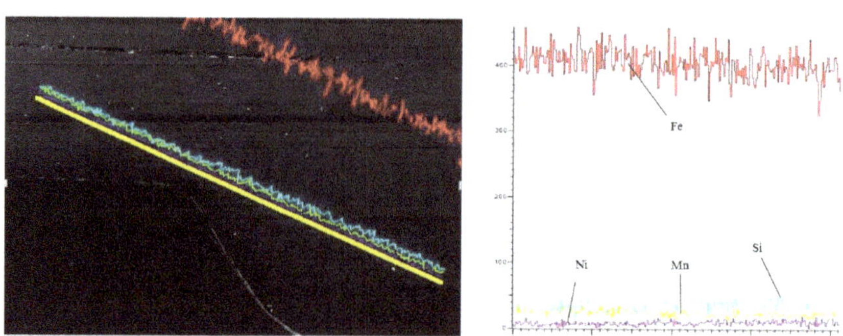

Figure 5. EDS detection results of welded joint.

3.2. Depth-to-Width Ratio of Welded Joints in Different Welding Parameters

The cross-sectional morphology of the welded joint observed at a 20-fold metallographic microscope magnification is shown in Figure 6. It shows the measuring method of depth and width. The welding torch at a certain angle with the welded is to obtain a welded joint with well formability. After several pre-experiments in the study, the welding torch in the angle of 70° is the best. The welding line energy, weld depth, weld width, and depth-to-width ratio in different wire feeding speeds and welding speeds are shown in Table 2, in which the welding line energy and depth-to-width ratio are calculated according to Formulas (1) and (2).

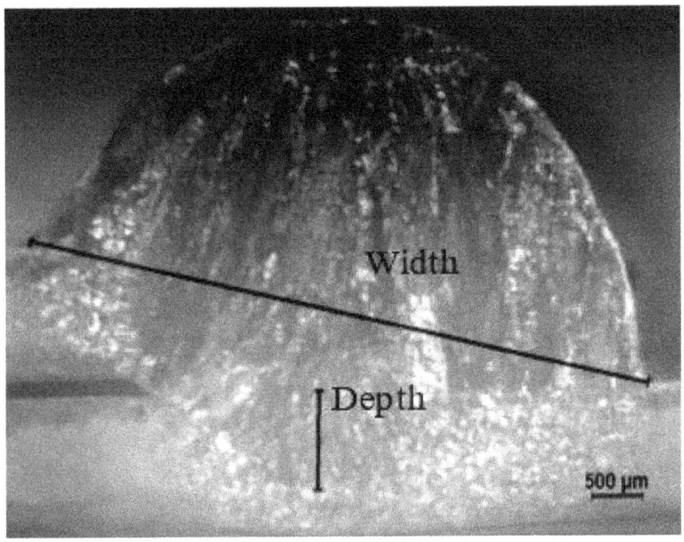

Figure 6. Section morphology of welded joint.

Table 2. Welding line energy and depth-to-width ratio in different welding process parameters.

Wire Feeding Speed/m/min	Welding Speed/mm/min	Welding Current/A	Welding Voltage/V	Line Energy/J/cm	Weld Depth [1] µm	Weld Width [1]/µm	Depth-to-Width Ratio
3	500	59	19	1345	515	4323	0.11
4	500	75	19.7	1773	708	5549	0.12
5	500	92	20.5	2263	1303	6512	0.20
4	350	75	19.7	2533	1515	6768	0.22
4	400	75	19.7	2216	837	6325	0.13
4	450	75	19.7	1970	799	5900	0.13
4	550	75	19.7	1612	437	5279	0.08
4	600	75	19.7	1478	350	4711	0.07

[1] The average confidence interval of the weld depth is ±10 µm and that of the weld width is ±50 µm.

When the welding speed is 500 mm/min, the single factor method is used to study the variation of weld forming parameters in the wire feeding speed of 3 m/min~5 m/min. The results show that within the parameter range of wire feeding speed from 3 m/min to 5 m/min, the weld forming effect is best at 4 m/min. When the wire feeding speed increases from 3 m/min to 5 m/min, the welding current and welding voltage rise accordingly, and the welding line energy increases from 1345 J/cm to 2263 J/cm. The weld depth and width are increased with the rise in welding line energy. When the wire feeding speed is 5 m/min, the welding depth-to-width ratio is the best. However, the line energy in the wire feeding speed of 5 m/min is too fast to manufacture welded joints without burn-through defects. The forming effect in the wire feeding speed of 5 m/min is not ideal. Therefore, based on the wire feeding speed of 4 m/min, the next step is to study the influence of the welding speed in the range of 350 mm/min to 600 mm/min. The welding line energy decreases from 2533 J/cm to 1478 J/cm when the welding speed increases from 350 mm/min to 600 mm/min. The welding depth-to-width ratio decreases from 0.22 to 0.07. When the welding speed is higher than 350 mm/min, the forming effect of welded joints is well. There are no obvious welding defects on the surface. Therefore, penetration and burning loss are manifested when the welded joint in the wire feeding speed of 5 m/min.

4. Discussion

4.1. Microstructures of HAZ in Different Wire Feeding Speeds

In the CMT + P mode, the welding speed of 500 mm/min is controlled and the effects of welding speed on the microstructures are shown in Figure 7. The HAZ is only a tiny area in the welded joint. However, failure of general materials occurs in this area. It is of practical significance to study the microstructures and grain size of the HAZ in welded joints. The HAZ of welded joints is divided into a complete phase transition zone and an incomplete phase transition zone according to the degree of the phase transition. The coarse-grained zone in the HAZ is a complete phase transition zone, but the grain size of the coarse-grained zone is larger than the other zone. Thus, the coarse-grained size is one of the reasons for the failure. Figure 7a,b, respectively, show the distribution of the HAZ at the wire feeding speed of 3 m/min and 5 m/min. The distribution range of the HAZ in the wire feeding speed of 5 m/min is more expansive than that in the wire feeding speed of 3 m/min. The heat input is increased with the rise in wire feeding speed, which leads to a broader HAZ range. Compared with the conventional CMT technology, the pulse can amplify the peak current to strengthen the penetration of the welder. Therefore, the heat input is reduced by increasing the welding speed to increase the grain size. At the same time, production efficiency is also improved.

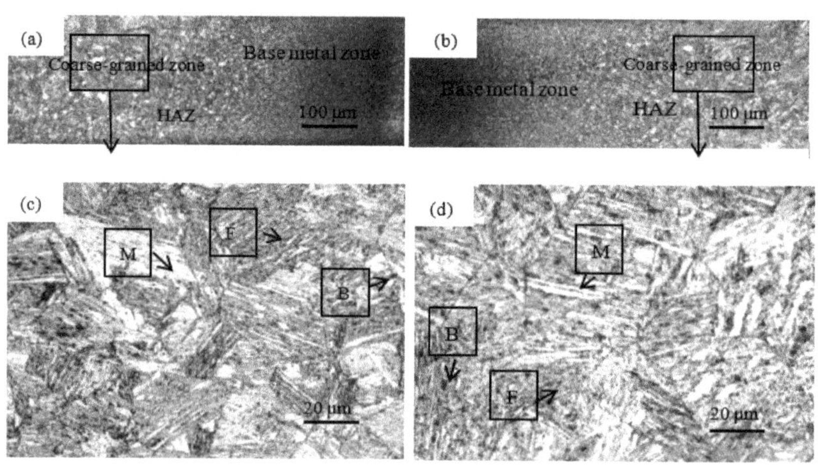

Figure 7. Microstructures of welded joints in different wire feeding speeds: (**a**) HAZ in the wire feeding speed of 3 m/min, (**b**) HAZ in the wire feeding speed of 5 m/min, (**c**) the coarse-grained zone in the wire feeding speed of 3 m/min, (**d**) the coarse-grained zone in the wire feeding speed of 5 m/min.

As shown in Figure 7c,d, the microstructures of the coarse-grained zone are martensite and ferrite. The temperature of the melting pool exceeds the temperature of Ac3, which results in the dendrite remelting of the welded metal. The original microstructures are crystallized to form martensitic after rapid supercooling, which is consistent with the research [36] that adopts the laser to weld the DP980 dual-phase steel. However, the difference is that the DP980 dual-phase steel welded by the CMT + P technology also generates a small amount of bainite. The heat input of the CMT + P technology results in the formation of bainite after a small amount of remelted austenite through shear transformation and short-range diffusion. The grain size of the coarse-grained zone was measured by the "intercept method" in GB/T 6394-2017. When the wire feeding speed is 3 m/min, 4 m/min, and 5 m/min, the grain size of the coarse-grained zone is 31.90 μm, 39.84 μm, and 50.93 μm. It is found that the grain size is directly proportional to the wire feeding speed. The faster the wire feeding speed is, the higher the heat input per unit time is. Therefore, the more sufficient the grain growth time is, the coarser the grain size in the coarse-grained zone is.

4.2. Microstructures of HAZ in Different Welding Speeds

In the CMT + P mode, the wire feeding speed of 4 m/min is controlled and the effects of welding speed on the microstructures are shown in Figure 8. The range of the HAZ in the welding speed of 600 mm/min is narrower than that in the welding speed of 350 mm/min. The heat input decreases accordingly to narrow the HAZ range. The temperature input into the HAZ also reaches the temperature of Ac3. The coarse-grained zone is composed of martensite, ferrite, and a small amount of bainite. When the welding speed increases from 350 mm/min to 600 mm/min, the martensite grain size in the coarse-grained zone decreases from 45.48 μm to 35.73 μm in Figure 9. It indicates that the grain size is inversely proportional to the welding speed. Due to the reduction of the heat input, it leads to the shortening of the grain holding time. The martensitic grains have no time to grow under the rapid cooling, which results in the formation of refined martensitic grains. Therefore, the influence of welding speed on the grain size is significant. The grain size of the coarse-grained zone is effectively refined by increasing the welding speed in CMT + P mode. The total amount of the heat input is decreased by increasing the welding speed. The pulse plays a role in amplifying the peak current and optimizing the welding arc. With the decrease of the heat input, the relative undercooling rate increases. The nucleation rate of microstructures in the coarse-grained zone increases and its growth rate slows down. Therefore, the grains of the coarse-grained zone are refined to improve the performance of the welded joints.

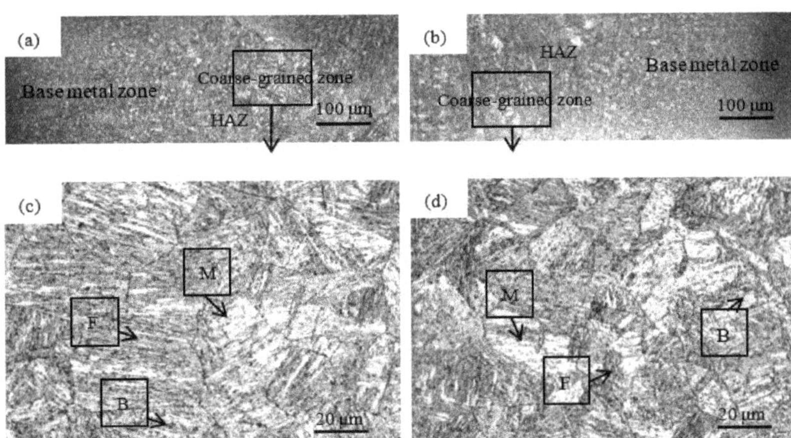

Figure 8. Microstructures of welded joints in different welding speeds: (**a**) HAZ in welding speed of 350 mm/min, (**b**) HAZ in welding speed of 600 mm/min, (**c**) the coarse-grained zone in welding speed of 350 mm/min, (**d**) the coarse-grained zone in welding speed of 600 mm/min.

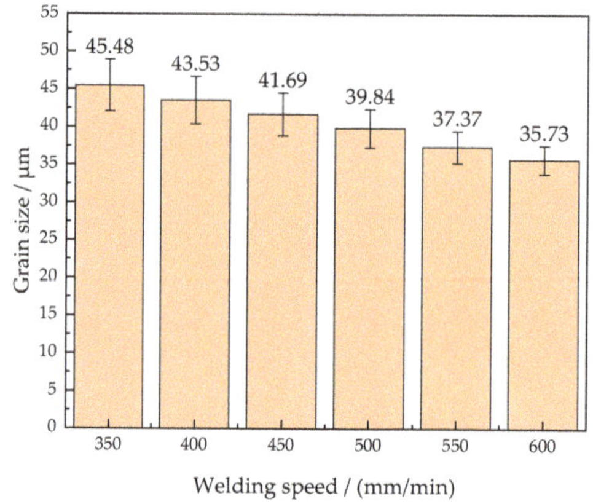

Figure 9. Grain size of the coarse-grained zone in different welding speeds.

4.3. Microhardness of Welded Joints in Different Welding Parameters

The microhardness and microstructures of the welded joints under different welding speeds are shown in Figure 10. In addition to the coarse-grained zone analyzed above, there is also an incomplete phase transition zone called the softened zone in the HAZ. The welding temperature in the softened zone is between Ac1 and Ac3, which is the ferrite transformation condition. However, the softened zone is far from the welding heat source. Due to the unstable heat input, the size and distribution of the ferrite structure are not uniform. Thus, the comprehensive mechanical properties of the softened zone are poor.

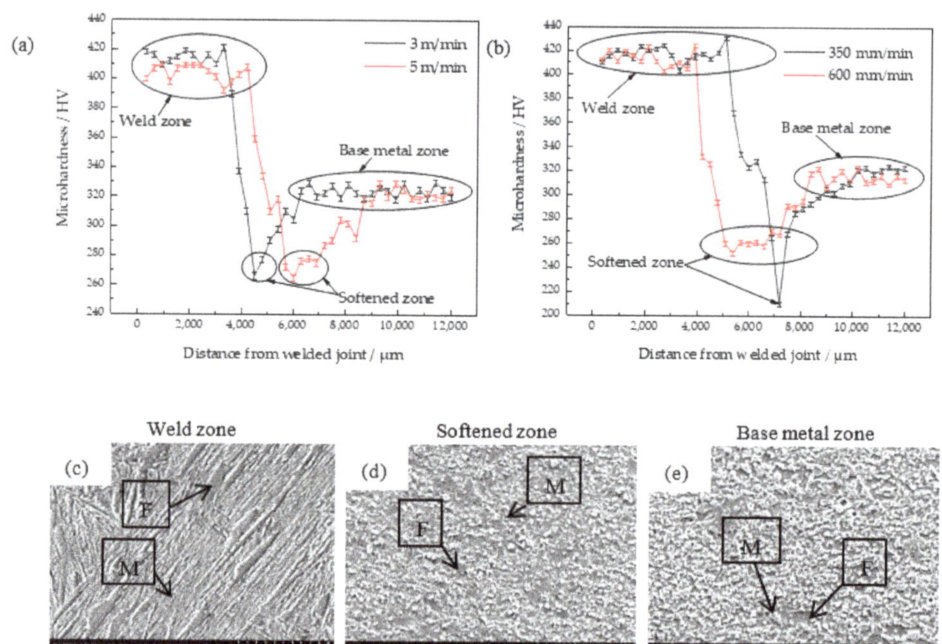

Figure 10. Microhardness and microstructures of welded joints: (**a**) microhardness in different wire feeding speeds, (**b**) microhardness in different welding speeds, (**c**) microstructure of the weld zone, (**d**) microstructure of the softened zone, (**e**) microstructure of the base metal zone.

Figure 10a shows that the distance of the softened zone in the wire feeding speed of 5 m/min is wider than that in the wire feeding speed of 3 m/min. The wire feeding speed is too fast, which results in a high total heat input and a sizeable thermal diffusion range. The HAZ deviates from the maximum temperature of the weld zone and the cooling speed is gradually accelerated. However, the temperature in this process is still higher than that of Ac1 for ferrite precipitation. Thus, more ferrite is formed in the softened zone.

Similarly, the heat input per unit of time is stable when the wire feeding speed is constant. The faster the welding speed, the shorter the residence time of the welding torch. Thus, the faster the undercooling rate, the more prominent the softened zone. As shown in Figure 10b, the welded joint in the welding speed of 350 mm/min has serious welding penetration. Due to the excessively high welding heat input, the microhardness in the softened zone decreases sharply. As the welding speed increases to 600 mm/min, the microhardness in the softened zone tends to be stable. Combined with the analysis of the line energy in Section 3.2, the welding line energy changes by 918 J/cm with the evolution of the wire feeding speed. At the same time, the line energy changes by 1055 J/cm with the evolution of the welding speed. It shows that the welding speed is the main factor affecting the heat input.

Figure 10a,b show that the microhardness of the welded joints is distributed in a high-low-medium trend. The high hardness area is the weld zone, which is formed by the filler. Because of the steady heat input, the microhardness in the weld zone is the highest. The average microhardness in the area is 420 HV and up to 430 HV. The weld zone composed of the lath martensite is shown in Figure 10c. Because the lath martensite is harder than the island-like martensite, its anisotropic has a stronger ability to hinder the movement of dislocations. Hence, the weld zone of the welded joint has a high microhardness.

The low microhardness area is the softened zone. The unstable martensite in the base metal is dissolved to undergo austenite transformation by heating cycles. After supercooling, parts of the new martensite are formed from the austenite. At the same time, the ferrite precipitates from another part of austenite. Due to the reduction of the original martensite, the microhardness of the softened zone decreases. The average microhardness in the softened zone is 250 HV. The softened zone composed of ferrite and island-like martensite is shown in Figure 10d. Part of the martensite in the base metal is transformed into ferrite after a short austenitizing process of heating, holding, and cooling. The remaining austenite is rapidly undercooled and transformed into the new martensite. Due to the enrichment of the ferrite in the softened zone, the microhardness drops sharply with the reduction of the martensite. It is the main reason for the failure of the welded joints.

The medium microhardness area is the base metal. The average microhardness is 320 HV, which is not strongly affected by the heat input. The base metal zone composed of ferrite and martensite is shown in Figure 10e. The martensite grain size in the base metal zone is larger than that of the softened zone, while the ferrite structure in the base metal zone is relatively reduced. It makes the base metal have higher hardness than the softened zone.

In conclusion, the expansion of the softened zone can be avoided by reducing the heat input. If the softened zone is reduced, the ferrite content in the softened zone will be decreased. That is, the microhardness can be improved by reducing the wire feeding speed or increasing the welding speed appropriately.

4.4. Tensile Strength of Welded Joints in Different Welding Parameters

Tensile strength of welded joints under different welding parameters is shown in Figure 11. It shows that the wire feeding speed of 4 m/min and the welding speed of 550 mm/min are the optimum process parameters. The optimum tensile strength is 973 MPa. Referring to Table 2 above, the line energy is 1612 J/cm at this time. The line energy in the optimum process parameters is reduced by 921 J/cm compared to that in the welding speed of 350 mm/min, and the tensile strength in the optimum process parameters is increased by 9.15%. With the increase of the welding speed, the weld depth and weld width are respectively reduced by 1078 μm and 1489 μm. It can be observed that the reduction distance of the weld width is greater than that of the weld depth. With the increase of the welding speed, the welding heat input decreases in the welding process. However, the weld depth is increased by the action of the pulse when the weld width is 5279 μm. The weld width in the optimum process parameters is lower than that in the low welding speed, which indicates that the HAZ width is smaller. At the time, the average grain size of the coarse-grained zone in the HAZ is tiny, which is 37.37 μm.

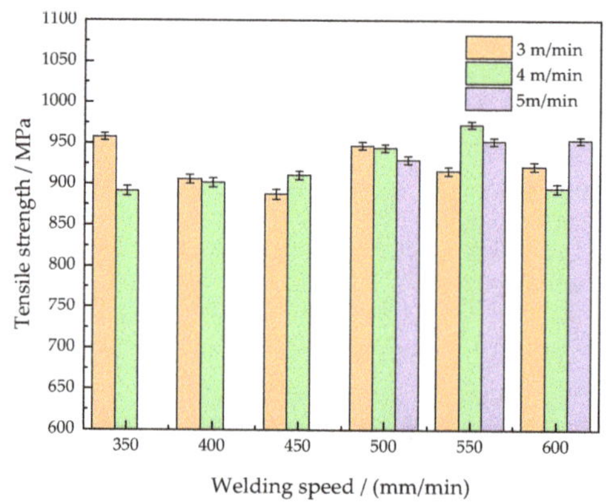

Figure 11. Tensile strength of welded joints in different welding parameters.

When the welding speed increases to 600 mm/min at the wire feeding speed of 4 m/min, the tensile strength of the welded joint suddenly decreases to 895 MPa. Because the welding speed is too fast to approach the optimal line energy, this results in the narrowing of the weld bead and the poor bonding strength of the welded joint. However, when the welding speed exceeds 600 mm/min, the welded joints have obvious welding defects of incomplete fusion.

All the tensile strength of the welded joints under different welding parameters in Figure 11 is higher than 85% of the tensile strength of the base metals, which are qualified welding products. It indicates that welded joints with excellent tensile properties can be produced by the CMT + P welding technology. The average tensile strength of the welded joints is 925 MPa under different welding parameters. The tensile strength of the dual-phase steel welded by the CMT + P welding technology is 1.4 times higher than that of the DP1000 dual-phase steel welded by the electron beam welding technology [15]. Because the CMT + P welding technology utilizes the arc as the heat source, it has the advantages of a stable welding arc, low heat input, and no spatter in the welding process. Moreover, the pulse current can improve the welding heat input and arc stability. Increasing the welding speed is a good idea to neutralize the increase of the heat input, which not only optimizes the mechanical properties of welded joints but also improves the welding efficiency.

In conclusion, adopting the CMT + P technology to weld the DP980 dual-phase steel can obtain welded joints with excellent mechanical properties. The pulse enhances the stability of the arc, which is suitable for high-efficiency welding of thin plate materials. Taking the optimal tensile strength as standard, the wire feeding speed of 4 m/min and the welding speed of 550 mm/min are the optimal welding parameters. Its welding line energy is about 1612 J/cm.

The fracture locations of the base metal and welded joint are shown in Figure 12. The fracture of the base metal occurs along the direction of the ferrite microcracks, and the fracture morphology is relatively flat. The fracture of the welded joint mainly occurs in the softened zone. Both of them are gray-black, cup-cone-shaped, and have no metallic luster. However, the fracture of the welded joint has a more significant tear ridge.

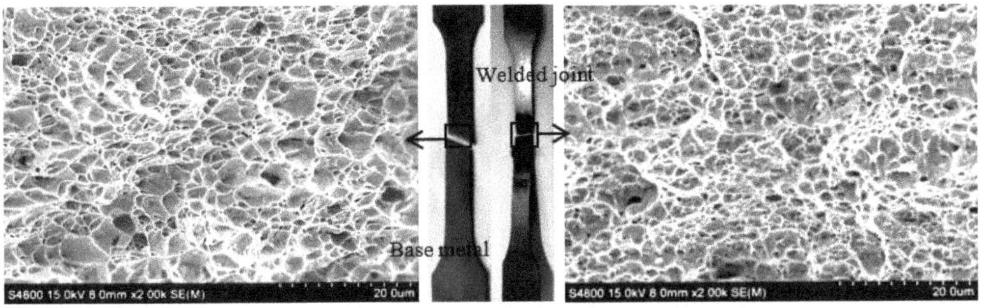

Figure 12. Fracture locations and morphology of the base metal and welded joint.

Compared with the ductile ferrite, the hard martensite has a stronger capacity to resist fracture and is less prone to fracture in the weld zone. Contrasted with the uniformly distributed ferrite in the base metal, the ferrite is enriched in the softened zone of the welded joint. The ferrite occurred in the softened zone consists of the ferrite without phase transformation and a small amount of the ferrite formed after austenite recrystallization. The excessive ferrite results in plastic deformation after stretching and the ductile fracture finally occurs in the HAZ.

According to the SEM morphology of the base metal, the microstructures of the fracture are evenly distributed. There are a large number of tensile equiaxed dimples in the fracture. Under the conditions of the unidirectional tensile, a three-dimensional stress concentration of the ferrite grain is caused, and microcracks are generated at the phase boundary between the ferrite and the martensitic [37]. As the cracks expand and merge, tiny dimples appear. The formation of fracture dimples indicates that the fracture mode of the base metal is microporous aggregation fracture. There are also dimples in the fracture of the welded joint. However, the size of the dimples in the welded joint is smaller than that of the base metal. It means that the welded joint has no larger buffer space than the base metal when plastic deformation occurs. Therefore, the welded joint has less ability to resist ductile fracture than the base metal.

5. Conclusions

(1) By using the CMT + P 70° lap welding method and the ER120S-G wire to weld the DP980 dual-phase steel with the thickness of 1.2 mm, the microstructures and properties of the welded joints are excellent. In the welded joints, Si and Ni are the main elements added. The martensite, ferrite, and a little bainite are the primary microstructures. By increasing the wire feeding speed from 3 m/min to 5 m/min, the average grain size of the coarse-grained zone increases from 31.90 μm to 50.93 μm. By increasing the welding speed from 350 mm/min to 600 mm/min, the average grain size of the coarse-grained zone decreases from 45.48 μm to 35.73 μm.

(2) The average microhardness of the weld zone is 420 HV, which is significantly higher than that of the base metal. The peak microhardness is 430 HV. The minimum microhardness occurred in the softened zone of the HAZ is 250 HV. The welding speed is the main factor affecting the range of the softened zone. By increasing the wire feeding speed or decreasing the welding speed, the range of the softened zone becomes wider.

(3) The wire feeding speed of 4 m/min and the welding speed of 550 mm/min are the optimum process parameters, when the tensile strength of the welded joint is 973 MPa. The tensile fracture is located in the softened zone and the fracture mode is a ductile fracture. In the CMT + P process, the effect of process parameters on the welding depth is greater than the welding width. The welding line energy for optimal process parameters is 1612 J/cm, the weld depth is 0.437 mm, and the weld width is 5.279 mm.

Author Contributions: Conceptualization, Y.L. and Z.L.; methodology, Y.L., Z.L. and Y.C.; software, Z.L., A.L. and X.L.; validation, Y.L. and C.H.; formal analysis, Z.L., Y.C. and A.L.; investigation, Y.L., Z.L. and A.L.; resources, Y.L. and C.H.; data curation, Z.L. and Y.C.; writing—original draft preparation, Y.L. and Y.C.; writing—review and editing, Y.L., Z.L., A.L. and X.L.; visualization, C.H.; supervision, Y.L. and C.H.; project administration, Y.L.; funding acquisition, Y.L. and C.H. All authors have read and agreed to the published version of the manuscript.

Funding: This study was financially funded by the Support Program for Innovative Talents in Universities of Liaoning Province (LR2019042), and the Key R & D Project of Liaoning Province (2020JH2/10100011).

Institutional Review Board Statement: Not applicable.

Informed Consent Statement: Not applicable.

Data Availability Statement: Not applicable.

Acknowledgments: We sincerely appreciate the support provided by Shenyang University in the preparation of this manuscript.

Conflicts of Interest: The authors declare no conflict of interest.

References

1. Xu, L.X.; Wu, H.B.; Wang, X.T. Influence of microstructural evolution on the hot deformation behavior of an Fe–Mn–Al duplex lightweight steel. *Acta Metall. Sin.* **2018**, *31*, 389–400. [CrossRef]
2. Hamid, A.; Morteza, S.; Rahmatollah, E.; Ehsan, G. Void formation and plastic deformation mechanism of a cold-rolled dual-phase steel during tension. *Acta Metall. Sin.* **2020**, *33*, 299–306.
3. Kuang, C.F.; Zheng, Z.W.; Wang, M.L.; Xu, Q.; Zhang, S.G. Effect of hot-dip galvanizing processes on the microstructure and mechanical properties of 600-MPa hot-dip galvanized dual-phase steel. *Int. J. Miner. Metall. Mater.* **2017**, *24*, 1379–1383. [CrossRef]
4. Zhou, H.; Liu, J.S.; Zhou, D.W.; Tao, T. Effect of aluminum foil on microstructure and temperature field of laser fusion welding joint of DP590 duplex steel /AZ31B magnesium alloy. *Trans. Nonferrous Met. Soc. China* **2020**, *30*, 2669–2680. [CrossRef]
5. Kucukomeroglu, T.; Aktarer, S.M.; Ipekoglu, G.; Cam, G. Investigation of mechanical and microstructural properties of friction stir welded dual phase (DP) steel. In *IOP Conference Series-Materials Science and Engineering*; IOP Publishing: Bristol, UK, 2019.
6. Deng, J.; Ma, J.; Xu, Y.; Shen, Y. Effect of martensite distribution on microstructure deformation behavior and mechanical properties of duplex steel. *Acta Metall. Sin.* **2015**, *51*, 1092–1100.
7. Dai, Q.F.; Song, R.B.; Fan, W.Y.; Guo, Z.F. Strain hardening behavior and mechanism of DP1180 duplex steel under high strain rate deformation. *Acta Metall. Sin.* **2012**, *48*, 1160–1165. [CrossRef]
8. Nie, W.J.; Shang, C.J.; Guan, H.L.; Zhang, X.B.; Chen, S.H. Microstructure regulation and anti-deformation behavior analysis of ferritic/bainite (F/B) duplex steel. *Acta Metall. Sin.* **2012**, *48*, 298–306. [CrossRef]
9. Tian, Y.; Shen, J.; Hu, S.; Gou, J.; Cui, Y. Effects of cold metal transition mode on the reaction layer of wire and arc additive-manufactured Ti-6Al-4V/Al-6.25Cu dissimilar alloys. *J. Mater. Sci. Technol.* **2021**, *74*, 35–45. [CrossRef]
10. Zhang, Q.; Zhang, P.; Yu, Z.S.; Yan, H.; Shi, H.; Wu, D.; Li, S.W.; Tian, Y. Microstructure and properties of an Al 6061/galvanized plate fabricated by CMT welding. *J. Wuhan Univ. Technol.* **2020**, *35*, 937–945. [CrossRef]
11. Liu, Z.Q.; Zhang, P.L.; Li, S.W.; Wu, D.; Yu, Z.S. Wire and arc additive manufacturing of 4043 Al alloy using a cold metal transition method. *Int. J. Miner. Metall. Mater.* **2020**, *27*, 783–791. [CrossRef]
12. Cam, G.; Ipekoglu, G. Recent developments in joining of aluminium alloys. *Int. J. Adv. Manuf. Technol.* **2017**, *91*, 1851–1866. [CrossRef]
13. Cui, J.J.; Dong, D.Y.; Wang, Q.; Li, Y.G. Failure behavior of DP780 dual phase steel resistance spot welded joints under dynamic load. *J. Mech. Eng.* **2021**, *57*, 70–79.
14. Deng, C.Y.; Liu, C.; Liu, X.G.; Gong, B.M.; Wang, D.P.; Xu, Q.J. Mechanical properties and microstructure analysis of DP800 dual-phase steel laser welded joint softened zone. *J. Tianjin Univ.* **2021**, *54*, 311–317.
15. Sezgin, C.T.; Hayat, F. The microstructure and mechanical behavior of TRIP 800 and DP 1000 steels welded by electron beam welding method. *Soldag. Inspeção* **2020**, *25*, 1–12. [CrossRef]
16. Huang, J.Q.; Xue, L.; Huang, J.F.; Zou, Y.; Niu, H.L.; Tang, D.Y. Arc behavior and weld performance of CMT welding under high pressure. *Acta Metall Sin.* **2016**, *52*, 93–99.
17. Chen, Q.H.; Lv, X.Q.; Xu, L.Y.; Jing, H.Y.; Han, Y.D. Microstructure and properties of CMT + P welded joints of P92 steel. *Trans. China Weld. Inst.* **2018**, *39*, 110–114, 133–134.
18. Kim, J.; Kim, J.; Pyo, C. Comparison of mechanical properties of Ni-Al-Bronze alloy fabricated through wire arc additive manufacturing with Ni-Al-Bronze alloy fabricated through casting. *Metals* **2020**, *10*, 1164. [CrossRef]
19. Silvayeh, Z.; Gtzinger, B.; Karner, W.; Hartmann, M.; Sommitsch, C. Calculation of the intermetallic layer thickness in cold metal transfer welding of aluminum to steel. *Materials* **2018**, *12*, 35. [CrossRef]

20. Brezinova, J.; Dupon, M.; Vojtko, M.; Vina, J.; Draganovska, D. Application of cold metal transfer welding for high pressure die casting mold restoration. *Metals* **2019**, *9*, 1232. [CrossRef]
21. Cao, R.; Chang, J.H.; Zhu, H.X.; Mao, G.J.; Xu, Q.W.; Shi, Y.; Chen, J.H.; Wang, P.C. Investigation of wire selection for CMT plug joining Mg AZ31-to-galvanized steel. *J. Manuf. Processes* **2018**, *32*, 65–76. [CrossRef]
22. Zheng, P.F.; Lu, Q.H.; Zhang, P.L.; Yan, H.; Zhao, J.; Shi, H.C. Effect of vibration on microstructure and fatigue properties of 6082 CMT-welded joints. *Trans. Indian Inst. Met.* **2021**, *74*, 3217–3225. [CrossRef]
23. Wu, W.; Xue, J.X.; Zhang, Z.H.; Ren, X.H.; Xie, B. Process optimization on multilayer morphology during 316L double-wire CMT + P deposition process. *Metals* **2019**, *9*, 1334. [CrossRef]
24. Xie, B.; Xue, J.X.; Ren, X.H.; Wu, W.; Lin, Z.B. A comparative study of the CMT + P process on 316L stainless steel additive manufacturing. *Appl. Sci.* **2020**, *10*, 3284. [CrossRef]
25. Lu, L.H.; Zhang, L.H.; Sun, W.C.; Shi, Y.; Fan, D. Design and experimental analysis of special power supply for double arc pulse MIG welding. *Trans. China Weld. Inst.* **2018**, *39*, 53–57, 131.
26. Shu, F.Y.; Lv, Y.H.; Liu, Y.X.; Xu, F.J.; Sun, Z.; Peng, H.E.; Xu, B.S. Finite element simulation of matrix softened phenomena of CMT + P narrow gap welding joints. *Trans. Nonferrous Met. Soc. China* **2014**, *24*, 1830–1835. [CrossRef]
27. Plangger, J.; Schabhuttl, P.; Vuherer, T.; Enzinger, N. CMT additive manufacturing of a high strength steel alloy for application in crane construction. *Metals* **2019**, *9*, 650. [CrossRef]
28. Wang, X.R.; Wu, Y.; Pan, H.; Yao, C.W.; Huang, J. Microstructure and softening of advanced high-strength steel QP1180 lap joints welded with CMT. *Mater. Lett.* **2021**, *287*, 129282. [CrossRef]
29. Zhang, Z.H.; Wang, X.N.; Ye, S.; Sun, Q.; Liu, Z.G.; Di, H.S. Microstructure and properties of laser wire welding joints of aluminum-silicon coated steel. *Trans. China Weld. Inst.* **2021**, *42*, 18–25, 97–98.
30. Zhang, Y.; Sun, L.L.; Tang, J.C.; Tang, M. Effect of high silicon fly ash active agent on liquid metal surface tension and TIG weld of Q235 steel. *J. Weld.* **2018**, *39*, 92–96, 133.
31. Zhang, X.F.; Li, J.X.; Wan, Y.X.; Wu, X.J.; Huang, Z.Y. Effect of annealing process on microstructure and properties of cold rolled Medium Manganese Aluminum Low density Steel. *Mater. Rev.* **2021**, *35*, 16099–16103.
32. Cao, Y.; Zhao, L.; Peng, Y.; Ma, C.Y.; Tian, Z.L.; Zhong, M.L. Effect of heat input on microstructure and mechanical properties of manganese steel joint during laser welding. *China Laser* **2018**, *45*, 83–91.
33. Cheng, B.; Wu, D.J.; Zhang, C.; Chai, D.S.; Ma, G.Y. Secondary phase transition mechanism of laser wire filled nickel-based alloy /304 stainless steel and its effect on intergranular corrosion properties. *Trans. Nonferrous Met. Soc. China* **2021**, *31*, 715–725. [CrossRef]
34. Yu, L.; Cao, R. Research status of welding cracks in nickel-based alloys. *Acta Metall. Sin.* **2021**, *57*, 16–28.
35. Zeng, Y.; Wu, X.C.; Xia, S.W.; Zuo, P.P. Effect of Nickel element on thermal stability of new die casting die steel. *Mater. Rev.* **2017**, *31*, 72–75.
36. Chen, X.M.; Wang, X.N.; Yan, Q.; Wang, G.S.; Hu, Z.R.; Zhang, M.; Chen, C.J. Effect of nickel foil on microstructure and properties of DP980/A6061 dissimilar laser welding joints. *China Laser* **2018**, *45*, 36–43.
37. Cai, H.J.; Hu, J.F.; Song, R.B.; Wang, L.; Yu, S.C.; Dai, Q.F. Microstructure mechanism of plastic deformation of 1200MPa grade cold rolled dual-phase steel at High strain rate. *J. Mech. Eng.* **2016**, *52*, 23–29. [CrossRef]

MDPI AG
Grosspeteranlage 5
4052 Basel
Switzerland
Tel.: +41 61 683 77 34

Materials Editorial Office
E-mail: materials@mdpi.com
www.mdpi.com/journal/materials

www.ingramcontent.com/pod-product-compliance
Lightning Source LLC
Chambersburg PA
CBHW040828250926
51850CB00007BA/1284
9783725822607